PRIMATE DIVERSITY

PRIMATE

DIVERSITY

Dean Falk UNIVERSITY AT ALBANY, SUNY

W. W. NORTON & COMPANY   NEW YORK·LONDON

Editor: John Byram
Project Editor: Mary Kelly
Copy Editor: Mary Babcock
Associate Managing Editor: Jane Carter
Director of Manufacturing: Roy Tedoff
Editorial Assisant: Paul Fyfe
Text Design: Jack Meserole
Illustrations: John McAusland, John Guyer, UG
Maps: Carto-graphics

The text of this book is composed in Trump Mediaeval with display set in Trajan
Composition by: UG / GGS Information Services, Inc.
Manufacturing by: Quebecor, Hawkins
Cover Illustration: Mori Sosen. Monkey and Wasp, late 18th-early 19th Century.
Hanging scroll. Los Angeles County Museum of Art, Etsuko and Joe Price Collection.

Library of Congress Cataloging-in-Publication Data

Falk, Dean
    Primate diversity / Dean Falk.
        p.   cm.
    Includes bibliographical references (p.   ) and index.
    ISBN 0-393-97428-6 (pbk.)
    1. Primates.   I. Title.
QL737.P9F3   2000
599.8—dc21   99-36427

W. W. Norton & Company, Inc., 500 Fifth Avenue, New York, N. Y. 10110
    www.wwnorton.com
W. W. Norton & Company Ltd., 10 Coptic Street, London WC1A 1PU
1 2 3 4 5 6 7 8 9 0

Dedicated to my four favorite juveniles:
Eve Penelope
Helen Dean
Judith Jane
Kylene Morris
granddaughters all

CONTENTS

OUR COUSINS: THE CHIMPANZEES

13

THE EARLIEST HOMINIDS

14

PREFACE

I decided to write this book because, over the years, I have been frustrated in my efforts to find an up-to-date textbook that meshes well with the sequence and order of topics and species that are covered in my college-level course on primates. Many people helped produce this book. First and foremost among them is my wonderful, thorough, gentle, hard-working editor at W. W. Norton, John Byram. John has a golden eye for the printed word and for meaningful content, and he has spent many hours improving both of these in *Primate Diversity*. I am also grateful to Mary Babcock for her hard work as copy editor, and to Paul Fyfe for tirelessly locating photographs and arranging permissions to include them. Along these lines, Bill McGrew generously provided a number of superb photographs of chimpanzees. I also thank Curt Busse, Michelle Sauther, and authorities at the National Park Zoo for providing photographs; John Guyer for preparing lovely illustrations; and Art Sansone for writing the Review Questions answer key.

I have tried to base this book on the latest primatological theory, taxonomy, and discoveries. If I have met this goal to some extent, it is because I have benefited from constructive and detailed critiques of earlier versions of various chapters by some of the top primatologists and evolutionary biologists in the world: Pamela Ashmore, Thad Bartlett, Marina Cords, David Frayer, Paul Garber, Kenneth Glander, Kevin Hunt, Jay Kaplan, Carol Lauer, Lois Lippold, John Mitani, James Moore, Pascale Sicotte, Russell Tuttle, Patricia Whitten, Anne Zeller, and Adrienne Zihlman. I also wanted *Primate Diversity* to convey some of the excitement experienced by primatologists as they observe their subjects in tropical forests, swamps, or savannas—and sometimes even in museums and zoos. I thank the following scientists for providing wonderful personal accounts of special moments in the field: Thad Bartlett, Sue Boinski, Marina Cords, Paul Garber, Sharon Gursky, Ellen Ingmanson, John Oates, Jane Phillips-Conroy, Melissa Remis, Peter Rodman, Alfred Rosenberger, Donald Sade, Dan Shillito, and Eleanor Sterling. In reading these accounts, I gained new respect for the kind of person who willingly gives up the creature comforts that most of us take for granted in order to study and help conserve primates in their wild habitats. I think you will too.

INTRODUCTION

WHERE PRIMATES LIVE AND HOW THEY MOVE
HOW PRIMATES ARE CLASSIFIED
THE PRIMATE PATTERN

You are probably reading this book because you have just signed up to take a course on primates. If so, I have good news for you. Odds are, you are going to love the subject and may even end up thinking that this class is one of the best you have ever taken. Primates (including us) are complex, highly intelligent animals that are at times incredibly beautiful. Although this alone makes them worth studying, there are other reasons for devoting an entire course to primates. For one, our fascinating nonhuman primate cousins provide an opportunity to learn more about our basic primate natures. We can also use these studies to speculate about the evolutionary origins of our humanity because we are genetically closer to our primate relatives than to any other creatures on Earth.

Over long periods of time, many primate species have come and gone for a variety of reasons, including natural disasters such as floods, fires, droughts, and storms. Unfortunately, today more than ever before, nonhuman primates are at risk of perishing, not because of natural disasters, but because of the ongoing human activities that encroach upon, or alter, the places in which they live. Most of the world's primates live in a variety of forested regions. These areas are rapidly being cleared for agricultural and commercial uses (e.g., logging or mining), or to make way for urbanization stimulated by pressures from expanding human populations. Forests have also been destroyed by civil war and (in Vietnam) bombing. In other areas, man-made arboricides and insecticides reduce and contaminate the plants and insects that some primates eat. The late renown astronomer, Carl Sagan, spoke eloquently about our planet as a lonely speck in the cosmic dark, and described Earth as a pale blue dot. He (1994:9) also wrote of "our responsibility to deal more kindly with one another, and to preserve and cherish the pale blue dot, the only home we've ever known." We should keep in mind, however, that it's not just our home—other primates (and indeed other animals) live here too.

Figure i-1 The postures that primates typically assume can literally vary by 180°.

Primates are an incredibly varied group of mammals. They range in average size from the 30-g (about an ounce) pygmy mouse lemur to the 160-kg (around 350-lb) male gorilla. Their habitats include deserts, swamps, rain forests, and snowy mountains. The postures that primates typically assume can literally vary by 180° (Fig. i-1). Some mate while suspended upside down from branches, and others can ingest whole meals while hanging upside down by their feet. The human primate, on the other hand (as the reader well knows), spends much of its time standing upright. To get from place to place, different primates may spring; take gigantic leaps; hop on the ground; swing by the arms, legs, and (in some cases) tails; or simply walk along the ground on all four limbs or on two legs. A few primates hibernate during the cool season, while others like yourself may go skiing or curl up with a book in front of a fireplace. Primate diets include different combinations of insects, saps, fruits, leaves, eggs, mosses, honey, nuts, meat, and (occasionally) each other, or even a little dirt. Some primates do not socialize much, while others are definitely party animals. In fact, primate social groups, and associated reproductive behaviors, occur in a bewildering variety of patterns. Care of offspring may be casual or intense, with the relative participation of mothers, fathers, and others varying greatly between different species (and sometimes between groups within a species). As detailed in the following chapters, the day-to-day social interactions of nonhuman primates approach those of humans in their degree of complexity, subtlety, and political machinations.

To illustrate the diversity of living primates, representative species are described in portraits in the chapters that cover a large number of primates (Chapters 3 to 9). Because some primates are much better known than others, some portraits are detailed while others (which are perhaps best described as written "snapshots") are not. These descriptions are not meant to recount all that is known about any particular species, but are intended to provide you with a general feel for each animal. (Readers who would like more information should refer to the references at the end of each chapter, and the more extensive list at the back of the text.) Portraits are not included in the more focused chapters that are devoted to only one genus of ape (Chapters 10 to 13) and early hominids (Chapter 14). (In a sense, these chapters are, in and of themselves, large portraits.) For Chapters 3 to 13, descriptive information for a number of species is summarized in chapter tables and is intended to aid the reader in grasping an overview of each species and making comparisons between them.

Despite the myriad variations described for primates, many species resemble each other in being lively, playful, and curious—in short, traits that reflect a high degree of intelligence. High intelligence is a product of intricately evolved, sophisticated brains. For this reason (and because I study brain evolution), each chapter is accompanied by a brief Neural Note. The 14 Neural Notes are presented in a logical sequence that builds on itself and illustrates important basic concepts regarding evolution of the brain and intelligence in primates, including people.

WHERE PRIMATES LIVE, AND HOW THEY MOVE

Primates are basically tropical animals. Millions of years ago, the climate was warmer and tropical forests existed over larger portions of the Earth's surface than is the case today. In keeping with this, the fossil record indicates that primates were also more widespread than contemporary nonhuman primates, which, with few exceptions, are restricted to a broad belt that encircles the world between the Tropics of Cancer and Capricorn (Fig. i-2). Approximately 90% of today's non-

Figure i-2 Most living nonhuman primates inhabit tropical regions that surround the equator and encircle the Earth.

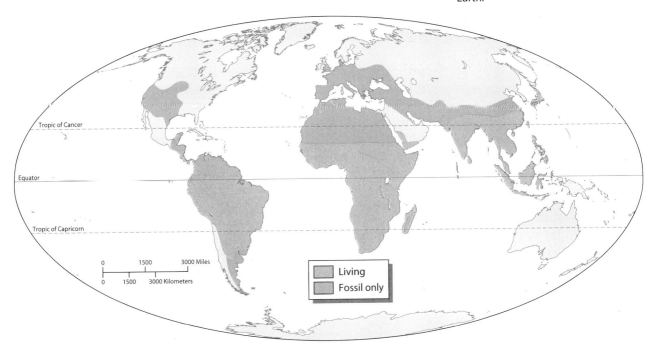

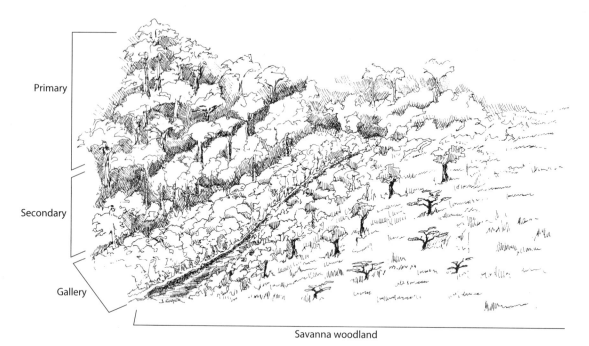

Primary

Secondary

Gallery

Savanna woodland

Figure i-3 Some of the more common primate habitats are pictured here.

human primates continue to live in a variety of forested habitats, including undisturbed or **primary forests**, regenerating or **secondary forests**, and **gallery forests** that are located along water (Fig. i-3). Some of the African monkeys (see Chapters 8 and 9) and apes (see Chapters 12 and 13), on the other hand, have shifted away from forests and have become adapted to grassland plains known as **savannas**.

The specific environment in which a primate lives constitutes its **habitat**. Primates that live in tropical forests time-share their habitats in order to obtain food and other resources. For example, many species are able to live in the same general area by reducing competition through various mechanisms including feeding at different times of the night or day, focusing on different types of food (e.g., insects, leaves, fruit) or different parts of the same plants, or foraging in different levels or stories in the canopies of trees (Fig. i-4). Insects tend to be found in shrubs of the lower level or **understory**, fruits are distributed at all levels, and leaves are especially plentiful in the lower part of the upper story. The lower and middle stories provide a closed (as opposed to open) canopy that is relatively easy to navigate, and it is here that most rain forest primates spend the majority of their time eating, traveling, and sleeping.

Primates that live in different habitats or levels of the canopy tend to differ in their locomotor patterns. The five main types of primate locomotion, outlined by John Fleagle of the State University of New York at Stony Brook, include (a) vertical clinging and leaping, (b) arboreal quadrupedalism, (c) terrestrial quadrupedalism, (d) suspension, and (e) bipedalism. **Vertical clinging and leaping** is seen in many arbo-

Figure i-4 Although nonhuman primates tend to time-share their environments, most rain forest primates spend the majority of their time doing so in the closed canopy of the under and middle stories.

real prosimians that have especially long hind limbs, hands, and feet used for leaping upright between vertical trunks that are found mostly in closed canopies (Fig. i-5a). **Quadrupedalism**, or moving on all four limbs, takes two forms. Arboreal quadrupedalism is the most common type of locomotion across all primates, and is characterized by grasping hands and feet and a long tail for balancing (Fig. i-5b). Quadrupeds that are ground living or **terrestrial**, on the other hand, have shorter fingers, toes, and tails. Their arms are also relatively long compared to the legs, especially in the knuckle-walking African apes (Fig. i-5c). Suspension also comes in a variety of forms. For example, primates such as howler monkeys or orangutans suspend their bodies by both arms and legs as they climb in trees and bridge intervening

Figure i-5 The five basic types of primate locomotion are illustrated by the vertical clinging and leaping of tarsiers (**a**), arboreal quadrupedalism of the Diana monkey (**b**), terrestrial quadrupedalism of the knuckle-walking gorilla (**c**), brachiation of the gibbon (**d**), and occasional bipedalism of the chimpanzee (**e**).

gaps. Another form of suspensory locomotion known as **brachiation** is seen in spider monkeys and gibbons that use arm-over-arm swinging to move through the trees (Fig. i-5d). Finally, although at times a number of primates walk upright on two legs, called **bipedalism** (Fig. i-5e), only humans habitually engage in this form of locomotion. Much more will be said about habitats and locomotion in the following chapters.

How Primates Are Classified

There are over 200 living species of primates in the world today. As discussed in Chapter 1, the fossil record shows that many other primate species lived during the past but are no longer represented; that is, they are **extinct**. Figure i-6 illustrates an arrangement of the living primates that is known as a **taxonomic classification**. This classification is hierarchical; each category includes all of those below it. For example, the genus *Homo* belongs to the family Hominidae, which joins the apes in the superfamily Hominoidea. Similarly, the Hominoidea merges with Old World monkeys (Cercopithecoidea) in the infraorder Catarrhini. The latter joins the New World monkey Platyrrhini to form the suborder Anthropoidea, which is one of the two main divisions of the order Primates. A **genus** is composed of one or more biological **species** whose members, by definition, are capable of interbreeding and producing viable offspring. People belong to the species *sapiens*, which belongs to the genus *Homo*. The Latin name *Homo sapiens*, which literally means "wise man," therefore supplies two pieces of information.

Thus, at each level of Figure i-6, divisions are recognized based on the extent to which groups or taxa resemble each other, and these distinctions become finer as one moves toward the bottom of the chart. The divisions are also based on descent because animals tend to resemble each other to the extent that they are genetically related. The exact composition of the various levels in Figure i-6 depends on the particular morphological, physiological, and genetic features that were compared in different primates. For this reason, the classification shown here is only one of a number of possible ways in which primates can be classified.

The classification used in this text is based on a traditional division of the order Primates into two suborders, the **prosimians** and the **anthropoids**. When most people think of primates, they usually envision an anthropoid or higher primate, in other words, a monkey, ape, or human. The prosimians or premonkeys, on the other hand, include lemurs, lorises (including bushbabies), and tarsiers who look nothing like typical monkeys (or apes). In fact, some of them have the general appearance of rodents, cats, or possibly even teddy bears! Some zoologists have adapted an alternative classification, owing to some taxonomic disagreement (see Chapter 3). Instead of recognizing prosimians and anthropoids as two suborders of primates, this alternative classification is based on one suborder called the **Strepsirhini**, which contains lemurs and lorises, and a second suborder, the **Haplorhini**, which represents tarsiers, monkeys, apes, and humans. Because other features related to the teeth, the skeleton below the neck, and the shape of the nostrils align tarsiers with lemurs and lorises, and because of the implications of recent discoveries of extremely old fossils that may be ancestral to living tarsiers, this text uses the traditional classification shown in Figure i-6.

As should be clear by now, the divisions in Figure i-6 are based on various features that separate different groups of primates. But what do all these groups have in common? In other words, what makes a primate a primate? Unfortunately, there is no simple answer to this question. No one feature sets primates apart from other mammals, and most features that are typical of primates are found to be lacking in at least a few primate species that are exceptions. The best that one can do is to enumerate a pattern of features that together describe the typical primate. Some of these features are found in the skeleton and therefore have the potential for fossilizing (unlike soft parts of the body or behavior). These bony traits are of particular interest to **paleontologists** who study fossils since they can be used to determine whether newly discovered specimens represent primates. Ten of the

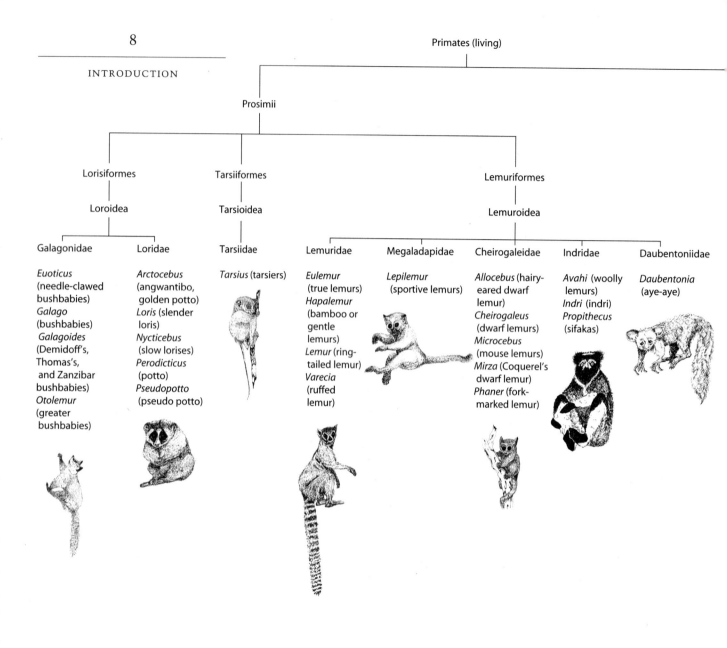

Primates (living)

Prosimii

Lorisiformes

Loroidea

Galagonidae

Euoticus
(needle-clawed
bushbabies)
Galago
(bushbabies)
Galagoides
(Demidoff's,
Thomas's,
and Zanzibar
bushbabies)
Otolemur
(greater
bushbabies)

Loridae

Arctocebus
(angwantibo,
golden potto)
Loris (slender
loris)
Nycticebus
(slow lorises)
Perodicticus
(potto)
Pseudopotto
(pseudo potto)

Tarsiiformes

Tarsioidea

Tarsiidae

Tarsius (tarsiers)

Lemuriformes

Lemuroidea

Lemuridae

Eulemur
(true lemurs)
Hapalemur
(bamboo or
gentle
lemurs)
Lemur (ring-
tailed lemur)
Varecia
(ruffed
lemur)

Megaladapidae

Lepilemur
(sportive lemurs)

Cheirogaleidae

Allocebus (hairy-
eared dwarf
lemur)
Cheirogaleus
(dwarf lemurs)
Microcebus
(mouse lemurs)
Mirza (Coquerel's
dwarf lemur)
Phaner (fork-
marked lemur)

Indridae

Avahi (woolly
lemurs)
Indri (indri)
Propithecus
(sifakas)

Daubentoniidae

Daubentonia
(aye-aye)

Figure i-6   This hierarchical taxonomic classification is only one of many possible ways of classifying primates.

Primates (living)

Anthropoidea

Catarrhini

Platyrrhini

Cercopithecoidea

Hominoidea

Ceboidea

Cercopithecidae

Cebidae

Atelidae

Colobinae

Cercopithecinae

Hylobatidae

Pongidae

Hominidae
Homo
(human)

Callimico
(Goeldi's
monkey)
Callithrix
(marmosets)
Cebuella
(pygmy
marmoset)
Cebus
(capuchin
monkeys)
Leontopithecus
(lion tamarins)
Saguinus
(tamarins)
Saimiri
(squirrel
monkeys)

Alouatta
(howler
monkeys)
Aotus (owl
monkeys)
Ateles
(spider
monkeys)
Brachyteles
(woolly
spider
monkey)
Cacajao
(uakaris)
Callicebus
(titi
monkeys)
Chiropotes
(bearded
sakis)
Lagothrix
(woolly
monkeys)
Pithecia
(sakis)

Colobus
(black-and-
white colobus
monkeys)
Nasalis
(proboscis
monkey)
Presbytis
(leaf-
monkeys)
Procolobus
(red and
olive colobus
monkeys)
Pygathrix
(snub-nosed
monkeys)
Semnopithecus
(Hanuman
langur)
Simias (pig-
tailed monkey)
Trachypithecus
(langurs)

Allenopithecus
(swamp monkey)
Cercocebus
(mangabeys)
Cercopithecus
(guenons)
Erythrocebus
(patas monkey)
Macaca
(macaques)
Mandrillus (drill,
mandrill)
Miopithecus
(talapoin
monkey)
Papio
(baboons)
Theropithecus
(gelada)

Hylobates
(gibbons and
siamang)

Gorilla (gorilla)
Pan (chimpanzees
including
bonobos)
Pongo (orangutan)

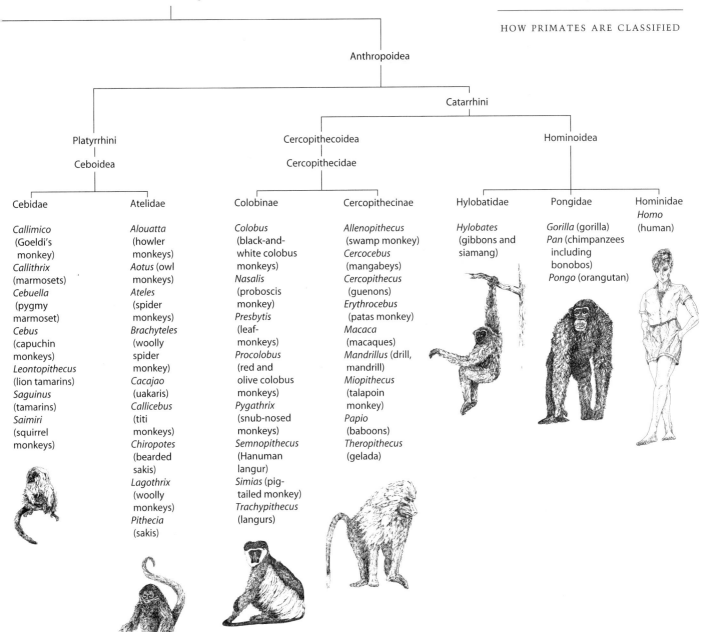

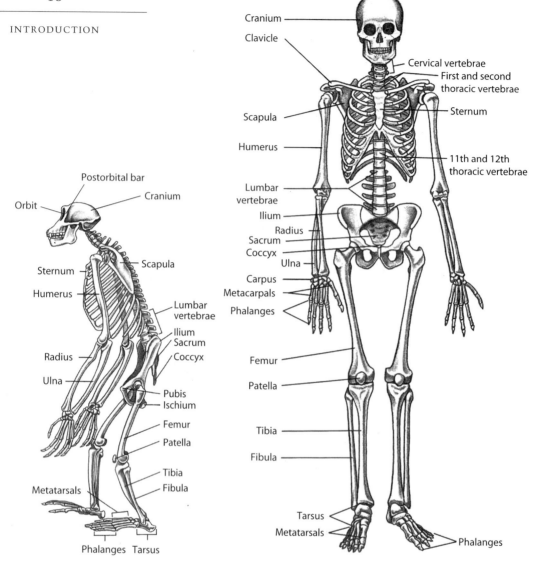

Figure i-7   The skeletons of a chimpanzee and a human illustrate some of the features that form part of the primate pattern.

best known features that constitute part of the primate pattern are listed below (Figs. i-7 and i-8). (Discussion of *why* these features evolved in primates is a subject of Chapter 1.)

THE PRIMATE PATTERN

BONY FEATURES

1. A shortened snout that contains at least three types of teeth

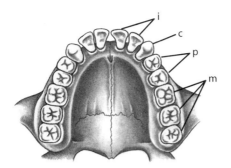

Figure i-8 Primate jaws contain at least three kinds of teeth. This human jaw contains four types of teeth: incisors (**i**), canines (**c**), premolars (**p**), and molars (**m**).

2. Eye sockets (orbits) that face forward and are protected on the side, toward the back, by bone (postorbital bar)
3. Three little bones of the middle ear housed within an outgrowth (petrosal bulla) of the skull, instead of being contained in a separate bone (or cartilage)
4. Collarbones (clavicles)
5. Fingernails and toenails instead of claws
6. Two separate bones in the forearm (radius and ulna) and leg (tibia and fibula)

OTHER FEATURES

7. Grasping feet (except humans) and hands, with mobility of thumbs and big toes as well as other individual digits
8. Tendency toward vertical posture
9. Trend toward longer lives with longer periods of infancy, childhood, and adulthood
10. Enlarged brains with increased areas for seeing and decreased areas for smelling

As we will see, some of these traits are developed more strongly in anthropoids than in prosimians. This is one reason why prosimians are sometimes called lower or primitive primates, while anthropoids are described as higher or advanced. Another reason is because specimens that look like prosimians have been discovered earlier in the fossil record than others that look like anthropoids. As discussed further in Chapter 1, for a variety of reasons paleontologists believe that living monkeys, apes, and humans once had ancestors that generally resembled some living prosimians. Of course, so did the modern prosimians themselves that, although frequently used as models for our ancestors, are just as much with us today as are the monkeys and apes.

Clearly, humans have long been fascinated with our primate cousins. They have also been ambivalent. Some nonhuman primates

1. In what ways are primate habitats being encroached upon by humans?

2. In what parts of the world do primates now live? Was this always true?

3. What are some of the ways in which primates that live in rain forests compete for food?

4. Name and discuss the five basic forms of primate locomotion.

5. You discover two groups of primates that appear somewhat different physically and that live at separate ends of an island. What information do you need to determine whether they belong to the same or different biological species?

6. What is your genus and your species?

7. Explain why the taxonomic classification presented in Figure i-6 is hierarchical.

8. How does a classification that recognizes the suborders of Strepsirhini and Haplorhini differ from that shown in Figure i-6?

9. You discover a nearly complete fossilized skeleton (including the skull) in extremely old deposits in Wyoming, and are hoping it represents a primate. What features do you look for?

have been revered as sacred (Hanuman langurs, crab-eating macaques); others, as profane crop-raiding pests that deserve to be shot, poisoned, trapped in steel snares, or clubbed to death. Nonhuman primates have been killed because they are regarded as evil (aye-ayes), or trapped as pets because they are supposed to bring good luck (rock macaques). Horribly, human beings continue to collect a variety of body parts of nonhuman primates for various purposes—and the list is entirely too long: Eyes of some species are collected for love-charms and medicine; skeletons, skulls, hands, feet, and heads of various primates are sought as trophies and charms; and teeth serve as ornaments. Gallstones of a number of species are used to make medicine and magic, and skins of other species become drumheads or ceremonial wear. Furs and pelts are used to decorate bridles and bows and to make clothing, and (ludicrously) the tails of a number of species are prized as dusters.

As if this were not enough, man's inhumanity to primates goes even further. *H. sapiens* all too often views other primates as resources that are available for human disposal. For example, according to Frances D. Burton of the University of Toronto, some species are literally impressed into labor to remove lice from the heads of humans (pygmy marmosets), pick coconuts (pigtailed macaques), and entertain tourists by passing hats to collect coins (capuchins) or providing photo opportunities (gibbons). In many areas where primates are legally protected, enforcement is insufficient. Consequently, young primates continue to be trapped and exported to foreign countries (including the United States) for the pet trade, for biomedical research, and to stock zoos. Tragically, for every infant captured, a number of **conspecifics** (frequently including their mothers) are reportedly killed. Nonhuman primates are sometimes hunted by humans for sport, and in many countries they are hunted for food.

What is the answer to this terrible situation that threatens the very existence of so many species in our order? Clearly, breeding programs and conservation efforts that are under way (frequently in countries that contain wild primates) should be supported, as should educational programs that are designed to stimulate public awareness about potential threats to nonhuman primates. If they are to work, conservation efforts in developing countries must take the local economic conditions into account, and be of practical benefit to the people who live there. To paraphrase Sagan, we need to take better care of the little blue dot and the forests that provide homes for many wonderful plants and animals. As you will hopefully agree by the end of this book, primates are wonderful, fascinating creatures. It would be a shame if we became the only ones left.

FURTHER READING

BURTON, F. B. (1995) *The Multimedia Guide to the Non-human Primates.* Scarborough, Ontario: Prentice Hall Canada.

FLEAGLE, J. G. (1994) Primate locomotion and posture. In S. Jones, R. Martin and D. Pilbeam (eds) *The Cambridge Encyclopedia of Human Evolution*. Cambridge: Cambridge University Press, pages 75–79.

SAGAN, C. (1994) *Pale Blue Dot: A Vision of the Human Future in Space*. New York: Random House.

REVIEW EXERCISES

10. What additional nonbony features round out the primate pattern?
11. Explain why living prosimians cannot be the ancestors of monkeys and apes.
12. Describe how human treatment of nonhuman primates has impinged on their existence. What, if anything, can be done about it?

1 THE BASICS OF PRIMATE EVOLUTION

INTRODUCTION

The very first primates probably coexisted with the great dinosaurs that roamed the planet long ago during the **Mesozoic era** (Fig. 1-1). The fossil record indicates that primates proliferated during the following **Cenozoic era**, which began about 65 million years ago (mya). However, the first primates to evolve (**basal primates**) did not resemble today's monkeys, apes, or even prosimians. For one thing, they had not yet become adapted to living in trees (**arboreal**) but instead were ground living or terrestrial. Basal primates probably resembled extant mammals known as shrews or hedgehogs; in other words, they looked something like little mice, with pointed snouts and whiskers. According to Richard Kay of Duke University and his colleagues, comparison of a wide range of fossil and living primates suggests that basal primates (for which the fossil record is unknown) probably had relatively large eyes that faced more to the sides than forward. They were also likely to have been timid, antisocial creatures that moved in a scurrying nervous manner. Furthermore, their sense of smell must have been keen, and they seem to have had a taste for insects (a fact that can be gleaned from the fossilized remains of teeth, as shown in Box 1).

The inferred combination of small bodies, large eyes, and a preference for insects indicates that these ancient primates had one characteristic for which we as their descendants should be particularly grateful—they slept during the day and were active at night; that is, they were **nocturnal**. As you are probably aware, dinosaurs, and indeed many other forms of plant and animal life, became extinct rather abruptly at the end of the Mesozoic era. Many scientists believe that this huge die-off was the result of a catastrophic event unleashed by a

16

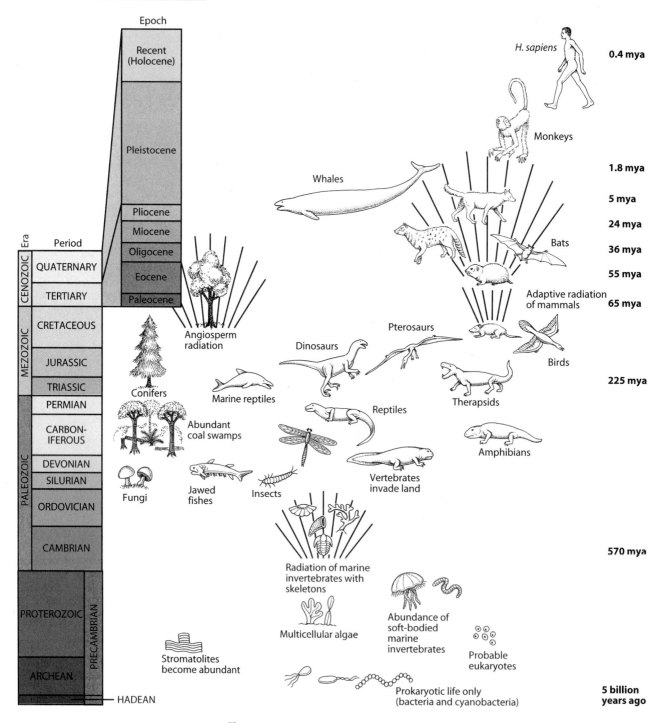

Figure 1-1 This chart of geological time is hierarchical, with eras divided into periods that are further divided into epochs. The approximate beginning in millions of years (mya) is noted for all of the eras and for the epochs during which primates evolved.

Box 1

How Do We Know What Extinct Primates Ate?

As discussed in Chapter 2, understanding the food preferences of the earliest primates provides important clues about their nutritional requirements, social organization, and general way of life. Fortunately, feeding patterns leave their stamp on fossils, and paleontologists are able to make reasonable guesses about food consumption from the sizes and shapes of a species's different teeth, as well as from estimates of its overall body size.

Imagine for a moment that you are an eater of insects. How would you go about it? One way would be to locate and quickly grab an insect. Because your prey is tiny, there is no need to bite into it with your front teeth as you might when eating something like an apple. Instead, you simply pop it into your mouth, crunch it up, and swallow. This form of eating relies heavily on the back teeth, or **premolars** and **molars**. In insectivores, the back teeth are relatively large compared to the front ones and have especially steep, pointed surfaces with sharp crests for shearing and crushing.

On the other hand, the front teeth are particularly important for fruit eaters. Biting into, peeling, or tearing fruit requires relatively large, wide **incisors**. The front teeth just to the outside of the incisors, that is, the upper and lower **canines**, provide sharp cutting edges as they shear past each other when the mouth is closing. The back teeth also differ dramatically from those of insectivores, having low and blunt surfaces that are useful for pulping fruit. In some frugivores, the molars are shaped into two simple transverse ridges.

The molars of leaf eaters resemble those of insectivores because fibrous plants, like insects, require shearing before swallowing. However, as illustrated in Chapter 2, insectivores and folivores can be distinguished on the basis of body size. Although insects are an excellent high-energy food, they are tricky to find and to catch. A big primate would have difficulty finding enough insects to live on. Leaves, on the other hand, are in great enough supply to sustain even the largest apes. For these reasons, living insect eaters tend to be considerably smaller than folivores. Clues from teeth and body size provide evidence for assessing the diets of primates that have long been extinct.

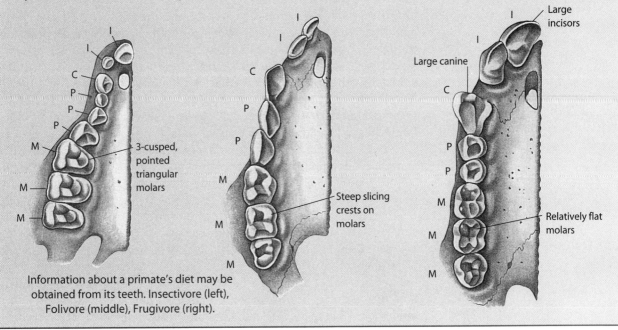

Information about a primate's diet may be obtained from its teeth. Insectivore (left), Folivore (middle), Frugivore (right).

10-km asteroid or comet that slammed into the Earth's surface near the Yucatan Peninsula at the end of the Cretaceous period. According to this "impact theory," bombardment of Earth by one or more large objects caused raging global fires and poisonous rainfall, and shredded the ozone layer. Pulverized debris formed a dust cloud that circled the globe and blocked out sunlight for months if not years. The ensuing dark cold "winter" caused extinction of many plants that depended on ultraviolet radiation, as well as the animals that fed on them. As a result, approximately 75% of all species of life (including all of the dinosaurs) ended at the division between the Cretaceous and Tertiary periods (Fig. 1-1). (This episode in Earth's history is called the **K-T extinction event**, *K* standing for the German word for "Cretaceous," and *T* for "tertiary.") An important trait that probably helped basal primates survive this terrible event was the fact that, as nocturnal animals, they were already adapted to living in the dark.

The K-T event was a major factor that contributed to the subsequent origin of a wide variety of mammalian species during the Cenozoic era, not just those that were primates. Because the event destroyed a good deal of life, resources became available for supporting new species. As a result, descendants of those timid little primates that survived the K-T extinction event flourished along with the descendants of other mammalian survivors. In fact, mammals were so successful during the Cenozoic that they ultimately gave rise not only to the species that live today, but also to many others that became extinct along the way.

"NATURE RED IN TOOTH AND CLAW"

Biologists have long wondered about the exact processes that result in a proliferation of new species over time (called **adaptive radiation**). Although researchers quibble about where and at what pace new species originate, most agree that the theory of **natural selection** that was developed in the mid-1800s by Charles Darwin and Alfred Russel Wallace is extremely useful for understanding the mechanisms that are responsible for the appearance of new species. Both Darwin and Wallace had read a famous essay about poor people who were unable to feed their children, and became fascinated with the consequences of limited food supplies for animals in general. It was this that led them to independently formulate the same theory that is based on the idea of natural selection.

The theory of natural selection is elegant in its simplicity: Species give birth to more offspring than can be supported by available resources. Certain individuals with favorable behaviors, anatomies, or physiological responses are more likely to take advantage of these limited resources and to successfully reproduce relatively large numbers of offspring. For example, potentially favorable traits might include better balance, specialized neural tissue for interpreting sounds, immunity to specific diseases, instinctual defense of territory, special enzymes that aid in the digestion of particular foods, or nurturing parental behavior. Not only do such characteristics permit some individuals to survive for longer and reproduce more than others, but also they are frequently genetically based and therefore may be inherited by offspring. This is how certain features are naturally selected in future populations over many generations, while other features get weeded out. Animal breeders, on the other hand, practice what may be termed *unnatural selection* when they deliberately breed stock for certain characteristics, rather than letting nature take its course.

If the forces of natural selection continue for long-enough periods of time in groups that have become separated by geographical boundaries (often living in very different environments), once-similar populations are likely to differ to such an extent that matings between them would no longer result in viable offspring. At this point, the populations have become distinct biological species. Although adherents of such a **biological species concept** believe that, by definition, interbreeding cannot occur between members of the two species, some primatologists prefer a less-rigid **ecological species concept** that recognizes the role of natural selection in shaping new species in different environments, but without the strict requirement that interbreeding between the two species be physically impossible. What is remarkable about the theory of natural selection is that, at the time of its formulation, the genetic mechanisms whereby traits are passed from parents to offspring were not well understood. In other words, Darwin and Wallace formulated their theory without knowing about genes. Today, of course, we do know about genes and, as described later, this information as well as other kinds of data are used not only to study and document primate evolution, but also to help understand the interactions of living primates with each other and their environments.

The Primate Fossil Record

One of the most important sources of information about primate evolution is the fossil record. Figure 1-2 shows the main groups of living primates on the right and indicates on the left some important primate fossils underneath the names of the epochs during which they lived (compare this figure with Fig. 1-1). Because of limitations in the fossil record, paleontologists frequently resort to educated guesswork when formulating their interpretations. An example is provided by the shadowing in Figure 1-2, which represents the best guess of paleontologist Robert Martin of the University of Zurich-Irchel about the evolutionary tree that connects the fossils with the living primates.

The fossil record is tricky to interpret for a number of reasons. First, it takes a good deal of luck for an animal (or plant) to become a fossil in the first place. It must die under circumstances that allow it to become deposited in the kinds of sediments that foster a process in which organic substances are replaced by minerals (i.e., parts of the

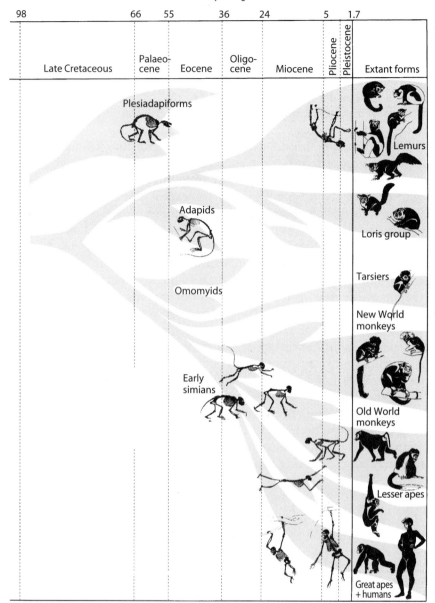

Millions of years ago

98 66 55 36 24 5 1.7

Late Cretaceous | Palaeo-cene | Eocene | Oligo-cene | Miocene | Pliocene | Pleistocene | Extant forms

Plesiadapiforms

Lemurs

Loris group

Adapids

Tarsiers

Omomyids

New World monkeys

Early simians

Old World monkeys

Lesser apes

Great apes + humans

Figure 1-2 Robert Martin's evolutionary tree shows the main groups of living primates and one possible interpretation of their evolutionary relationships.

individual slowly turn to rock). This happens regularly in only certain kinds of environments in specific parts of the world. Furthermore, usually only small hard portions of individuals become fossilized. For this reason, the fossil record of primates consists mostly of teeth and small fragments of bone. (Indeed the rare find of a fossilized skull or partial skeleton is cause for celebration.) Although the skull and teeth are particularly useful for identifying the exact species of a fossil, one really needs to observe the **postcranium** to assess details of body size and limb proportions that are related to how an early primate moved. On the other hand, although a fossilized portion of the pelvis or foot bone might indicate something about locomotion, without associated parts of the skull one might not know what species one is observing! For these reasons, paleontologists must proceed cautiously in their evolutionary interpretations while keeping in mind that some new discovery may overturn even their best guesses.

PALEOCENE PRIMATES (65 TO 55 MYA)

To study primate evolution, it is important to keep in mind the past geographical locations of the continents. During the early part of the Mesozoic era, the Earth contained one supercontinent known as Pangaea. Over time, the Earth's crust broke into separate continents that began drifting with respect to each other. This process continues in modern times, as anyone who has experienced an earthquake can affirm. Because of **continental drift**, the geography of the world during the Paleocene epoch appeared much different from the geography today (Fig. 1-3). For example, India was its own continent and North and South America were not connected. The distribution of primates changed with the drifting of continents as the Cenozoic era progressed, a fact that has important implications for understanding the details of where and under what climatic conditions certain groups of primates originated.

As mentioned earlier, discoveries of previously unknown fossils sometimes stimulate paleontologists to revise their overall view of primate evolution. This happened during the early 1990s in the case of Paleocene fossils from North America and Europe known as **plesiadapiforms**, which were traditionally thought to represent the oldest

Figure 1-3 This map represents the coastlines as they existed 60 mya during the Paleocene.

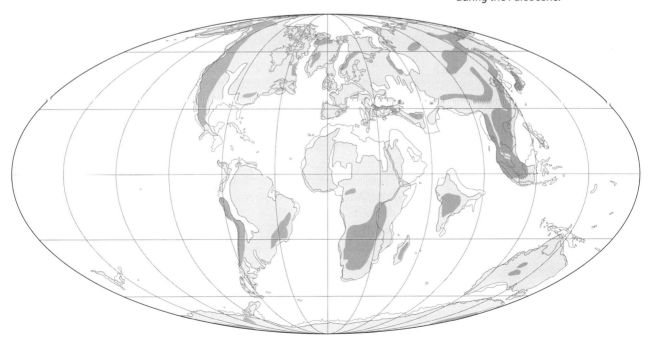

known primates (see Fig. 1-2). These fossils lack typical primate features including a postorbital bar and forward-facing eyes but have other primate-like characteristics such as an auditory bulla. Thus, plesiadapiforms appear to be remains of primitive or archaic primates that were possible links between early mammals and the earliest ancestors of modern prosimians. Consequently, they have been identified as the first-known major radiation of primates, or part of the "first wave" of placental mammals that became arboreal.

Because plesiadapiforms were represented mostly by jaws, teeth, and smashed skulls but precious few postcranial fossils, their lofty status as the so-called first wave of primates was ripe for a crash. That crash may have come in the form of the first intact plesiadapiform skull and some previously unstudied fossils of the finger and wrist bones that were found in Wyoming in 1990. These new finds led some workers to conclude that plesiadapiforms had a number of nonprimate features including a membrane between its hind limbs and forelimbs that could be used to glide when traveling from tree to tree. Today, some paleontologists would boot the plesiadapiforms out of the order Primates altogether; others caution that the evidence for a specific link between plesiadapiforms and nonprimates is open to question.

EOCENE PRIMATES (55 TO 36 MYA)

If the Paleocene plesiadapiforms were not primates, then when and where did the earliest known primates develop? Fossils that might help to answer this question are dated to the Eocene epoch. By this time, North America and Europe had begun to drift apart (Fig. 1-4), and overall gradual warming resulted in generally warm tropical habitats. Fossil primates have been found in Eocene deposits in North America, Europe, Africa, and China.

Two groups of Eocene primates have received particular attention. Fossil evidence regarding the forms of the teeth and the sizes of the body and eyes indicates that **omomyids** were tiny nocturnal primates that ate insects and fruit. They appear to have given rise to modern tarsiers and anthropoids, either independently or through a common ancestor. Until recently, most omomyids were found in North America and Europe. However, in 1994, teeth that are nearly identical to those of modern tarsiers were discovered at a site in southeastern China that may date to 45 mya. *Tarsius* therefore seems to be associated with the oldest known fossil record of any primate. However, more fossil material is needed to determine whether the body and head of the Chinese species (*Tarsius eocaenus*) also resemble those of modern tarsiers. If it turns out that tarsiers that look like modern forms inhabited Chinese forests 45 mya, then *Tarsius* is a dramatic example of a living fossil or **structural ancestor**. (Living primates, of course, cannot actually be fossils or ancestors. It's simply that they are thought to resemble their earlier predecessors.)

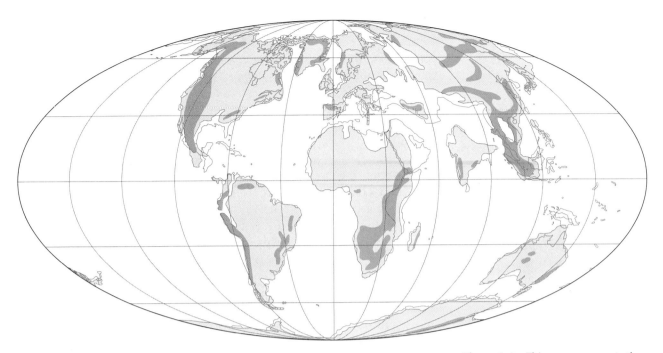

Figure 1-4 This map represents the coastlines as they existed 45 mya during the Eocene.

The second group of well known Eocene fossils are the **adapids**, which were also abundant in North America and Europe. These species were larger than the omomyids, and probably preferred eating fruit and leaves to insects. They may have been **diurnal** (i.e., sleeping at night and active during the day), although the fossil evidence regarding the relative size of the eyes is difficult to interpret in these primates. Some workers believe that adapids gave rise to living lemurs and lorises; others think they were related to these forms but not directly ancestral (i.e., they were distant cousins). Because of their bones and teeth, adapids and omomyids have been recognized as the earliest-known primates of modern aspect (or "second wave") and are believed to be related either directly or indirectly to living prosimians.

The ancestors of the first monkeys (early anthropoids or **simians** that eventually gave rise to living monkeys, apes, and humans) also seem to have evolved by Eocene times. Some of the oldest evidence is again from deposits in southeastern China that may be 45 million years old. *Eosimias sinensis* (Fig. 1-5, color plate) is known only from partial mandibles, but details of its jaw and teeth distinguish it from omomyids, adapids, and living prosimians. Instead, it appears to be a relatively primitive form that may be linked to the origin of early anthropoids. However, the continent on which higher primates first evolved is not clear because another fragmentary species that may be on the anthropoid line (*Algeripithecus minutus*) lived in Africa at about the same time that *Eosimias* lived in China.

Although either an African or an Asian origin seems possible for simians, one thing is for sure. There is a huge element of luck when it comes to finding fossils and it is extremely unlikely that those recov-

REVIEW EXERCISES

1. Who besides Charles Darwin developed the theory of natural selection?
2. Explain some of the reasons why it is difficult to interpret the fossil record.
3. What can postcranial remains reveal that a fossilized skull cannot?
4. Has the geography of the world remained constant over time?
5. How has the "first wave" of Paleocene plesiadapiforms recently been reinterpreted? Why did this happen?
6. Which living primate may have the oldest-known direct fossil record?
7. To date, the earliest recognized fossil record of primates that clearly resemble living prosimians is from which epoch?
8. What are simians, and in which epoch do their earliest-known fossils appear?

ered for any species happen to be the oldest in their lineage. As Martin points out, because fossils represent such a small portion of any given population, the times of origin of different branches of the primate tree are probably underestimated. Martin even goes so far as to suggest that primates of modern aspect may have originated in the Cretaceous period (before the dinosaurs died out).

The Arboreal Theory

The Eocene primates were well on their way to developing the features that comprise the primate pattern. A theory that accounts for *why* this pattern evolved in our ancestors was formulated during the beginning of this century by British anatomists G. Eliot Smith and F. Wood Jones. According to this **arboreal theory**, the primate pattern resulted from natural selection for features that permitted early primates to live in trees. Collarbones, fingernails and toenails, two bones in the forearm, and grasping hands and feet permit stability of the shoulder, increased mobility of the arm, and good grasping and manipulative ability of the hands and feet. These features are useful for moving safely through trees without falling. A shortened snout is associated with a decreased dependence on the use of the mouth as a grasping organ and the nasal region to sense smell. It is also correlated with movement of the eyes from the sides to the front of the face so that the two visual fields overlap. This latter feature facilitates three-dimensional depth perception or **stereoscopic vision**. All of these traits are associated with changes in increasingly large brains that facilitate hand-eye coordination (good for feeding) as well as other neurologically based motor skills. And, of course, primates with bigger, more complex brains need longer developmental stages in which to program these brains. In fact, of the 10 traits chosen to represent the primate pattern (see Introduction), only the petrosal bulla remains unaccounted for by the arboreal theory. If Smith and Jones were right, then humans have retained the primate pattern because our ancestors once lived in trees!

However, the arboreal theory has not gone uncriticized. Matt Cartmill of Duke University emphasizes that nonprimate arboreal mammals such as opossums, tree shrews, and squirrels lack the short face, close-set eyes, and enlarged brains that supposedly characterize arboreal life. If the primate pattern is not primarily related to life in the trees, what is its adaptive significance? Cartmill's answer is the **visual predation hypothesis** (or bug-snatching theory). Based on comparisons with chameleons, marsupials, and cats, Cartmill suggests that the grasping and visual abilities of the earliest primates were selected in conjunction with seeing and catching insects in either the undergrowth or the trees of tropical forests.

Unlike Cartmill, Katharine Milton of the University of California, Berkeley, not only accepts but extends the arboreal theory to speculate more about the early causes of anthropoid brain evolution. She believes that the worldwide spread of forests during the late Cretaceous period paved the way for a shift from an insectivorous diet in the first arboreal primates to reliance on a variety of edible plant parts. Natural selection enhanced the efficiency of foraging for plants by refining the locomotor, grasping, manipulative, cognitive, and visual abilities (including color vision) needed to harvest them. In particular, Milton notes that because quality items such as fruit trees are rare and appear seasonally in scattered patches in tropical forests, big brains with good memories were especially useful for mapping their locations and remembering when they would ripen. Milton's suggestion is particularly interesting in light of the **expensive-tissue hypothesis**, formulated by Leslie Aiello of University College London and Peter Wheeler of Liverpool John Moores University. Because the brain and guts consume much more of the body's energy than do many other tissues (i.e., they are metabolically more expensive to sustain), this hypothesis suggests that a trade-off occurred in which the development of enlarged brains was based on a commensurate reduction in gut size that, in turn, required a shift to a higher-quality diet.

Whatever one's views on the arboreal theory are, two things are certain: (1) Arboreal ancestors of the living monkeys, apes, and humans had already evolved by the Eocene epoch. (2) By this time, some primates had already shifted away from the insectivorous diet of their ancestors and were relying heavily on plant food.

OLIGOCENE PRIMATES (36 TO 24 MYA)

By the end of the Oligocene epoch, cooler drier climates in Europe and North America had contributed to the extinction of local populations of prosimians. Parts of Africa and Asia remained warm and tropical, however, and here diurnal anthropoids flourished, apparently pushing the diminishing numbers of prosimians into nocturnal niches. A landmark event that occurred during the Oligocene was the appearance of the earliest known monkeys in South America. Since the continents had almost reached their present positions by this time (Fig. 1-6), the source of these oldest known fossil New World monkeys is mysterious. Because fossil anthropoids are known from the Oligocene of Africa but not from North America (where there is, however, a record of more primitive primates), some researchers think that the first New World primates may have crossed the Atlantic on rafts of floating vegetation that originated in Africa. An alternative view is that ancestral anthropoids existed in North America despite the lack of a fossil record, and that they haphazardly arrived in South America (which was an island continent at that time) as a result of multiple raftings between various islands on uprooted trees (island hopping). Although both of these hypotheses have their proponents, neither one is very satisfying because it is difficult to imagine primates living for very long, or in sufficient numbers to found new populations, on floating rafts of vegetation. Nevertheless, it is clear that by the Oligocene,

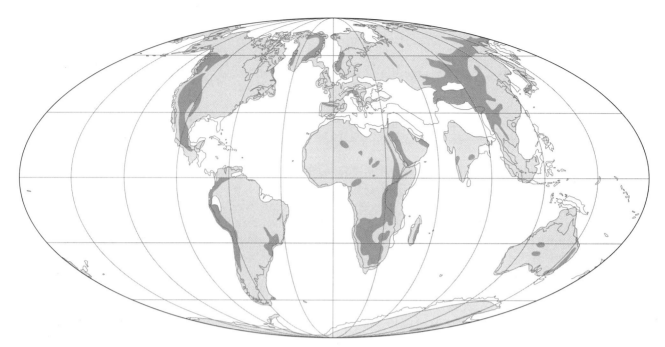

Figure 1-6 This map represents the coastlines as they existed 30 mya during the Oligocene.

Old and New World monkeys had diverged to forever go their separate ways.

It is not known how widespread primates actually were in the Oligocene because most fossils are from one site, the Fayum Depression located to the southwest of Cairo in Egypt. Today these badlands are hot, wind blown, and dusty. But during the Oligocene, the Fayum Depression was a tropical, swampy forest on the shores of an ancient sea that teemed with animal life (Fig. 1-7). Elwyn Simons of Duke

Figure 1-7 Although the Fayum Depression is now a desert, it was a swampy forest during the Oligocene.

How Primate Brains Are Distinguished from Those of Other Mammals

The topic of brain evolution is of crucial importance for understanding what, if anything, makes primates unique among mammals. This subject may be approached in two ways: The first is to compare brains of different living animals with those of primates. Because paleontologists believe that primates may have evolved from mammals that resemble today's insectivores (i.e., moles, hedgehogs, and shrews), brains from these species have traditionally been used in comparative studies. Using this technique, the late Leonard Radinsky summarized several trends that seem to have occurred during primate evolution. Two of the most important trends that Radinsky noted were an increase in the size of the brain compared to body weight (a ratio known as **relative brain size**), and a related increase in the size of the outside part of the brain or **neocortex** (sometimes called the **cerebral cortex**).

Radinsky was also a pioneer in the second approach for studying brain evolution, which focuses on studying casts of the insides of fossilized braincases, called endocranial casts, or **endocasts**. These casts of the interior of the skull show the general size and shape of the brains of extinct (and living) primates, and frequently reveal important details about the surface of the brain including which if any parts are relatively enlarged. For example, fossil endocasts led Radinsky to conclude that the earliest record of modern-sized frontal lobes is from an endocast of *Proconsul* (18 mya), which also appeared advanced in other respects such as having a relatively large brain.

This brings us back to the question of what, if anything, are the unique features of primate brains? Unfortunately, it is next to impossible to identify any features of the brain and endocasts that clearly distinguish primates from all other mammals. Instead, as is the case for the entire primate pattern, an overall pattern of various neurological features (each of which occurs in some other mammals) appears to typify most primates. The typical primate brain includes features that are associated with enhanced motor and visuospatial skills. Interestingly, these neurological features are sometimes more strongly developed on the right or left side of the brain, particularly in the anthropoids. One primate that is distinguished from all other mammals in some ways is *Homo sapiens*, which has the greatest relative brain size of any mammal (about three times as large as would be expected in an ape of similar body weight). An associated feature, of course, is that *H. sapiens* engages in certain unique activities such as symbolic speech.

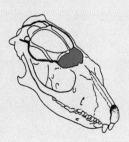 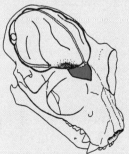 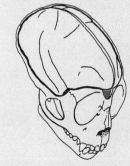

Relative brain size (brain size divided by body size) is three times larger in the galago (center) than the hedgehog (left) and three times larger in the monkey (right) than the galago.

Figure 1-8 The largest Fayum primate was *Aegyptopithecus zeuxis*, which is represented by many fossilized bones including these crania. (Photograph courtesy of Elwyn Simons.)

University has excavated this site since the 1960s, and his efforts have revealed a diversity of arboreal quadrupedal primates with varying amounts of leaping abilities. Details of the skulls and teeth indicate the presence of at least four genera of anthropoids and two of prosimians. However, because the Fayum primates do not exactly resemble modern species, it is not clear whether any of them, in fact, represent an actual ancestor.

A case in point is provided by what may be the best known Fayum primate, *Aegyptopithecus zeuxis*. Represented by several skulls (Fig. 1-8), jaws, teeth, and limb bones, *Aegyptopithecus* was the largest Fayum primate, weighing up to 8 kg (18 lb). Although this short-limbed, heavily muscled, diurnal primate had the same number and general kinds of teeth as modern Old World monkeys and apes, the teeth are primitive and hard to classify as either monkey-like or ape-like. Its small skull appears monkey-like in some respects, and its brain as reproduced on an endocast (see Neural Note 1) reflects features that are intermediate between prosimians and anthropoids. Some workers, such as Simons, consider *Aegyptopithecus* to be ancestral to the Miocene apes that eventually gave rise to living apes and humans (**hominoids**). A recent analysis by Brenda Benefit and Monte McCrossin of Southern Illinois University, however, suggests that the low cranial vault, long snout, and tilted face of this primate may have been present in the common ancestor of all Old World higher primates including monkeys, apes, and humans (**catarrhines**) (Fig. 1-9).

MIOCENE PRIMATES (24 TO 5 MYA)

As monkeys continued to evolve in the New World (see In the Field), by the early Miocene epoch a variety of apes had evolved in the tropi-

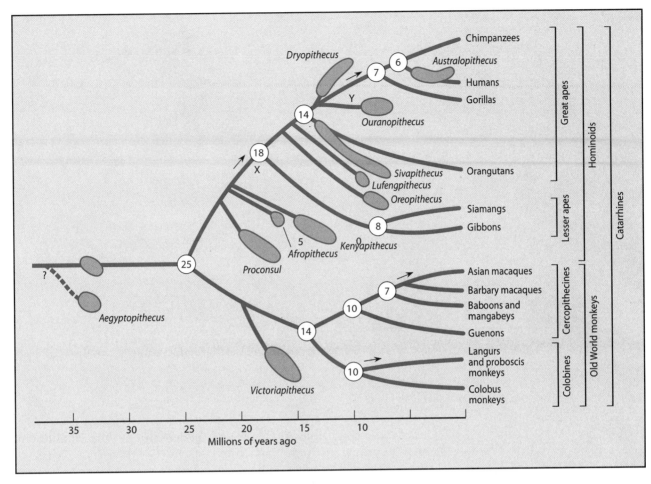

Figure 1-9 This evolutionary tree of fossil and living Old World primates suggests that *Aegyptopithecus* manifests characteristics that were present in the common ancestor of Old World monkeys, apes, and humans.

cal forests and woodlands of eastern Africa (Fig. 1-10). Unlike living bonobos, chimpanzees, gorillas, and orangutans, the earliest recognized apes had many postcranial features in common with quadrupedal monkeys. What initially distinguished ancient apes from monkeys were several dental features that reflect a more frugivorous than folivorous diet, for example, an extra fifth cusp on the lower molars that arranges the cusps into the shape of a Y (known as the **Y-5 pattern**, Fig. 1-11). At the beginning of the Miocene, the apes were much more abundant and varied than the Old World monkeys, probably because of the high availability of fruit. However, during the middle to late Miocene, the climate became cooler and drier as tropical forests were replaced by woodlands and a wide variety of apes spread to Europe and Asia. In response to a changing diet, the molars of apes became larger with thicker enamel for chewing hard gritty plant food.

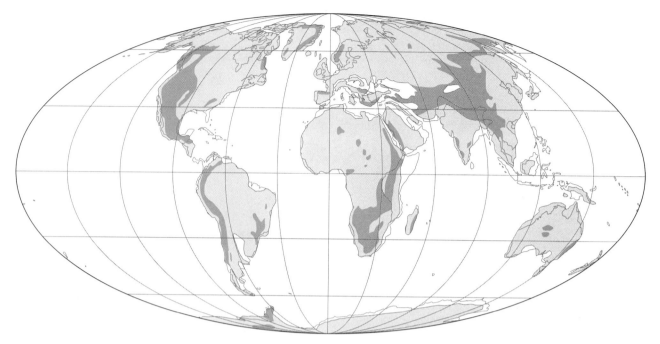

Figure 1-10   This map represents the coastlines as they existed 12 mya during the Miocene.

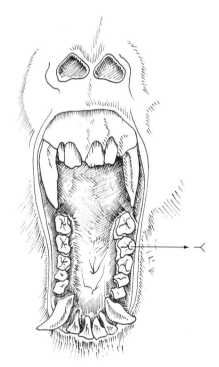

Figure 1-11   In the Y-5 cusp pattern of the lower molars of fossil and living apes, the Y rests on its side, nestled among the five cusps.

Under these conditions, the once prolific apes (more than 30 different types according to John Kappelman of the University of Texas at Austin) declined to a few relic species while the more folivorous monkeys flourished.

The best known Miocene ape is *Proconsul*, which lived about 18 mya in eastern Africa (Fig. 1-12). Through archaeological detective work that spanned 30 years, Alan Walker of Pennsylvania State University and his colleagues have recovered most of its skeleton and have even acquired enough individuals of all ages to chart its growth. *Proconsul* consisted of two or more species that exhibited a unique combination of apelike and monkey-like features: apelike big toe and lack of tail; monkey-like wrists and limb proportions. It was probably a cautious, slow-moving arboreal quadruped. *Proconsul* had a relatively large brain and a special feature of the skull (the frontal sinus) that is found in living African apes and humans. However, it lacks other special features that would either link it with the living apes or exclude it from their ancestry. Some workers therefore believe that *Proconsul* is generalized enough to be an ancestor of apes and humans.

All in all, the Miocene apes came in a huge variety of species, some of which were large bodied and lacked tails. After this spectacular proliferation of apes, however, nearly all of them became extinct by the end of the Miocene, with only one lineage surviving to give rise

Figure 1-12 *Proconsul* is shown here as it might have appeared 18 mya.

REVIEW EXERCISES

1. How does the arboreal theory account for the traits that comprise the primate pattern?
2. Explain how Cartmill's bug-snatching theory differs from the arboreal theory.
3. What is the importance of big brains with good memories for Milton's version of an arboreal theory?
4. Are you more persuaded by Cartmill's or Milton's theory? Is it possible to reconcile the two theories?
5. What is the New World and when did primates first get there? How might they have gotten there?
6. Name and describe the best known Fayum primate. During which epoch did it live?
7. Discuss two ways in which the topic of brain evolution may be studied.
8. What general trends characterized primate brain evolution?
9. When did the earliest-known apes proliferate?
10. What did the earliest-known apes eat, and how was their diet reflected in their molars?
11. Did the bodies of the earliest apes appear more like those of living great apes or living monkeys?
12. Name and describe the best-known Miocene ape.

to living apes and humans. Except for *Proconsul* and *Kenyapithecus* (see Fig. 1-9), most of the fossil apes have special features that are missing in the living apes, which therefore are unlikely to be their direct descendants. Another possible exception is *Sivapithecus* from Turkey and Pakistan, which may have a fairly close evolutionary link with modern orangutans (see Fig. 1-9). Paleoanthropologists continue to debate the identity of the one lineage that preceded living apes and humans and, so far, the fossil record for living chimpanzees, bonobos, and gorillas remains a complete blank (see Fig. 1-9).

PLIO-PLEISTOCENE PRIMATES
(5 MYA TO 10,000 YEARS AGO)

The time span of Plio-Pleistocene primates actually consists of two epochs, the Pliocene, which lasted from 5.0 to 1.8 mya, and the subsequent Pleistocene (or Ice Age), which continued until the beginning of

Good Things Come in Small Packages

When it comes to the fossil record, the "field" often extends into the museum where important finds are brought to be cleaned, pieced together, and interpreted. Alfred Rosenberger, one of the world's leading authorities on New World monkeys, was instrumental in discovering that platyrrhines represent a single adaptive radiation that has continued without the constant replacements that characterized the evolution of Old World monkeys. Below, he describes a fateful day that contributed to this important discovery—a day in which the field came to him in a hotel room! In his own words:

ALFRED ROSENBERGER IS ONE OF THE LEADING EXPERTS ON FOSSIL AND LIVING NEW WORLD MONKEYS.

I had been waiting for my colleagues to arrive from Colombia for more than a week. This hotel room was not the likeliest place for us to hold a miniconvention, but there's no schedule to the process of discovery. Surely, many of the major steps were planned: where and how to go about looking for fossils of particular interest; setting up an approach to study the stuff if it came; devising questions or tests to narrow the range of possible answers to basic questions—What is it? How does it relate to living species? What are its adaptations relative to extinct forms and living relatives? What does it tell us about broader evolutionary matters? But those rare moments of insight can occur just about anywhere and anytime once the stage is set. So there we were arguing back and forth, hoping to divine meaning from a stone, or at least set up a pathway to search for serious clues when we hit the museum the next day.

On the table lay a small, nondescript cardboard box, just the right size to hold a pair of earrings. My friend and colleague Tak casually picked it up and handed it to me. There was no longer any question they had found something interesting. I unwrapped the rubber band, flipped off the box-top, and set the thing back down. Glistening toward me, jewel-like, was a pearly gray, perfect row of mouse-sized teeth. They were set naturally each in its place, seated in an off-white, matte jawbone, altogether about the length of my little finger. Lodged in one corner of the box, almost hidden beneath a protective cotton cushion, was a second piece, a fingernail-sized hunk of weathered bone. I picked up both and turned over the odd one to find a stubby tooth fixed underneath. I looked up to see the four of us beaming, from ear to ear. Two pieces of the same individual: the left side of a lower jaw, all teeth preserved; the matching third upper molar of the same and a chunk of

the Recent epoch (the Holocene) that started 10,000 years ago. By the beginning of the Plio-Pleistocene, global temperatures were falling as increasing amounts of water became locked up in ice sheets (Fig. 1-13). A drier climate caused equatorial forests to shrink, giving way to patchy grasslands known as savannas. There is virtually no fossil record of apes in Africa and a very scant one in Asia during the Plio-

cheekbone. Geology had kept this specimen secret for nearly 15 million years. Tak Setoguchi and his team from the Primate Research Institute of Japan had begun to unlock its mystery by discovering it in the sand. Now, together as a group, it would be our happy task to unravel the puzzle of its meaning and share it with the world.

We soon determined that our fossil was an extinct owl monkey, making this new fossil the partner of another discovered in Argentina, both being the fossils most closely related to the living owl monkeys. This we learned from two lines of anatomical evidence that converged on only one possible explanation. For the teeth and mandible—the canine, premolars, molars, and jaw shape—all resembled the *Aotus* pattern, to be sure. But to establish phylogenetic ties, general similarities fall into the category "Okay, but not enough." What was really important was the scooplike, tall and wide lower incisors of the specimen. They were more similar to modern owl monkeys than anything we had ever seen before in the fossil record. And we knew that for the entire adaptive radiation of the New World monkeys, *Aotus* stands out in the anatomy of its whole upper and lower snout. Looking at it from the mandibular teeth, a scooplike dental pattern is the foundation for that odd owl monkey snout which, until that day, seemed truly unique to *Aotus*.

We named the species *Aotus dindensis*, after the living genus and the place where the fossil was discovered. In looking more broadly at the second line of evidence, outside the teeth, we grew even more confident of our interpretation. We established that the chunk of cheekbone securing that tiny third molar on one surface re-vealed the gentle curvature of an eye socket on its flip side. That smooth concave depression, and its unbroken inner edge, bore a striking resemblance to the lower corner of an *Aotus* orbit, which is overly large to accommodate big, big eyeballs. Large eyes, a nocturnal adaptation, are the most obvious characteristic setting *Aotus* apart from other modern platyrrhine monkeys. Tall, wide incisors are less easy to understand functionally. But here, combined in a single fossilized package, were two *Aotus*-like conditions with no obvious structural interconnection. Why would they co-occur in two different species? The simplest explanation was that our *A. dindensis* was part of the same phylogenetic line.

Three months in the hot desert, a couple of hours hovering over unburied treasure in a hotel room, months poring through specimens in the museum and pictures and measurements in the lab, more months writing up results and waiting for them to be scrutinized by other scientists as a rite of passage toward publication—these were all inescapable steps on the road to discovery. It is the way of modern science, as exemplified in the field of paleontology. Now, more than 10 years since the process of revealing this fossil began, the process continues. We test, challenge, refigure hypotheses, and test again. Our road led to one interpretation; other researchers see the evidence and implications differently. Controversy is a common plot in the story of many fossils because we have so little to go on, and figuring out what happened in the past is a difficult thing to do. But the best ideas survive. Whether they are cooked up in a dinner tent, an airport hotel room, or a silicon microprocessor.

Pleistocene. However, Old World monkeys continue to appear in the fossil record, and a few fossils of New World monkeys have also been found. By the beginning of the Plio-Pleistocene, a new type of primate had evolved out of apelike stock. Unlike apes, however, these first hominids, which would one day give rise to humans, are represented in the Plio-Pleistocene fossil record, possibly because they displaced

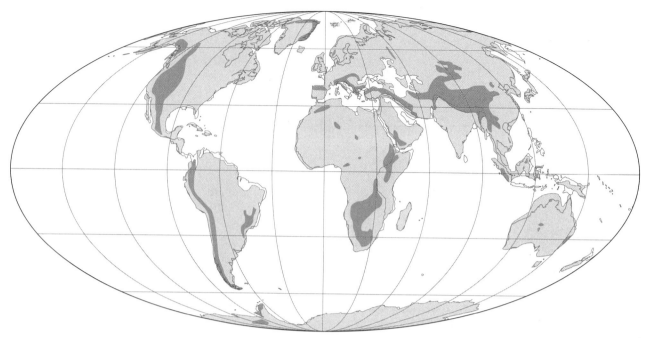

Figure 1-13 This map represents the coastlines as they existed 5 mya during the Plio-Pleistocene.

apes from nonforested regions that are most likely to yield fossils. Because early hominid evolution is best discussed with reference to the living primates, this subject is postponed until Chapter 14.

USING LIVING SPECIES TO STUDY EVOLUTION

So far, our brief survey of primate evolution has been based strictly on fossils, the subject matter of paleontology. Individual specimens are dated by using various scientific techniques that estimate the ages of associated rocks or sediments or, in some cases, by relying on already established dates for contemporaneous species. However, it is often difficult to infer the precise dates when species originated (or died out) from the spotty fossil record. Another problem with paleontology is that interpretation of evolutionary relationships must rely on superficial similarities and differences between specimens, which can be misleading if one is interested in biological speciation. Fortunately, other methods based on comparisons of living species permit researchers to make educated guesses about the evolutionary histories of primates.

Molecular Techniques

Although information about genetic relationships cannot be determined directly from the fossil record (with rare exceptions), biochemical or molecular methods that rely on living species can provide information about their evolutionary histories, or family trees. These techniques are based on comparisons of actual genes (e.g., using DNA sequencing or the less-specific method of DNA hybridization) or on the proteins produced by genes (e.g., by amino acid sequencing or the more-general technique of electrophoresis). As the reader may be aware, these molecular techniques not only are applied to evolutionary studies, but also may be used to ascertain the likelihood of parentage or to gather evidence for criminal investigations.

Molecular biologists base their evolutionary analyses on **molecular clocks**. That is, they assume that genetic changes or **mutations** in certain genes have occurred randomly and at constant rates over long periods of time. Thus, a greater accumulation of differences in their DNA implies that two living species have been separated for a longer period of time than would be the case if they had fewer mutations separating them. Those few cases in which the fossil record provides reasonable information about the actual time of divergence of two groups are used in conjunction with molecular comparisons to determine actual mutation rates for particular genes. These molecular clocks may then be applied to comparative genetic data to determine the divergence patterns and associated dates in a variety of species.

Depending on the particular genetic material or protein that is being compared (for various reasons, **mitochondrial DNA** [mtDNA] found in cell structures called mitochondria is currently the molecule of choice), these methods can be powerful. Indeed, some researchers assert that molecular techniques succeed where the fossil record has not; that is, that they have the potential for revealing how closely related living species are to each other and how long ago they diverged. Others, among them paleontologists, vociferously disagree. One thing, however, is for sure—since taking root in the 1960s, the field of molecular biology has had a major impact on our understanding of primate evolution. This is illustrated by three apparent success stories:

A first example concerns the fossil record of our own family, the hominids. In the 1960s and 1970s, many paleontologists believed that 14-million-year-old jaws and teeth from a fossil formerly called *Ramapithecus* represented the earliest-known hominid. Molecular biologists (from the University of California, Berkeley) thought that this was impossible because their comparative studies indicated that hominids did not even exist as a separate group until approximately 5 mya. After years of squabbling between the two camps, the molecular and fossil evidence finally slid into place with discovery of more fossils (particularly of the face). The new material caused most workers to remove *Ramapithecus* from hominids and instead group it with *Sivapithecus*, an ape that is likely to have evolutionary ties with living orangutans.

Another more recent example that shows the power of combining molecular with fossil evidence concerns the hotly debated identity and geographical origin of the common ancestor of gorillas, chimpanzees, and humans. Caro-Beth Stewart of the University at Albany, State University of New York and Todd Disotell of New York University integrated their study of DNA from living monkeys and apes with evidence from the fossil record. The simplest evolutionary tree that is consistent with their analyses is shown in Figure 1-9. When Stewart and Disotell added geographical information to the pot, they concluded that at least one African ape migrated to Europe or Asia about 19 mya (X in Figure 1-9) where it proceeded to give rise to other apes including the ancestors of gibbons and orangutans. About 10 mya (Y in Figure 1-9), one of these Eurasian apes migrated back to Africa and founded the lineage that eventually led to the African apes and humans. Because it is based on molecular, fossil, and geographical data, the Stewart-Disotell hypothesis is very exciting and worthy of future testing.

Molecular studies are also shedding light on other details of primate evolution. For example, anthropologists have long pondered the exact branching patterns among the gorillas, chimpanzees (including bonobos), and hominids. Although a few workers still contend that the three lineages separated at one point in time, recent mtDNA sequencing studies by Maryellen Ruvolo of Harvard University support an earlier branching date for gorillas (8 to 10 mya) than for the ancestor of chimpanzees and bonobos from hominids (6 mya). If so, of the two African great apes, chimpanzees including bonobos are the most closely related to humans (see Fig. 1-15), a conclusion that is supported by cladistic studies (noted below) as well as other molecular studies.

Other Methods

Another method for studying living primates is based on a concept of grades that should be familiar to anyone who has attended primary school in the United States. A **grade** is a group of species (or children) that appear similar in their general level of organization (or education). In theory, a series of living primates may be used as a comparative model for generally approximating the sequence of evolutionary stages that occurred during the Cenozoic. Figure 1-14 is a classic illustration of this concept. The organization of primates into grades is somewhat flexible. For example, most workers believe that lorises, galagos, and lemurs belong to the same general lemur grade shown in Figure 1-14. Similarly, New and Old World monkeys are usually placed within the same monkey grade.

However, as you will recall, New and Old World monkeys had very long separate evolutionary histories. Thus, although grades are based on similarities in appearance, one must exercise caution and take fossil information into account when using grades to assess the general trends that occurred during primate evolution.

Despite the drawbacks of classifications that are based on grades, many workers continue to compare living primates to determine more accurate evolutionary trees. However, instead of identifying a broad set of generalized similarities (i.e., grades), some investigators rely on the more-limited specialized (**derived**) features to sort species

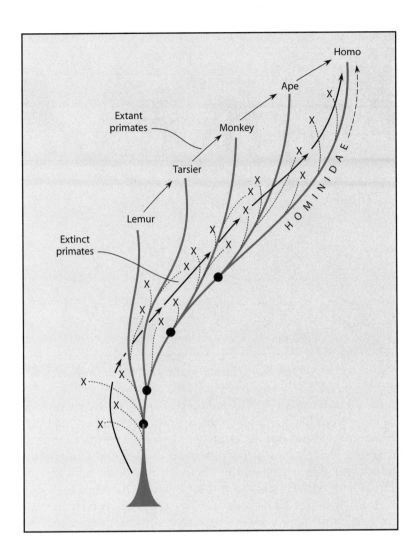

Figure 1-14 This diagram shows a series of living grades which suggests a general trend of evolutionary development that may have characterized primate evolution. The lemur grade contains lemurs, lorises, and galagos; that for monkeys includes New and Old World monkeys. The large dots indicate ancestors at the base of each grade.

into groups or branches known as **clades**. For example, Figure 1-15 is a cladogram that is consistent with a good deal of anatomical and genetic (see below) evidence in living apes and humans. This cladogram suggests that gorillas branched off from the lineage leading to humans before the ancestor of bonobos and chimpanzees did. The cladogram does not contradict the idea that a general ape grade (which includes both gorillas and chimpanzees) preceded the hominid grade during the course of evolution. Instead, it provides more specific details. Thus, both cladists and gradists compare living species to make inferences about their evolutionary histories (although this practice is more explicit with the former than the latter). The difference is that cladists analyze shared special features of living primates (or other organisms) and tend to ignore overall general likenesses.

In this book, the different groups of living primates are covered in an order that proceeds from left to right along the primate classification chart (see Fig. i-6); that is, galagos, lorises, and tarsiers are studied

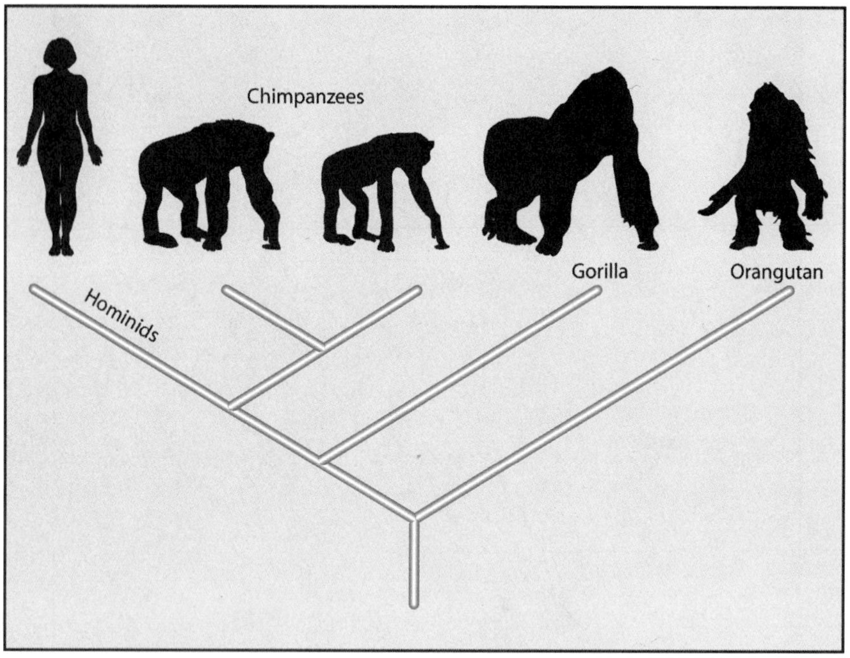

Figure 1-15 This evolutionary tree, which is based on comparative molecular studies, illustrates one view regarding the possible branching pattern among great apes and hominids.

first, then lemurs, New World monkeys, Old World monkeys, apes, and finally hominids. This sequence is the traditional one for surveying primates and does not necessarily reflect the evolutionary branching patterns that led to today's primates. It is very important to remember the point made earlier about grades, because of the prevalent but questionable assumption that primates are more advanced (or less primitive) as one progresses step-by-step along the traditional sequence beginning with prosimians and ending with humans. All of the living primates have evolutionary histories that reach back to the beginnings of the Cenozoic (and farther). Therefore, they have all advanced in time, but each species has evolved its own unique combination of characteristics.

Approaches using grades and clades are useful, especially when complemented with additional information from the fossil record. Although it is incorrect to view prosimians at the left of Figure i-6 as less advanced than anthropoids on the right, this book is organized according to the traditional sequence of grades for two reasons. First, prosimians occur before anthropoids in the primate fossil record; and second, certain features of the primate pattern are more developed in anthropoids than in prosimians. In fact, in the final analysis it appears (perhaps surprisingly) that the sequence of grades shown in Figure i-6 actually does roughly approximate the general unfolding of primate evolution that happened over vast amounts of time. But why should this be true? The answer is that scientists have traditionally organized categories of organisms into hierarchies that reflect increasing complexity, although often without any conscious notion of evolution. Thus, an assumption of general development is implicit as one moves across the primate grades, just as one thinks of a sixth-grade child as being more educated than a fourth grader. (However, one must keep in mind that although sixth- and fourth-grade children have attained different levels of education, it would be a mistake to assume that all sixth-grade children are more intelligent than fourth graders, or to assume that every adult attended each grade of school.)

Because classifications like that in Figure i-6 appear to approximate (if only roughly) the general sequence of events that occurred over an unimaginably long duration of primate evolution, certain living species may be used as models for various stages of ancestral development. In general, these models (called structural ancestors, **living fossils**, or **living links**) have few of the specializations that characterize many living species and therefore appear to have changed very little during their evolution. That is, structural ancestors seem to be relatively generalized or **primitive** compared to many of their living contemporaries. They are not really ancestors, but they provide a good guess as to how the early ancestors of primates may have looked and even behaved.

The Tree Shrew: Structural Ancestor of All Primates?

Take, for instance, the tree shrew, which comes in a variety of species that live in forests in South and Southeast Asia (the best known genus is *Tupaia*). This little animal is so peculiar that different zoologists have classified it variously with primates, insectivores (moles, hedgehogs, and shrews), rabbits, flying lemurs, and bats, and more recently have resorted to placing it in its own order of mammals (called Scandentia). Whatever its correct classification, many workers agree that the tree shrew provides a good structural ancestor for primates because it has retained a large number of features thought to have been present in the earliest basal primates. Some species inhabit shrubs and low bushes, while others scurry nervously along the ground or move back and forth between terrestrial and arboreal habitats. Superficially, tree shrews appear to be small squirrel-like animals, with long bodies and tails, slender wet muzzles, eyes on the sides of the head, and digits that have limited grasping ability and claws instead of nails (Fig. 1-16). They eat fruit and insects, and are rather antisocial. For example, absentee mothers have multiple newborns that they leave in nests for up to 2 days before returning to them to nurse. If placed in a cage, males are extremely aggressive and will even fight to the death. (As the late anthropologist Carleton Coon once quipped, tree shrews exhibit uninhibited human behavior because they are gluttonous, wildly irate, and libidinous.)

Despite the nonprimate appearance and antisocial behavior of tree shrews, certain features led earlier workers to classify them with primates. These include a primate-like middle ear, ring of bone around the eyes, visual area of the brain, and partial mobility of the big toe

Figure 1-16 Although the tree shrew is no longer viewed as a primate, morphologically it is believed to resemble the common ancestor of all primates. It may thus be termed a structural ancestor.

1. What epochs make up the Plio-Pleistocene, and when did it begin and end?
2. What new type of primate evolved out of ape-like stock during the Plio-Pleistocene?
3. What structures do molecular biologists compare in different living species and how do molecular clocks work?
4. Do paleontologists and molecular biologists always agree? Give some examples to illustrate your answer.
5. Recent mtDNA sequencing studies suggest that, of the two African great apes, the ——————— is most closely related to humans.
6. Explain the concept of grades. Do grades of living primates correspond exactly to evolutionary branching patterns? Why or why not?
7. Contrast grades with clades. What is a cladogram?
8. What is the significance of derived features, and which methodological approach relies exclusively on them?
9. What methodology do gradists and cladists have in common?

and thumb. As you will recall, the first primates that lived before the beginning of the Cenozoic probably resembled tree shrews anatomically and socially and, like them, were not yet committed to an arboreal way of life. This is why workers who disagree on how to classify the tree shrew concur that it makes an excellent structural ancestor for primates. It should be noted, however, that all but one species of tree shrew are diurnal and, in this fundamental respect, they do not appear to resemble the ancestors of basal primates, which were probably nocturnal. This illustrates the important fact that, like other species, structural ancestors experienced at least some change during their own long evolutions. An additional important method for studying living primates—the field study—will be one of the subjects of the next chapter.

SUMMARY

As we have seen, the study of primate evolution is fraught with certain problems. For one thing, the fossil record only samples a small portion of the past. Because primate fossils include mostly jaws and teeth, their interpretations are subject to radical change when new material is discovered. This happened recently with the plesiadapids (now thought by some to be related to certain gliding mammals rather than primates), and the former *Ramapithecus* (which has now been grouped with *Sivapithecus* and may be a close relative to the ancestor of orangutans rather than humans). Interestingly, in both instances predictions made from molecular rather than fossil studies turned out to be right. Postcranial fossils are also important for studying evolution because they can reveal much about a species's general body size and form of locomotion (e.g., whether it was arboreal, bipedal, etc.). However, interpreting postcrania is fraught with its own problems. For example, without an associated skull, it is often difficult to know the exact species represented by postcranial remains. Another problem with primate paleontology is that the fossil history of the primates closest to humans (chimpanzees, bonobos, and gorillas) is unknown. Finally, small samples of fossils result in incorrect estimates about the times of divergence of major lineages. When this situation is compensated for with statistical techniques, numerous branches of the primate family tree appear to be older than indicated by previous estimates. Although the field of molecular biology remains controversial, it is gaining acceptance as a useful approach for estimating the branching patterns and dates of the evolutionary tree that ultimately gave rise to the living primates. Another avenue for studying primate evolution entails comparing living species to speculate about general evolutionary trends (via grades) or to construct hypothetical evolutionary trees (based on clades). Although the concepts of grades, clades, and structural ancestors are useful for studying primate evolution, taxa are organized in this book in an order that is consistent with the concept of grades. A portrait for the tree shrew

(*Tupaia*) illustrates a good candidate for the structural ancestor of all primates despite the fact that zoologists continue to debate the details regarding its correct classification. The merging of paleontology and molecular biology has had a healthy impact on our present understanding (which will surely change) of primate evolution.

FURTHER READING

CONROY, G. C. (1990) *Primate Evolution.* New York: W. W. Norton.

GIBBONS, A. AND E. CULOTTA (1997) Miocene primates go ape. *Science* **276**:355–356.

KAY, R. F., ROSS, C. AND B. A. WILLIAMS (1997) Anthropoid origins. *Science* **275**:797–804.

WEINER, J. (1994) *The Beak of the Finch.* New York: A. A. Knopf.

REVIEW EXERCISES

10. Define and describe structural ancestors. Name a good candidate for the structural ancestor of primates.
11. What primate features are exhibited by tree shrews? What nonprimate traits do they have?
12. Are tree shrews securely assigned to primates?

2

METHODS AND THEORY FOR STUDYING LIVING PRIMATES

INTRODUCTION

Primatology has undergone a revolution during the past 25 years. Today, both the questions that primatologists ask and the analytical framework used to interpret data have shifted away from the more-or-less purely descriptive methodology that characterized the early field studies of the 1960s. Instead, the new primatology is grounded in analyses that explore the relationship between social behavior, on the one hand, and the genetic contributions of individuals and groups to future generations (or **reproductive fitness**), on the other. For primates, as with other organisms, the name of the evolutionary game is to reproduce, and natural selection favors any behavior that allows an individual or group to outreproduce its rivals. Therefore, these behaviors are viewed as mechanisms or **reproductive strategies** that allow the genes of particular individuals (or copies of those genes in their relatives) to be perpetuated into the future. Not surprisingly, such strategies may entail relationships with mates, offspring, other kin, or even nonrelatives. For example, the various ways in which primates select mates is of interest, since competition for and success at acquiring mates influence the number and quality of an individual's offspring. Once offspring are produced, there are various styles of parental nurturing, or **investment** in the young, that differ between males and females, and these too affect not only whether or not particular offspring survive, but also the amount of time and energy each parent has to devote to other reproductive efforts. Individuals may also influence their reproductive fitness by more subtle means such as supporting close relatives that carry copies of some of their genes (**kin selection**), or establishing reciprocal relationships with other unrelated individuals (e.g., by grooming or protecting them) that ultimately contribute to the general well-being and reproductive potential of both participants.

Because primatologists are keenly aware that reproductive strategies occur within environmental contexts rather than in isolation, they study **ecology**, which focuses on the relationship between species and their particular environments. The subfield of **socioecology** is concerned less with interactions between individuals, and more with identifying correlations between wide social behaviors (including grouping patterns) and certain environmental factors. Socioecologists therefore explore how factors such as group size and composition, population density, ranging patterns, territoriality, preferred foods, and predators limit species and their interactions with others. This is not to say that socioecologists ignore reproductive fitness. Far from it! For instance, a number of workers are currently engaged in documenting the various ways that males and females of the same species (i.e., **conspecifics**), or adults of different species, harvest energy (food) from the environment and convert it into a variety of activities related to reproduction (e.g., competition for mates, bearing and nursing young). Some of the most important methods used by these researchers are described in the following section.

FIELD METHODS

Imagine yourself as a primatologist who is about to go to some far-off tropical place to study a particular species of primate. Before you get on the plane, you should know what research has already been done on your subject or focal species so that you do not reinvent the wheel, and your project should be clear. What questions are you asking or what gap(s) in the current state of knowledge do you hope to fill? How will you go about it? What problems will you face? How will you be sure that your observations are typical for your species rather than due to mere chance? Fortunately, the steps for proceeding are well established, straightforward, and logical.

Upon arriving in the field, your first task is to find your subjects and to **habituate** them to your presence. Depending on how shy the particular primates are, this may take a good deal of time and patience. With persistence, however, your subjects should cease to be alarmed by your presence and (ideally) should begin to go about their business as if you were not even there. Now your real work begins. The next step is to learn to recognize each individual primate by sight. Scars, birthmarks, battle wounds, distinctive coloration, hairdos, and ways of moving may be useful. Many primatologists name their subjects, and these names often reflect some distinctive aspect of appearance. Learning to identify all of the individual primates is likely to be one of your most difficult and frustrating accomplishments in the field.

Once you know who is who, it is time to begin making and recording observations in a predetermined and systematic manner. The idea is to sample the behavior of the species in as representative and unbiased a way as possible. The most widely used method is **focal animal sampling**, described by Jeanne Altmann of the University of Chicago in 1974. With this method, one or more animals that have been targeted for study (e.g., adult females) are observed in a predetermined

Figure 2-1 With notepad, watch, and binoculars, primatologist Cheryl Knott and her assistant Tadyn record observations of orangutans in Borneo.

order for a given length of time. With clipboard, stopwatch, and binoculars in hand, the primatologist (Fig. 2-1) locates and observes the activities of the first focal animal on the list, and records them either in a diary format or on a checksheet for a predetermined amount of time. When that period is over, the primatologist then moves on to the next individual and repeats the whole procedure. (Recently, some workers have foregone checklists in favor of entering data directly into laptop computers.) The focal sampling approach provides unbiased data describing the rate as well as pattern of social interaction for each animal. The recorded behaviors are chosen to reflect the particular interests of the observer and the questions to be addressed. These behaviors might include such activities as feeding, travel, rest, grooming, and play. Questions that might be asked include whether males of a particular species fight more than females or whether animals groom close relatives more than distant relatives or nonrelatives. To avoid results that are biased because of the time of day or year, the entire procedure should be repeated at different predetermined times of the day and (ideally) during different seasons.

Focal animal sampling is good for studying rates and patterns of interactions of a relatively small number of animals. However, if the investigator wishes to learn about the behavior of the whole group, another method that increases the number of animals observed must be used, and consequently the length of time that can be spent on any individual decreases. One such technique is called **instantaneous** or **scan sampling**. Here, the observer records the current behavior of every animal in the group (or subset of the animals in the group) as rapidly as possible. The objective is not to describe sequences of behavior but rather to document the activity state of each animal. These samples are generally performed at frequent, preselected intervals (such as once per minute for small groups and once per 3, 5, or 10 minutes for larger groups). As with focal sampling, care must be taken to avoid biases due to time of day and season of the year. When done correctly, scan sampling effectively provides a snapshot of what the group is doing at any given time of day, and allows calculation of the percent of time animals spend in different activities.

The methods just described are widely used for observing primates in their natural habitats. However, they are ideals and as Dorothy Cheney of the University of Pennsylvania and her colleagues (1987:6) note, "All primatologists are familiar with the mysterious process that causes subjects to disappear from sight shortly after a focal sample on them has begun"! What to do then? The observer tries a third method of **ad libitum (or ad-lib) sampling**. In other words, the primatologist wings it by recording whatever animals or behaviors are of interest at the time. Although ad-lib sampling is not representative sampling in the statistical sense, it is nevertheless very important for recording behaviors that are rarely observed in some species such as copulations or tool use. Moreover, ad-lib samples can provide useful

---

So You Want to Be a Primatologist. . .

Jane Phillips-Conroy is a well-known primatologist from Washington University in St. Louis. Every year she conducts field research in Africa or South America, where she continues her long-term biomedical and behavioral research on a variety of Old and New World monkeys. When asked what advice she gives to students who want to become primatologists, Jane replied that there is a special romance about primatology, in particular the African great apes, and many undergraduates hope to become future Jane Goodalls or Dian Fosseys. If not that, they see themselves as becoming university professors or professionals in conservation-oriented organizations such as the World Wildlife Fund. This is what Jane tells her students:

JANE PHILLIPS-CONROY IS SHOWN HERE IN THE PLACE SHE MOST LIKES TO BE—IN THE FIELD! (PHOTO COURTESY OF JANE PHILLIPS-CONROY.)

"In all of these occupations, a Ph.D. is needed. Undergraduates who want experience preparatory to graduate school often volunteer for lower-level technical positions in animal care facilities, rehabilitation centers, or as field assistants (usually self-financed) on field projects. Once they get to graduate school, we find that primatology students frequently take 7 years plus to get their Ph.D. when it involves fieldwork: 2 years of coursework, then another 2 years for grant writing and getting into the field, then 1½ years in the field, and an equal amount of time writing and finishing the degree when they get back, if they're lucky. Graduate programs are cutting back on support and many students now incur debt against an uncertain employment future. For this reason, I recommend that, along the course of their education, students learn skills in at least one allied area such as gross anatomy, statistics, or computer consulting, so that they will be more likely to market themselves. Other useful techniques can be learned in courses and practical experience in genetics, botany, osteology, and forensics.

information on dominance relationships among animals, as these are based on which animals win and lose fights. As primatologist Glen Hausfater has shown in his research on baboons, such fights are unlikely to be influenced by the sampling procedure used to record the events.

In addition to systematic behavioral observations, one can collect primate fecal samples for laboratory analysis (related to diet, seed dispersal, etc.), or record information about the available food sources (including plants) and other animals in the area. Primate vocalizations may also be sampled with a tape recorder for future laboratory analysis. Becoming a primatologist skilled in collecting scientific data is not an easy task (see In the Field). However, if carefully collected and computerized, an investigator's data can provide a gold mine of information that can be analyzed statistically to approach new questions for years to come.

So it's a long haul, and students need to know ahead of time whether or not they will be able to hack it. I always recommend they get into the field before actually embarking on their long dissertation project. A *pilot* study is really important because students can figure out whether or not it's for them before they've invested so much time, effort, and money. It also helps them perfect their data-gathering skills. It's important that students learn something about the difficulties of working with primates: Whether working with captive or free-ranging primates, there are potential health risks. For example, in the hands-on kind of stuff that I do, there is risk of exposure to pathogens that affect the animals. I've certainly contracted stuff from them like amoebas. We're now more cautious than ever when handling vervets, which carry diseases that can cause death in humans. Although I wore gloves and masks, I got a strange malady last summer after working with vervets and, I confess, I briefly thought I must be insane to do this! Even if you don't handle animals, there are risks of contracting malaria or other diseases, not to mention that fieldwork often takes place in politically conflicted countries that may be unstable and prone to violence. As foreigners, lining up the permits that are required to do research can be tricky and time-consuming. So fieldwork is not for the faint-of-heart or personally fearful. Students need to be motivated, flexible, and tolerant. Previous experience might offer a clue here. Do they like camping, hiking, and outdoor activities? Can they be adaptable in terms of cuisine? (Usually one eats a vegetarian diet, especially if there is no refrigeration.)

And then, after all this, students get out of the field, write their dissertations, and *there are few or no jobs*. This prospect is something they should anticipate and be prepared for, rather than be ambushed by it at this point. They must be flexible in the careers they will consider. They may end up as researchers in primate centers, or working on environmental enrichment for animals in research centers or zoos. Or they may become established researchers (almost always supported by grants) working on biomedical projects that employ nonhuman primates as models for human disease. These careers can be exciting, productive, and employ the skills they learned, but now in a new context. Few of them will become Jane Goodalls or Dian Fosseys. I tell my students that the bottom line is that primatology is a hard course that is long, physically demanding, and has no great job prospects. However, *if they love it after they have tried it and want it more than anything—no doubt about it, they should do it!"*

Figure 2-2 Nonhuman primates such as this lemur (a sifaka) from Madagascar are not conscious of their reproductive strategies. This mother carries her infant like a belt around her waist.

The theory that primatologists use to shed light on the mating, parenting, and other social behaviors of their subjects, in and of itself, could provide the subject matter for an entire book. However, only the concepts that are the most relevant for the present text (i.e., that apply to the subsequent discussions about various species) are sketched in this chapter. As already noted, a key premise is that individuals behave as if they are motivated to perpetuate their genes or copies of them that exist in close relatives. Thus, primatologists may think of individuals as gene machines, but primates themselves do not (Fig. 2-2). These points need to be emphasized because some reports lend themselves to the interpretation of greater conscious strategizing on the part of primates than can possibly exist. For the most part, it is safe to assume that primates are just doing what comes naturally. As you will see, because of the way evolution works, such an assumption in no way compromises the theories that clarify contemporary primatology. Because these theoretical constructs have a good deal of explanatory power, they are useful for formulating new questions and, in this sense, continue to guide the maturation of evolutionary biology, including primatology.

Males and females, of necessity, have different reproductive strategies because of the simple fact that only females conceive and grow offspring within their bodies (a process known as **gestation**), and then give birth to and nurse them. Minimally, males need only to provide sperm to reproduce (although some male primates devote a good deal of time to parenting). Females, on the other hand, must provide an enormous amount of physiological energy for gestation, nursing or **lactation**, and parenting (Fig. 2-3). Thus, the cost of reproduction, or parental investment, is much higher for females than males. For the most part, primates seem to follow a general rule that applies to many

Figure 2-3 Female primates such as this Barbary macaque with twins provide an enormous amount of physiological energy toward gestation, lactation, and parenting. (Photograph courtesy of Meredith Small.)

animals, formulated by Robert Trivers of Rutgers University: Members of the sex that invests less heavily in offspring compete for access to the more heavily investing sex. Thus, male primates usually compete aggressively with each other for sexual access to female primates, rather than the other way around.

PRIMATE SOCIAL SYSTEMS

Because males compete for access to females, the distribution of males and the types of social groups that primates form are constrained by the density and distribution of females. In other words, males go where the females are. Thus, groups with four or less adult females are likely to have only one adult male, whereas groups with more than four females usually have more than one male. Primatologists use some version of the following scheme to categorize typical social groups, known as **residence patterns**, that are based on ratios of adult males to adult females (Fig. 2-4). Not surprisingly, residence patterns are associated with particular mating systems. You should keep in mind, however, that these are ideals in the minds of primatologists; that is, species may utilize more than one pattern, or merge aspects of different patterns, depending on their ecological circumstances.

1. *One-male, multifemale groups.* Also called unimale groups (or sometimes harems), these groups consist of one breeding male, and several adult females and their young. Males thus have more than one sexual partner; they are **polygynous**. This type of social organization typifies gelada and hamadryas baboons, some howler monkeys, and some langurs.

2. *One-female, multimale groups.* Females that mate with multiple nonpolygynous males are **polyandrous**. In groups where one adult female mates with more than one male, males usually cooperate in parenting her offspring. Although fairly rare, such **cooperative polyandry** occurs among some New World monkeys.

3. *Multimale, multifemale groups.* These groups contain numerous adults of both sexes and their young. Both sexes mate promiscuously; that is, the mating pattern in these groups is **polygamous**. Depending on the species, one sex or the other generally migrates to another group before breeding, which minimizes harmful genetic defects due to inbreeding. This pattern is found in many Old World monkeys, some New World monkeys, and chimpanzees.

4. *All-male groups.* In some species, males reside in all-male groups before joining or forming a mixed-sex group in which to

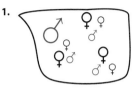

One-male, multifemale

One-female, multimale

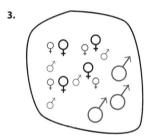

Multimale, multifemale

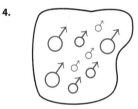

All-male

One-male, one-female

One-adult

Figure 2-4 Primate species use one or more of these six residence patterns, which are associated with particular mating systems. Smaller symbols represent immature group members.

REVIEW EXERCISES

1. Why aren't primatological studies as purely descriptive as they were in the past?
2. What is reproductive fitness?
3. Discuss reproductive strategies. Are nonhuman primates conscious of them?
4. What does socioecology add to studies of reproductive fitness and strategies?
5. Why do males and females have different reproductive strategies?
6. State the general rule formulated by Robert Trivers regarding parental investment and competition for mates.

Figure 2-5 Male-male competition for females results in relatively large bodies and canines, as seen in this male hamadryas baboon from Ethiopia.

Figure 2-6 The degree of sexual dimorphism is greater in species that form one-male, multifemale groups than in those that form multimale, multifemale groups, or monogamous groups.

breed. For most males, residence in an all-male group is temporary, but some individuals may return if they are driven out of a bisexual group. These groups commonly coexist with multimale, multifemale groups (e.g., savanna baboons) and with one-male, multifemale groups (e.g., some of the Hanuman langurs).

5. *One-male, one-female groups.* One adult male and female form a pair-bond and live with their immature offspring in a territory that they defend. Mating is ideally exclusive between the partners or **monogamous**. All gibbons and a few species of prosimians and New World monkeys live in these groups.

6. *One-adult systems (solitary).* This social organization consists of an adult female and her immature offspring, on the one hand, and lone males, on the other. Males and females usually interact only for purposes of mating, and each sex may have multiple partners. Some prosimians and orangutans are solitary.

Male Strategies

Sexual selection is a form of natural selection that focuses on matings. One type, **intrasexual selection**, involves competition between members of the same sex for mates and, for reasons discussed already, the most common form of this particular reproductive strategy entails direct male-male combat for access to females. The ultimate result is selection for males with relatively large bodies and canines compared to females (Fig. 2-5), that is, a high degree of **sexual dimorphism**. As a rule, male competition and sexual dimorphism are most fierce in social groups that have a surplus of females, such as one-male, multifemale groups like those that characterize gorillas, some baboons, and howler monkeys (Fig. 2-6). (This rule, again, has interesting exceptions that will be discussed later, such as the orangutan [see Chapter 11]).

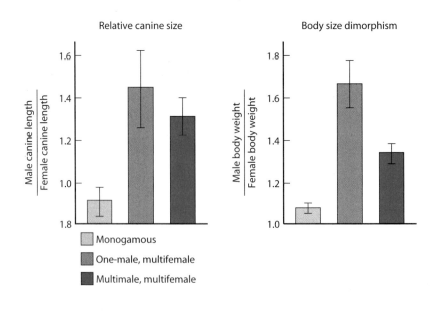

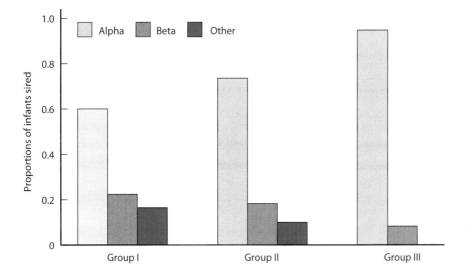

Figure 2-7 In three groups of long-tailed macaques, the highest-ranking (alpha) male fathered a larger proportion of infants than did the second-ranking (beta) and lower-ranking males.

Another reproductive strategy, **infanticide** or the killing of infants, is sometimes used by outside males that periodically drive adult males from their one-male groups, which they then take over. At first glance, the male proclivity for killing infants in newly acquired one-male, multifemale groups seems odd or even bizarre. However, as elaborated by Sarah Blaffer Hrdy of the University of California, Davis, the removal of nursing infants causes their mothers to resume ovulation and become sexually receptive, thus increasing the new resident's mating opportunities at the expense of his predecessor's reproductive fitness. (Infanticide is rarely engaged in by female primates.) This form of intrasexual selection is discussed in more detail in Chapter 7.

Some degree of sexual dimorphism also characterizes multimale, multifemale groups to a greater (e.g., some baboons) or lesser (e.g., muriquis) extent, and a certain amount of direct competition for females occurs between males in these species. For example, they may chase, threaten, bite, fight with, or generally hassle each other (e.g., during attempted copulations). Such interactions are sometimes kept in check by an established pecking order, or linear **dominance hierarchy** that reflects male fighting ability, the general ability to gain access to resources, and, at least in some documented cases, reproductive fitness (Fig. 2-7). **Dominance** is simply the ability of one individual to intimidate or defeat another in aggressive or competitive interactions. It should be noted, however, that dominance behavior is not always linear; pairs of individuals sometimes form friendly **alliances** that allow them to prevail over a third individual that is dominant to both under other circumstances.

Although their overall body and canine sizes are not as impressive as those of males that compete for exclusive access to multiple females, males that live in multimale, multifemale groups have their own anatomical specializations that are tailored to increase reproduc-

Figure 2-8 The large testes of this male muriquis monkey from Brazil is thought to be associated with sperm competition.

Figure 2-9 Male and female siamangs are monogamous and the two sexes produce different songs or calls, during which they inflate their throat sacs.

tive fitness. In these groups, females that are sexually receptive (or in **estrus**) mate with any number of males. Interestingly, males in these groups have large testes relative to those in other kinds of groups (Fig. 2-8), presumably as a result of **sperm competition**. This form of intrasexual selection is based on the assumptions that (a) larger testes produce more sperm than do smaller testes, and (b) males that deposit the greatest amount of sperm are more likely to impregnate females that mate with multiple partners during a short period of time than are their competitors. Sperm competition is discussed further in Chapter 6.

Males that live in monogamous groups (e.g., gibbons, Mentawai leaf-monkeys, spectral tarsiers, and indris), however, lack both the large anatomical weaponry of those that live in unimale, multifemale groups and the large testes of males that live in multimale, multifemale groups. How, then, do males in these highly territorial, monogamous species compete for and maintain their pair-bonds with their mates? The answer seems to be, for the most part, with their voices! Although these primates are not particularly dimorphic in body and canine size, mated pairs do differ in their vocalizations and, in fact, are known for their beautiful duets in which males and females sing different sex-specific parts (Fig. 2-9). Although both sexes defend their

territory and presumably each other, there is some indication that males in at least some of the gibbon species engage more in intrasexual vocal encounters than do females, for example, as they chase away intruders or enter vocal contests at the edges of their territories. This is discussed further in Chapter 10.

Female Strategies

Female reproductive strategies differ dramatically from those of males. These include utilizing resources (e.g., food) for reproduction, and competing with other females for those resources rather than for mates. Such competition, which is costly, is often accompanied by relatively stable, clear-cut dominance hierarchies among adult females in some species of capuchin monkeys, vervets, macaques, baboons, and mangabeys. Compared to male hierarchies (which are completely separate), female dominance relationships are often quite stable and predictable over time in these groups, with maternal rank generally transferred to daughters, but with younger sisters outranking their older sisters. As is the case for males, female rank is correlated with the ability to obtain resources such as choice food, as well as with reproductive success in some species (Fig. 2-10).

Females also use other reproductive strategies to a greater or lesser extent, depending on the species. For instance, they may be more or less selective in choosing potential fathers of their offspring (i.e., by engaging in nonrandom **female choice**), thereby having some unconscious control over the quality of the offspring in which they eventually invest. As described in portraits of primates throughout this text, a form of natural selection called **intersexual selection** occurs when a trait appears in one sex because it has been preferred by the other. This type of selection usually results from female (rather than male) choice. Choosy females may focus on any number of male characteristics such as likableness (i.e., being a good friend), high position in a dominance hierarchy, or physical attractiveness that may or may not

REVIEW EXERCISES

1. Why do you think male primates go where the females are?
2. Review the six types of residence patterns that typify primates.
3. Why do you think that primate residence patterns are associated with specific mating patterns?
4. Distinguish between sexual selection, intrasexual selection, and intersexual selection.
5. Discuss male reproductive strategies and the types of residence patterns that they are likely to be associated with.
6. Describe some female reproductive strategies. Why do these differ from those of males?
7. Why is female philopatry the rule for most social primates that are not monogamous?

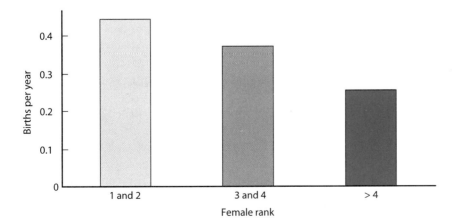

Figure 2-10 Among gelada baboons in Ethiopia, high-ranking females (low rank numbers) have a greater number of infants than do lower-ranking females.

be correlated with good health (e.g., having an impressive coat, colorful face [Fig. 2-11, color plate], or big nose).

A mother may also encourage other individuals to invest in her offspring by protecting, carrying, and sharing food with them. For reasons that are discussed later, such **allomothering** is more likely to occur among genetically related than among unrelated individuals, and stable associations among female relatives or **female-bonded kin groups** turn out to be quite common in multimale, multifemale groups in certain ecological settings. A closely but imperfectly correlated residence pattern in which females remain in the groups in which they were born, or **natal groups**, for life, while males emigrate is the rule for most primates that are not monogamous (e.g., ringtailed lemurs, squirrel monkeys, and macaques). Although male **philopatry** also exists in some species (e.g., chimpanzees), the reason why female-bonded groups and female philopatry are more common today is probably because females that, of necessity, invested heavily in offspring benefited more from mutual support systems during their evolution than did males that were less engaged in demanding parental activities.

KIN SELECTION AND ALTRUISM

Although much of this discussion has focused on competition, it is also important to emphasize that most of the nonsolitary primates are highly social or **gregarious** animals that frequently cooperate with each other in various ways. For example, primates sometimes share food and defend others during disputes. In groups that have dominance hierarchies, coalitions or alliances of individuals may join together and take precedence over a primate that ordinarily outranks each member of the coalition. Or, upon spotting predators, individuals may give warning calls that stimulate others to flee danger. Another important manifestation of friendly or affiliative behavior is social grooming, which is common among the gregarious primates. Groomers use their hands to pick through hair, removing (and sometimes eating) bits of debris as they go along (Fig. 2-12). Besides serving to keep coats clean and parasite free, grooming calms, soothes, and at times appeases recipients. Indeed, this activity has been colorfully described as a kind of social cement that bonds primates together.

Behaviors are **altruistic** if reproductive fitness is reduced in the animals performing them, while that of the recipients is increased. For example, an adult male red colobus monkey that manages to save an infant from the clutches of a chimpanzee but gets killed in the process (see Chapter 13) is behaving altruistically. Most altruistic behaviors are not this obvious, however. Primatologists tend to view food sharing and grooming as altruistic because they increase the resources or general well-being of recipients at the expense of time and energy that could be invested in the actors' own reproductive fitness. If the name of the game is to perpetuate one's genes, why do gregarious primates support others so extensively?

The answer appears to be that their support, while extensive, is also selective. Numerous studies reveal a strong tendency for primates to direct altruistic acts toward others that carry copies of their genes, that is, toward relatives. Thus, food sharing, grooming, and supportive coalitions occur most often among close kin. These ideas are formally quantified in the theory of kin selection, elaborated by biologist W. D. Hamilton. Although the mathematics of this theory are beyond the scope of this text, the gist of it is that altruistic behaviors toward relatives occur to the extent that the benefits experienced by the recipients outweigh the costs to the actors. Put another way, primates maximize the number of copies of their genes that are perpetuated, even at their own expense. Thus, the closer a relative, the more likely it is that an individual will behave altruistically toward that animal.

This rule is illustrated by a classic example: It is known that, on average, primate full siblings share one-half of their genes, while first cousins share one-eighth of theirs. Thus, quantitatively speaking, an individual's genetic contribution to future generations, in theory, would be matched by that of two brothers or eight cousins. As one famous geneticist is reported to have quipped, "I'd lay down my life for three brothers or nine cousins!" (The numbers change a bit in this example if the siblings have different fathers, as is the case for many primates living in the wild, but the principle remains the same.) Altruistic behaviors are predicted to occur if the ultimate effect is to increase a primate's **inclusive fitness**, or the sum of that individual's reproductive fitness and that added by altruistic acts directed toward kin. There are also many incidences in which primates behave altruistically toward others that are not kin. In these cases, individuals often

Figure 2-12 Two female olive baboons and their infants spend relaxed time grooming.

seem to have give-and-take relationships based on the proverbial saying "You scratch my back, and I'll scratch yours." Such **reciprocal altruism** has the potential for increasing the reproductive fitness of both participants, even if they are not related.

The subtleties involved in primate social behavior are frequently interpreted as indicating high intelligence. For example, in order for kin selection to work, primates must somehow distinguish between their relatives and other individuals. Similarly, reciprocal altruism and alliance formation imply that individuals not only keep track of past interactions, but also use this knowledge for their own ends. People frequently engage in complex behaviors like these, and those who do so most successfully are viewed as street smart, or even highly intelligent. Although it may seem inviting to attribute the social complexity seen in some monkeys and apes to high intelligence, we should proceed with caution. As noted earlier, nonhuman primates are not conscious of their reproductive strategies, although they nonetheless behave in ways that maximize reproductive fitness. Similarly, despite the fact that we might be tempted to **anthropomorphize** (i.e., attribute human-like characteristics to nonhumans) when it comes to interpreting the motives and thoughts that underlie primate social interactions, the level of intelligence demonstrated by higher primates is, in reality, a far cry from that seen in humans (see Neural Note 2).

ENVIRONMENTAL FACTORS: DIETS AND PREDATORS

As noted earlier, socioecologists explore the interplay between social life and environmental factors such as the kinds and distribution of available foods, or the intensity of pressure from predators. Stability of habitats is also an important consideration. At a general level, socioecologists have identified two different reproductive strategies used by animals living in stable versus unstable habitats: Where food supply is highly unpredictable, parents tend to produce large batches, or litters, of offspring that develop rapidly. Mortality is high, life is short, and parental investment is low in this strategy of **r-selection**, which is commonly seen in fish, rodents, and some birds. Most primates, on the other hand, are characterized by **K-selection** in which few offspring are born at once, and they develop slowly and are well cared for by one or both parents. Consequently, while fewer are born, more survive. Life spans are also longer and population sizes and food supplies remain relatively stable.

Despite these generalizations, it is worth noting that primates vary in the extent to which they are K-selected, and this variation correlates with body size. Thus, some of the smaller species such as mouse lemurs (see Chapter 4) have high birthrates, multiple births, rapid development, and short life spans. Larger-bodied species such as gorillas (see Chapter 12), on the other hand, have low birthrates, single births,

Just How Smart Are Monkeys and Apes?

Monkeys and apes are thought to be relatively intelligent compared to prosimians because they usually have larger brains (after body size is taken into account) and perform better on certain tests. Depending on the species, nonhuman primates can learn to varying degrees to discriminate shapes, colors, and locations in laboratory tests. Some can also determine, by sight or by touch, whether objects are the same or different, and are able to select the odd component from a group of three. Certain individuals in some species can work puzzles and manipulate objects, conspecifics, or people to acquire desired items. Nonhuman primates can remember goals and carry them out at later times, and some (such as vervet monkeys) have specific calls that stand for specific predators. Although prosimians do not appear to use tools in the wild, numerous species of monkeys and apes do. Some apes are even able to recognize themselves in mirrors, keep track of low numbers, and communicate with humans by using sign languages or symbols, as shown by the famous ape language experiments.

Many monkeys and apes use tools. Here, an adolescent female chimpanzee uses a twig to remove a loose tooth of an adolescent male, while an adult male looks over her shoulder. (Photo courtesy of W. C. McGrew.)

Although the above survey is informative to a degree, it gives no indication of how monkeys and apes think. Just what, if anything, is on their minds? Focusing particularly but not exclusively on monkeys (especially vervets), Dorothy Cheney and Robert Seyfarth of the University of Pennsylvania wrote a fascinating book on this subject. After a thorough review of the evidence, they concluded that monkeys lack a theory of mind; that is, they do not recognize their own knowledge (they do not know that, or what, they know) and therefore cannot attribute mental states to others. Consequently, monkeys do not engage in teaching, and learn only from passive observation. With the possible exception of chimpanzees, nonhuman primates rarely use deception as an apparent means to manipulate the perceptions of others, an activity at which humans excel (e.g., by telling lies). Finally, even the cleverest of the signing apes is without intellectual curiosity, as shown by their failure ever to ask questions about the world in which they live. This discussion is not meant to diminish the social complexity or many intelligent behaviors that characterize nonhuman primates, however. After all, it is from such substrates that the intellectual capabilities of our own species evolved.

slow development, and long life spans (Fig. 2-13). However, because most primates only produce one offspring at a time, the concepts of r- and K-selection are not as generally applicable as they are for many other animals. Nonetheless, it is interesting to note that prosimians that regularly produce multiple births live in habitats that are associated with seasonal fluctuations in food supply (see Chapter 4).

Figure 2-13 The largest living primates such as this gorilla weigh up to several hundred pounds and are K-selected. The mouse lemur is among the smallest living primates, weighing about 56 g (2 oz), and is relatively r-selected compared to the gorilla.

Although humans tend to eat an extensive variety of items (i.e., they are **omnivores**, as we saw in Chapter 1), many other primates feed more heavily on particular kinds of foods including insects (**insectivores**), fruits (**frugivores**), leaves (**folivores**), grains and seeds (**granivores**), and even saps and gums (**gummivores**). The smallest species of primates tend to be insectivorous and gummivorous, while the largest species are mostly folivorous (Fig. 2-14). Frugivores are in the middle. Part of the reason why diets correlate to a degree with body size is because small animals have relatively high metabolisms that require small amounts of high-energy foods. Insects fit the bill. Compared to insectivorous primates, however, the larger primates need more calories and therefore relatively larger quantities of food. For them, insects do not fit the bill, but leaves, being both plentiful and stationary, do. This relationship between body size and diet is formalized in **Kay's threshold**, which states that purely insectivorous primates can be distinguished from other primates because they usually weigh less than 500 g. Richard Kay also notes that purely folivorous primates always weigh more than 700 g because smaller bodies with higher metabolisms would not be able to sustain the lengthy digestion required for plant fiber. As described in subsequent chapters, the diets of many primates also correlate with interesting anatomical specializations of teeth, hands, and guts.

Home ranges are fixed areas that contain all of the sources of food and water and places to sleep that primates need to survive. Diets are correlated to some degree with population densities and ranging patterns of primates. For example, insectivores such as the slender loris (see Chapter 3) tend to feed alone because their food is usually dis-

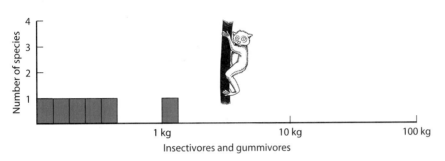

Insectivores and gummivores

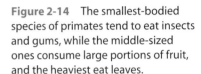

Frugivores (eating insects)

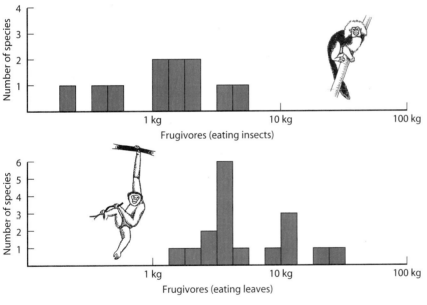

Frugivores (eating leaves)

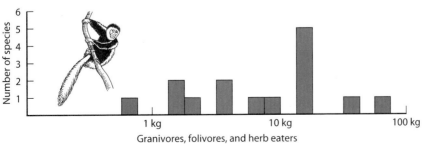

Granivores, folivores, and herb eaters

Figure 2-14 The smallest-bodied species of primates tend to eat insects and gums, while the middle-sized ones consume large portions of fruit, and the heaviest eat leaves.

Figure 2-15 Insectivores such as this slender loris tend to forage for food alone.

1. What is altruism, and of what benefit are altruistic acts to primates?
2. How is the quip "I'd lay down my life for three brothers or nine cousins" consistent with the theory of kin selection?
3. Why are most primates described as K-selected rather than r-selected?
4. Contrast the diets of the smallest and largest primates. Why do they differ?
5. What is the relationship, if any, between diet and ranging patterns of nonhuman primates?
6. Which do territorial primates *directly* defend—their land, their food, or their mates?
7. What kinds of animals prey on nonhuman primates?

persed and a larger number of individuals would not be able to find enough to eat in one place (Fig. 2-15). Depending on the season, scattered fruit trees that ripen at different times attract small subgroups of frugivorous primates like spider monkeys (see Chapter 6) that must forage over relatively long distances. Folivores such as howler monkeys (see Chapter 6), on the other hand, often remain in stable groups that travel within smaller home ranges than those of frugivores or insectivores because they have access to nearby leaves that are relatively abundant. In order for a number of different species to share a range, they have to use its resources differently, and primates are a good example of an order that can subdivide resources to provide niches for many species.

Depending on the species and environment, the ranges of different groups within a species may or may not overlap. Those species in which groups vigorously defend the boundaries of their ranges from conspecifics are **territorial**. Territorial species such as gibbons use various means to protect their home ranges, from spacing mechanisms based on loud vocalizations to out-and-out physical combat (Fig. 2-16). There is some debate as to whether territorial primates are simply guarding access to limited resources that are located in small-enough areas to defend, or whether they (especially males) are also preventing outsiders from approaching their mates. As we will see (in Chapter 10), both explanations seem plausible at times.

Predators are another environmental factor that constrains how primates live. Animals that sometimes prey on primates include wild cats (Fig. 2-17), snakes, birds of prey, crocodiles, and other primates including humans. The severity of predator pressure varies tremendously between species, and at times, between groups of conspecifics living in different habitats. Primatologists have long speculated that terrestrial primates that spend most of their waking hours on the ground are more exposed to predators than are arboreal primates that spend most of their active hours in trees. Although this is far from

Figure 2-16 These male hamadryas baboons are engaged in a territorial dispute.

proven, it would help to explain why terrestrial primates generally live in larger social groups than their arboreal relatives. The extent to which group living protects species from predators, as well as the degree to which predator pressure keeps population sizes in check, will be examined for a number of species in later chapters.

SUMMARY

Contemporary primatologists use sampling techniques to make systematic behavioral observations of primates, as well as laboratory and other collection methods to analyze data. For primates, as with other organisms, the name of the evolutionary game is to reproduce, and natural selection favors any behavior that allows an individual or group to outreproduce its rivals. Primates use various reproductive strategies to maximize the likelihood that they and their relatives will be represented in future generations. Because such strategies take place within environmental contexts, socioecological factors that constrain the lives of primates are also important (e.g., availability of food, ranging patterns, and predators). Six residence patterns characterize living primates and are correlated with particular mating systems.

Figure 2-17 One of the numerous animals that prey on primates, a leopard stalks off with a fresh baboon kill.

FURTHER READING

BOYD, R. AND J. B. SILK (2000) *How Humans Evolved*, 2d. ed. (Part Two: Primate Behavior and Ecology). New York: W. W. Norton & Company.

CHENEY, D. L. AND R. M. SEYFARTH (1990) *How Monkeys See the World*. Chicago: University of Chicago Press.

3

GALAGOS, LORISES, AND TARSIERS

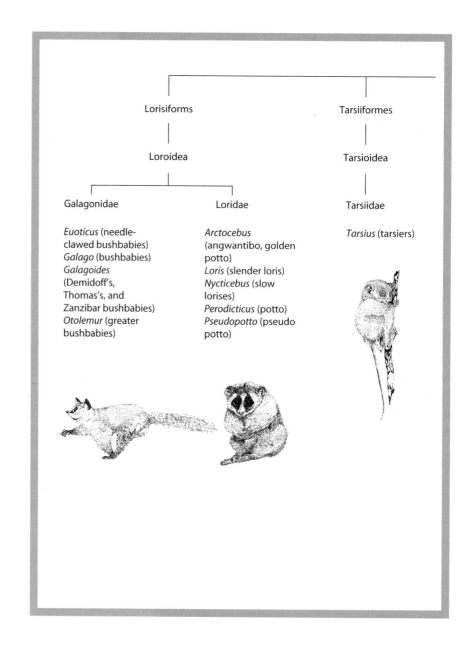

Lorisiforms

Loroidea

Galagonidae

Euoticus (needle-clawed bushbabies)
Galago (bushbabies)
Galagoides (Demidoff's, Thomas's, and Zanzibar bushbabies)
Otolemur (greater bushbabies)

Loridae

Arctocebus (angwantibo, golden potto)
Loris (slender loris)
Nycticebus (slow lorises)
Perodicticus (potto)
Pseudopotto (pseudo potto)

Tarsiiformes

Tarsioidea

Tarsiidae

Tarsius (tarsiers)

INTRODUCTION

From that initial moment long ago when the first little primates survived the K-T extinction event and scurried into the future, their descendants have been on the move. By at least 50 mya (and possibly well before), the adapid ancestors of lemurs and lorises had shifted from foraging for insects in the undergrowth to scrambling in trees with their grasping hands and feet in search of fruit and leaves. Meanwhile, the smaller omomyid ancestors of tarsiers and anthropoids had also become arborealists that moved in a scurrying-leaping fashion, but ones that continued to eat a large amount of insects. Early on, a dramatic event occurred when some of these little omomyids became diurnal. As the Cenozoic era progressed, apes prospered, some primates descended to the ground while others lounged in European oaks, and gorilla-sized monkeys roamed the African terrain. The more than 200 species of primates that exist today, confined mostly to three continents, represent only a fraction of the physical types that evolved during the Cenozoic. Yet even these few survivors of the perverse pulses of cooling and drying that almost seemed designed to bedevil tropical animals are enormously diverse. It is to them that we now turn.

Millions of years ago, prosimians inhabited parts of North America, Europe, Africa, and Asia. Because of competition with increasingly successful anthropoids, their geographical distribution eventually became severely restricted. Consequently, today's prosimians are native only to the island of **Madagascar** (and the nearby small Comoro Islands off of the east coast of Africa), sub-Saharan Africa, and parts of South and Southeast Asia (Fig. 3-1). Competition also caused prosimians either to remain in or to be driven into nocturnal niches, except on Madagascar and nearby small islands where some diurnal

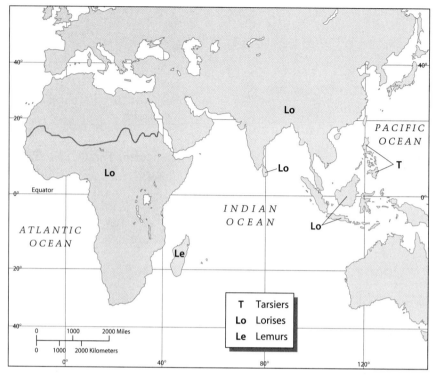

Figure 3-1 Today's prosimians occupy Madagascar, sub-Saharan Africa, and parts of South and Southeast Asia.

species were able to evolve, owing to a complete lack of monkey and ape competitors.

Figure i-6 divides prosimians into three major groups: lorises (including galagos) from Africa and Asia, tarsiers from Southeast Asia, and lemurs from Madagascar. Representatives from each of these groups are shown in Figure 3-2. The lemurs and lorises are generally regarded as belonging to the same natural grade. On the other hand, some workers place tarsiers in their own grade of prosimians, while others prefer to sort them with the anthropoids into a haplorhine group (see page 7). There is a certain irony in prosimians being thought of as lower primates. Although the name *premonkey* implies lack of advancement or primitiveness, prosimians are incredibly varied in their appearances and behaviors, partly because a wide array of specializations are found in different species. In fact, some prosimians such as the aye-aye (a lemur) are among the most specialized primates alive today. (Unfortunately, aye-ayes are so specialized that they could easily become extinct.) This chapter focuses on the general characteristics of prosimians and specifically on lorises, galagos, and tarsiers. Detailed discussion of lemurs, which constitute the most diverse group of prosimians, is postponed until Chapter 4.

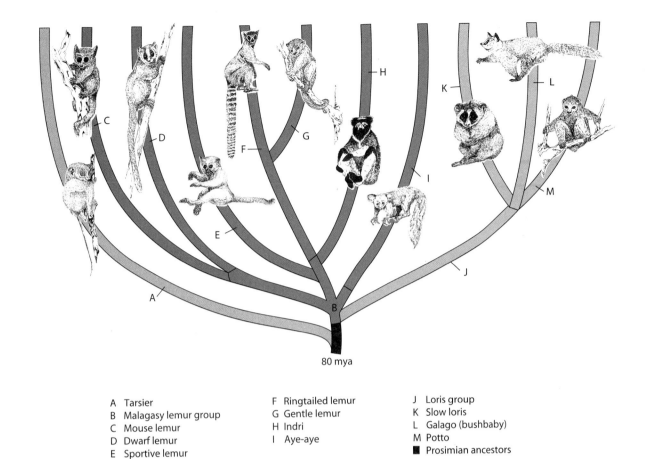

A Tarsier
B Malagasy lemur group
C Mouse lemur
D Dwarf lemur
E Sportive lemur

F Ringtailed lemur
G Gentle lemur
H Indri
I Aye-aye

J Loris group
K Slow loris
L Galago (bushbaby)
M Potto
■ Prosimian ancestors

Figure 3-2 Living prosimians are highly diverse in appearance, as illustrated by this family tree.

WHAT UNITES PROSIMIANS

Despite their variability, one can delineate a general prosimian pattern in much the same way that was done for primates in the Introduction. That is to say, while no one feature distinguishes prosimians from all other mammals, a set of particular traits typifies most prosimians (although there are exceptions for individual species). For example, the number of different kinds of teeth together with their configuration within the upper and lower jaws set prosimians apart from other primates (Fig. 3-3). Thus, most prosimians have a **dental formula** of 2.1.3.3./2.1.3.3., which means that each side of their upper (numerator) and lower (denominator) jaws contains (from front to back) two incisors, one canine, three premolars, and three molars. Because the numerator and denominator are the same in this formula, it is shortened to 2.1.3.3. This dental formula alone does not distinguish prosimians, as it is also found in most species of New World mon-

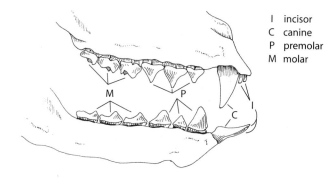

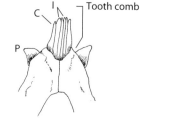

Figure 3-3 The prosimian dental formula is 2.1.3.3, and the incisors and canines of the lower jaw project together to form the tooth comb in all prosimians except tarsiers.

keys. However, in prosimians, the lower incisors and canines project forward into a **grooming** or **tooth comb** and this feature in conjunction with the dental formula is part of the prosimian pattern.

The Prosimian Pattern

BONY FEATURES (Figs. 3-3 and 3-4)

1. Dental formula of 2.1.3.3 with tooth comb in lower jaw
2. Lack of postorbital closure
3. Unfused metopic suture on top of skull
4. Unfused midline in lower jaw
5. Grooming claw on second toe of foot

OTHER FEATURES

6. Digits act together (powerful grip)
7. Long muzzle with moist nose
8. Immobile upper lip and inexpressive face
9. Tendency toward vertical clinging and leaping
10. Multiple pairs of breasts, frequent multiple births
11. Large eyes, most species nocturnal
12. Relatively great dependence on hearing and smell (use of scent)

A word of caution is necessary with respect to the always problematical tarsier. As detailed below, *Tarsius* is not well described by the typical prosimian pattern. Although it has a grooming claw on each foot (in fact it has two per foot), the tarsier lacks a dental comb and its skull has a postorbital plate of bone behind each eye. Also contrary to the prosimian pattern, *Tarsius* has a flattened snout and expressive lips and face. Meanwhile, other prosimian traits such as large eyes and an anatomy that is specialized for vertical clinging and leaping are exaggerated.

At first glance, the 12 features outlined in the prosimian pattern may appear to represent an unrelated list. Upon closer inspection, however, many of the features converge on three functional complexes that unite lemurs, lorises, and tarsiers. The first is related to the primitive and widespread prosimian characteristic of nocturnality. Most nocturnal prosimians have very large eyes (feature 11) that function to see in the dark rather than allow for detailed vision during the day. Thus, their retinas have a special light-reflecting layer called the **tapetum lucidum** that makes their eyes appear to glow in the dark like a cat's. The extent to which prosimians see colors and perceive depth is questionable. In general, it is believed that vision is not as important for nocturnal prosimians as it is for their diurnal relatives. Instead, the sense of smell appears to be crucial and is facilitated by a long wet snout region (feature 7). Hearing (feature 12) is also thought to be an important sense for nocturnal prosimians, as indeed it is for diurnal primates like ourselves when we tune in to strange noises if we awaken during the night.

A second complex that cuts across prosimians has to do with the strong tendency for primates in general to be extremely social: Because the social life of prosimians evolved within a nocturnal framework, olfactory communication (feature 12) is frequently emphasized

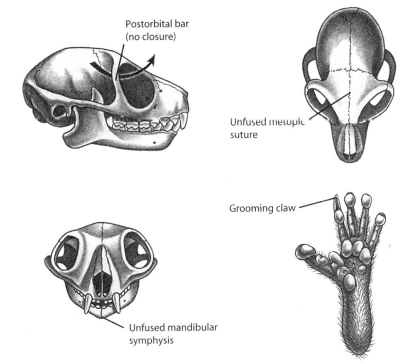

Postorbital bar
(no closure)

Unfused metopic
suture

Grooming claw

Unfused mandibular
symphysis

Figure 3-4 Bony features 2 to 5 of the prosimian pattern are illustrated.

The Neurological Basis of Smell: Olfactory Bulbs

According to the arboreal theory of primate evolution that was discussed in Chapter 1, selection for features that permitted early primates to live in trees corresponded with increased visual acuity, on the one hand, and a decreased dependence on the sense smell, on the other. However, this is not to say that the importance of the sense of smell has decreased in all living primates. Far from it, as this chapter shows. Because the social life of prosimians evolved within a nocturnal framework, olfactory communication is frequently emphasized over visual gestures, especially in nocturnal species. In fact, even those lemurs that are diurnal (see Chapter 4) have specialized scent glands that are used for communication. As illustrated by several remarkable examples in this chapter, the strong reliance on olfactory messages sets prosimians apart from most monkeys, apes, and humans. This distinction is even reflected in their brains.

The ability to smell depends on a phylogenetically old part of the brain, the olfactory bulbs, and we know from humans that this sense is tied closely to memory and emotions. It is not closely connected with areas of the brain that process language, however, which explains why people often have difficulty naming familiar smells. Humans have very small olfactory bulbs, and they are not much larger in monkeys and apes. Compared to anthropoids, however, prosimians have noticeably larger olfactory bulbs, a difference that is even more dramatic when one realizes that absolute and relative brain sizes are significantly smaller in the prosimians. Using the size of the olfactory bulbs as a clue, researchers studying endocasts (described in Neural Note 1) glean information about the relative importance of smell that characterized various species of fossil primates.

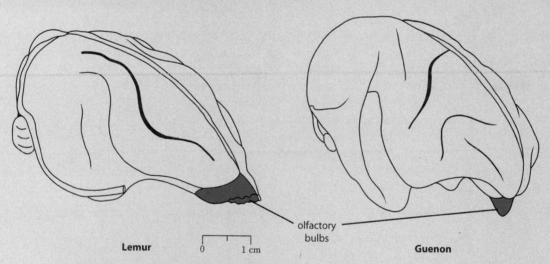

Lemur 0 1 cm

olfactory bulbs

Guenon

Prosimians have larger olfactory bulbs than anthropoids, as illustrated by these endocasts from a lemur and a guenon.

over visual gestures including facial expressions (feature 8). Many prosimians have specialized glands for producing strong-smelling chemicals (pheromones). These glands may be located on the head, chest, abdomen, legs, or anogenital region. Prosimians mark objects such as trees by applying their glandular secretions, feces, or urine (urine marking). Olfactory communication facilitates reproductive behavior, territorial defense, and recognition of individuals (see Neural Note 3). The prosimian specializations of tooth combs (feature 1) and **toilet claws** (feature 5) seem to have evolved as a grooming package that is useful for (among other things) applying and distributing scent to furry coats.

A third trait that cuts across many species in the three major groups of prosimians is a special form of locomotion called vertical clinging and leaping (feature 9). As described by the late British primatologist John Napier and Alan Walker, this form of moving may have evolved during early primate evolution. At any rate, prosimians that engage in vertical clinging and leaping include the tarsiers, numerous lemur species, and the fast-leaping loroids (galagos). These prosimians are all small to medium-sized primates that have an upright clinging posture when at rest in their arboreal habitats. When moving between trees, they use their two hind limbs in unison to propel themselves into long leaps (Fig. 3-5). On the ground, they frequently proceed bipedally in a series of leaps or quick hops. For example, bushbabies leap froglike from tree to tree, and hop on their hind limbs like a kangaroo when on the ground. Tarsiers and sifakas (a kind of lemur) may rotate their bodies as much as 180° in midflight, and land more or less

REVIEW EXERCISES

1. Name and briefly describe
   the three major groups of
   living prosimians.
2. Review the 12 features of
   the general prosimian pattern.
3. What three functional
   complexes are represented
   by many of the features in
   the prosimian pattern?
4. What are some of the ways
   in which prosimians use
   scent?
5. Describe vertical clinging
   and leaping. What is its
   possible evolutionary significance?
6. What is the formula for
   the intermembral index
   and how does it relate to
   relative hind-limb length?
7. What part of the brain
   processes smells? Is this
   region larger in the bigger-
   bodied primates? Why or
   why not?

upright on their hind feet when they reach their destination. Good grasping abilities (feature 6) are adaptive for clinging as well as for landing. Given their spectacular leaping abilities, it is not surprising that vertical clingers and leapers have greatly elongated hind limbs relative to the forelimbs. (Relative limb length is quantified with a ratio called the **intermembral index**, which is calculated as follows: length of the humerus + length of the radius $\times$ 100/length of the femur + length of the tibia. The lower the intermembral index, the longer the hind limbs.) Napier and Walker suggested that long hind limbs relative to the length of the vertebral column are an almost universal primate characteristic, and that the variety of locomotor patterns seen in modern anthropoids therefore resulted primarily from evolutionary adjustments in forelimb length.

GALAGOS AND LORISES

The loroids consist of two families, the Galagonidae, which includes the fast-leaping galagos (bushbabies), and the Loridae, which contains the slowly creeping lorises (see Fig. i-6). Bushbabies are found only in Africa, while their cousins, the lorises, live there and in Asia. Although their geographical range is much broader than that of lemurs (see Chapter 4), loroids comprise only about half as many species. Because they are all nocturnal, they have been able to avoid competition with diurnal monkeys and apes (and likely extinction). Where more than one species overlaps (i.e., are sympatric), they separate ecologically by utilizing different levels of the canopy or undergrowth (with smaller species lower). All loroids forage alone for fruits, flowers, leaves, gums, insects, or small mammals and birds, with slow lorises specializing in slow insects. Although males and females do not live in large social groups, their social systems can be surprisingly complex as illustrated later by the lesser bushbaby.

Galagonidae

In the wild, galagos make a surprising variety of vocalizations including some that are said to sound like the crying of a human baby (hence, their common name). Bushbabies are found only in Africa, where they are divided into four genera (see Fig. i-6) that contain approximately 11 species. There is currently some confusion about how to classify galagos, however; some groups that appear physically similar are beginning to be recognized as different species because they emit very different vocalizations and odors. An example of this confusion is the southern lesser bushbaby, described below, which recently has been removed from *Galago senegalensis* and placed in its own species, *Galago moholi*.

In all bushbabies, adult females have relatively small ranges that overlap those of related females, while adult males have considerably larger ranges that incorporate those of a number of females. This distribution is partly due to the fact that males leave the mother's terri-

Figure 3-6 The lesser bushbaby, *Galago moholi*, has a slender body, long bushy tail, and large mobile ears.

tory upon reaching sexual maturity, but females do not. Therefore, the small social groups that exist are descended through females, are **matrilineal**, consisting of mothers, daughters, sisters, and their offspring. As will be seen from the following portrait, bushbabies are excellent leapers because of their slender bodies, long bushy tails, and long hind limbs. Consequently, they are generally noisier and easier to spot than their slower-moving loris and potto relatives (described in the following section).

THE SOUTHERN LESSER BUSHBABY (*GALAGO MOHOLI*) The rat-sized lesser bushbaby is neither the smallest nor the largest of the galagos, averaging about 200 g (7 oz) (Table 3-1). Its fur is gray with tinges of yellow, and it has a short face with facial vibrissae. This species is also characterized by short fingers, a long thin tail, and large spoon-shaped ears that move independently and can be folded up (Fig. 3-6). The eyes are huge and the head can be rotated approximately 180° so that it is facing completely backward, a behavior that often occurs as it clings upside down by its feet! The hands are also capable of grasping, having nails and well-defined tactile pads that bear prints. Curiously, grasping is facilitated by moistening the hand and footpads with drops of urine. This habit of **urine washing** is widespread among lorises and galagos, as described by the classic primatologist Osman Hill:

> All lorisoids have a strange habit of washing their hands by micturating on them. One hand is passed over the genital area and wetted and then the

TABLE 3-1 Summary Statistics for Galagos, Lorises, and Tarsiers

	G. moholi (southern) lesser bushbaby	P. potto (potto)	N. coucang (slow loris)	T. spectrum (spectral tarsier)
Adult mean body weights				
Female	188 g (6.6 oz)	1.08 kg (2.4 lb)	1.20 kg (2.6 lb)	108 g (3.8 oz)
Male	211 g (7.4 oz)	1.02 kg (2.3 lb)	1.21 kg (2.7 lb)	126 g (4.4 oz)
Adult brain weight (g)	4.8	14.3	10	3.8
Maximum known life span (yr)	16	26	20	12
Nocturnal (N) or diurnal (D)	N	N	N	N
Tapetum	+	+	+	0
Olfactory communication[a]	uw, um	um, sm	uw, sm	sm, uw
Location (habitat)[b]	A: w, s, fe	A: pf, sf	SA: rf	SA: pf, sf, sc, mn, mf
Locomotion[c]	vcl, bp	sc	sc	vcl, aq
Diet[d]	i, g	f, g, i	f, i, g, e	i, v
Solitary foragers	+	+	+	0
Nests	+	0	0	+
Mean no. of offspring	1–2	1	1	1
Mouth carry	+	0	0	+
Seasonal breeders	+		0	+
Gestation (days)	121–125	194–205	191–193	181–194
Social organization[e]	dp	dp	dp?	mo, mm
Female dominance	0			
Territorial	+	+		+
Ratio of adult females to males				1:1
Infanticide/allomothering	+/+	+/		
Average group size	1–3	1–2		2–6
Density	31/km²		20/km²	30/km²
Home range	F 4.4–11.7 ha M 9.5–22.9 ha	F 6–9 ha M 9–40 ha		1–10 ha
Conservation status[f]	L	L	L	L

Note: Information in this table has been selected from many sources and should be viewed as tentative (i.e., subject to change in light of new information). Sources include Rowe, 1996 (including conservation statuses); Harvey et al. (brain weights), 1987; Bearder, 1987; Niemitz, 1984; and other references mentioned in the text. Abbreviations: +, yes; 0, no. Blank entries indicate that no information is available. Dietary items are listed in order of decreasing importance to the species.

[a]Olfactory communication: sm, scent marks; um, urine marks; uw, urine washes.

[b]Location (habitat): A, Africa; SA, Southeast Asia; fe, forest edge; mf, montane forest; mn, mangroves; pf, primary forest; rf, rain forest; s, savanna; sc, scrub; sf, secondary forest; w, woodlands.

[c]Locomotion: aq, arboreal quadruped; bp, bipedal hopping; sc, slow climber; vcl, vertical clinger and leaper.

[d]Diet: e, eggs; f, fruit; g, gums (sap); i, insects; v, small vertebrates.

[e]Social organization: dp, dispersed polygyny; mo, monogamy; mm, multimale, multifemale groups.

[f]Conservation status: L, lower risk.

two hands, or one hand and one foot, are rubbed together and again passed through the urinary stream for a final rinse. The origin and value of this curious habit are obscure, but it has been noted in some emotional states (e.g., alarm). (Hill, 1953:127)

G. moholi is widespread across central southern Africa (Fig. 3-7) and lives in habitats that range from forests to woodland-savannas. A vertical clinger and leaper with powerful long hind limbs, this species moves rapidly as it leaps up to 15 feet between trees, or hops kangaroo-like on the ground. *G. moholi* eats mostly insects and gums.

By radio tracking individuals from a savanna area in South Africa, Simon Bearder of Oxford Polytechnic and Robert Martin were able to trace social development in the southern lesser bushbaby and to discover details about its social system. Like the mouse lemurs described in Chapter 4, lesser bushbabies forage alone for insects at night and sleep during the day in small groups (up to six) in tree hollows or nests (but adult males frequently sleep alone). During mating season, males chase estrous females and, when mounting them, make loud calls that end in whistles. This species breeds seasonally (twice a year) and frequently produces twins that initially stay in the nest. Related females may rear their offspring in the same nest. As they begin to mature, mothers move their babies by mouth (like a cat) and park

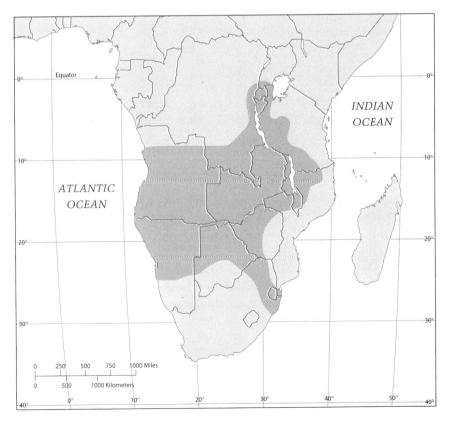

Figure 3-7 *Galago moholi* is widespread in central southern Africa, living in habitats that range from forests to woodland-savannas.

them (i.e., hang them up like a coat hanger) in trees near the nest while they forage.

Adult females and males establish independent territories, which they each defend with scent markings and vocalizations. However, unlike many lemurs, bushbabies lack female dominance over males. Territories of adult females may overlap (especially those of kin), but those of fully adult males do not. Males immigrate away from their area of birth and, when fully adult, establish territories that are larger than those of females. Each male's territory overlaps with that of one or more females, which he visits and mates with regularly. Such a mating system is known as **dispersed polygyny**. (Although polygyny technically refers to males mating with more than one female, estrous lesser bushbabies also mate with more than one male; i.e., they are promiscuous.) These relationships are quite stable, and it is possible that adults may even meet at night for occasional grooming or play. Thus, female territories are constrained by available food, while those of males are determined by the distribution of females. According to Simon Bearder of Oxford Polytechnic, bushbabies provide an ideal model for the structural ancestor of all primates because their constellation of a dispersed polygynous mating system, solitary foraging, partially gregarious sleeping habits, lack of female dominance, and male transfer between groups could have given rise to the wide range of more-specialized social systems that characterize many of today's primates.

Loridae

As shown in Figure 1-6, the lorids contain three genera of pottos (*Arctocebus*, the golden potto; *Perodicticus*, the potto; and *Pseudopotto*, the pseudo potto) and two genera of slow lorises (*Loris*, the slender loris; and *Nycticebus*, the slow loris). The pottos are from Africa, while the slow lorises live in southern Asia. The name *loris* derives from the Dutch word *laeris*, which means "clown." This name probably originated because of the humorously slow and cautious way that slow creepers such as the potto or slow loris move in their arboreal habitats. These species move silently through the dense forest cover, one limb at a time, as if they were hanging on for dear life (not an altogether unreasonable approach to height). They proceed with extreme care and freeze to avoid the attention of predators and prey. Lorids have smaller ears, shorter tails and hind limbs, and longer bodies than bushbabies. They also have remarkable hands and feet with incredibly strong grasps, in part due to their greatly reduced second digits—all the better for hanging on in the dark. Even youngsters have good grasps that are taken advantage of by their mothers, which park their single babies by hanging them on trees. In keeping with their low profile, the sense of smell is extremely important for lorises and pottos, while vocal communication is less so. Portraits are provided below for one African and one Asian lorid.

THE POTTO (*PERODICTICUS POTTO*) The potto is a curious, cuddly-looking creature (Fig. 3-8) whose name is taken from an African phrase meaning "softly softly." This name reflects the fact that pottos are slow-creeping nocturnal lorids that vocalize infrequently and have soft reddish-brown fur and stumpy bottle-brush tails. Males and females are approximately the same size (see Table 3-1), weighing from 0.8 to 1.6 kg (1.8 to 3.5 lb). (The recently discovered *Pseudopotto*, on the other hand, is smaller and has a much longer tail.) Pottos have the typical prosimian tooth comb and grooming claws on their second toes, and they have musk glands on their arms, genital regions, and other parts of their bodies that produce a skunklike odor. Other parts of their anatomy are very specialized. For example, their hands have extraordinarily strong grasps, partly because the second finger is reduced to a mere stub, allowing the first and third digits to act much like a pincer (Fig. 3-9). Pottos also have unusual vertebral spines that protrude at the back of their necks and are covered by thick skin, fur, and long hairs, forming a kind of hump or shield that is used under defensive (e.g., as protection when sleeping curled up) or aggressive circumstances. As vividly described by Pierre Charles-Dominique of the Museum of Natural History in Brunoy, France:

> In the event of an attack by, for example, an African palm civet, the potto turns toward it, head buried between its hands and presenting its shield. The aggressor's charges are dodged by sideways movements without the potto loosening the grip of its hands and feet on the branch. Then, straightening its body, and maintaining a clamp-like hold on the branch, the potto delivers fearful bites or a violent blow with its shield, toppling the predator to the ground, where it is difficult for the aggressor to find a route back to its prey. (Charles-Dominique, 1984: 37–38)

Pottos are arboreal quadrupeds that live in forests, wooded savannas, and plantations in Central Africa and along its west coast (Fig. 3-10). They climb slowly and use extreme caution as they wind their way through dense vegetation in which they are nicely camouflaged, partly because of their brownish coats. This form of concealed locomotion is a **cryptic feature** that serves to conceal the potto from potential predators. Unlike their bushbaby cousins, pottos do not leap or flee danger, although they sometimes drop to the ground when threatened by a predator such as a snake. Pottos eat fruit, gums, and insects, and, curiously, are capable of ingesting whole meals while hanging upside down by their feet. When other food is not available, they are able to consume poisonous millipedes, irritating caterpillars, and other foul-smelling and hairy insects that most primates would pass up. They usually forage alone for food at night, although Myrdene Anderson of Purdue University reports that they also explore their surroundings and engage in a certain amount of mutual grooming, fighting, and play.

Males have relatively large territories that overlap the smaller territories of a number of adult females and their young offspring. Unlike bushbabies, however, pottos are not characterized by small matrilin-

Figure 3-8 Because of its anatomy, the African potto is one of the more specialized lorids.

Figure 3-9 The potto's hands are capable of extremely strong grasps, partly because of the greatly reduced second digits.

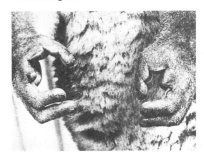

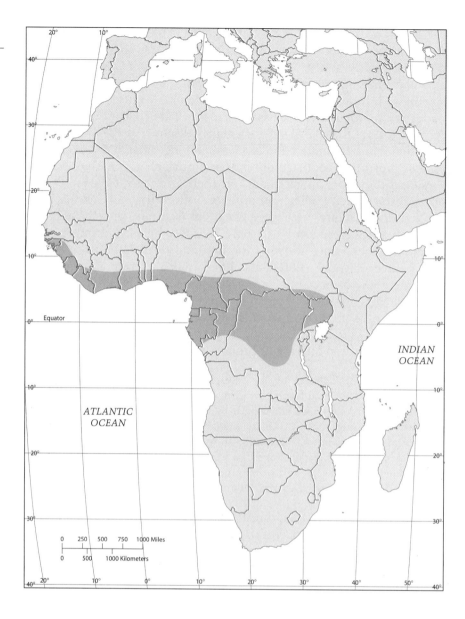

Figure 3-10 Pottos live in Central Africa and along its west coast.

eal social groups. Adult males and females with young sleep independently in thick foliage during the day. As noted, pottos have a strong odor generally, and give off a distinct fear scent when they are alarmed. Both sexes use urine to mark large branches in their territories, and use scent to mark members of the opposite sex. Pottos reach sexual maturity within the first year of life, and engage in dispersed polygyny, with males periodically visiting the home ranges of two or three estrous females that advertise their condition by producing whistling calls. According to Anderson, pottos use their neck shields in social, erogenous contexts, and they engage in ventral-to-ventral copulation instead of restricting themselves to the male ventral to female dorsal position. They are also reported to mate while suspended by their feet from a branch. Potto gestations take about 6 months and usually result in the birth of one or sometimes two infants. Potto mothers frequently park their young in trees when they go off to forage at night. Anderson notes that youngsters like to baby-sit their newborn siblings, which resemble white furry frogs.

Pottos are preyed on by snakes and small carnivores such as civets. Unlike their cousins the bushbabies, pottos prefer fight (as described earlier) to flight. However, given their druthers, pottos avoid both responses by engaging in slow, silent, concealed locomotion and freezing at the slightest sound or disturbance. According to Charles-Dominique (1984:37), "this petrified stance can be maintained for hours, so long as the danger lasts, and will frustrate even the most patient watcher!" Perhaps because of their cryptic habits, pottos are less threatened than some of the other primates. Nevertheless, their populations are decreasing in parts of Africa because of habitat destruction by humans.

THE SLOW LORIS (*NYCTICEBUS COUCANG*) Although all loroids are nocturnal, the slow loris looks, moves, and in other ways behaves differently than its cousins, the galagos (Fig. 3-11). To begin with, slow lorises are relatively robust, attaining the size of a small cat (averaging around 1.2 kg or 2.7 lb, see Table 3). (According to Radoslaw Ratajszczak of the Zoological Garden in Poznan, Poland, smaller individuals that weigh around 500 g are pygmy lorises and should be regarded as a separate species, *Nycticebus pygmaeus*.) The head is rounded, with large eyes and little ears that are hidden in the fur. A tail is usually absent. The trunk is stocky and the hind limbs are not particularly long since this species does not leap. The slow loris has an endearing owllike face, with pupils consisting of vertical slits in the middle of eyes that are encircled in dark rings. The eyes are separated by a white stripe, and a dark stripe runs down the middle of the top of the head and spine. There are at least four geographical variations of this basic face (Fig. 3-12), which leads one to reconsider whether loris is derived from

Figure 3-11 The head of the slow loris (*Nycticebus coucang*) is rounded, with large eyes and little ears that are hidden in the fur.

Figure 3-12 Slow lorises (*Nycticebus coucang*) are characterized by four basic types of facial markings.

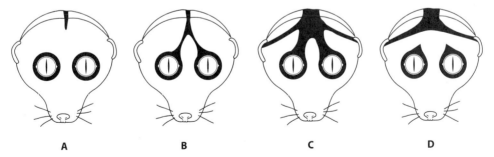

A   B   C   D

Geographical differences in the facial markings of slow lorises (*Nycticebus coucang*).
A   *N. c. bengalensis*
B   *N. c. tenasserimensis*
C   *N. c. coucang*
D   *N. c. borneanus*

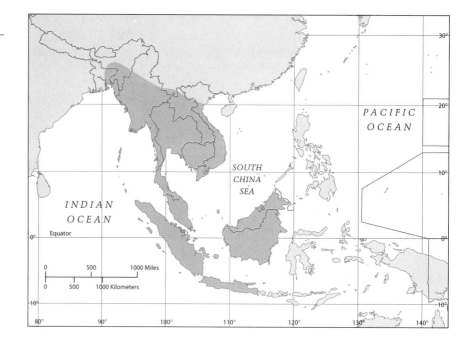

Figure 3-13 Slow lorises have a wide distribution in Southeast Asia.

1. Diurnal prosimians are found only in —————
   —————————————.
2. Where do bushbabies live and how do they move?
3. Discuss the anatomical specializations of the potto and their functions in this species's day-to-day life.
4. What anatomical adaptations does *Nycticebus coucang* have for moving?
5. What is baby parking and which primates do it?
6. Describe dispersed polygyny and name a species that engages in it.
7. Give two possible reasons why the word *loris* derives from a Dutch word that means "clown."
8. How do slow creepers deal with predators?

the Dutch word for "clown" because of the way it moves or because of its face paint.

Slow lorises are well represented across Southeast Asia (Fig. 3-13). They move among trees in tropical rain forests in a deliberate hand-over-hand climbing fashion that almost appears as if they are filmed in slow motion. As they progress, their bodies undulate in a side-to-side rhythmic motion, and they may alternate between moving under a tree limb (like a sloth) and proceeding along its top, in a spiral kind of motion. Above all else, *Nycticebus* is built for hanging on. The digits on the hands and feet are strong, the second digits are reduced, and the thumb and big toe (pollex and hallux) are widely opposable to the remaining digits—all of which makes for an extremely strong grasp. Special blood vessels in the limbs allow the muscles to remain contracted for long periods, for example, while the animal feeds suspended from the hind limbs. In fact, Hill (1953:132) observed that "after death, rigor mortis in the feet will keep the corpse suspended for hours and render its removal a matter of great difficulty"!

The slow loris subsists on insects, with an occasional egg or lizard, and will cheerfully accept fruit (bananas) and milk in captivity. Unlike galagos and most lemurs, *Nycticebus* does not seem to be a seasonal breeder and females give birth to only one offspring at a time. Infants are born relatively well developed and are capable of clinging to their mothers' bellies. Thus, they are not stashed in nests. In fact, this species does not sleep in tree hollows or leaf nests as galagos do. Instead, slow lorises prefer to curl up in a fork of a tree while gripping twigs with their hands and feet, where they remain inconspicuously sleeping during the day.

The five to six recognized species of living tarsiers (all belonging to the genus *Tarsius*) constitute the third major group of prosimians shown in Figure i-6. Confined to islands in Southeast Asia, these little nocturnal primates physically resemble the African bushbabies (see Fig. 3-2) and occupy a similar nocturnal predatory niche. However, of the prosimians, the tarsier has been particularly troubling to zoologists and this one genus (*Tarsius*) has caused some workers to accept a different taxonomic classification than the one presented in Figure i-6. According to their view, tarsiers do not belong with the prosimians because they share special features with anthropoids. As noted in the Introduction, this alternative classification contains one suborder (the Strepsirhini) that includes lemurs and lorises, and another (the Haplorhini) that represents tarsiers and anthropoids. The features that are shared by tarsiers, monkeys, apes, and humans include the following:

1. A nose region that is covered by hairy skin instead of naked wet skin
2. A mobile single upper lip in place of one that is anchored to the gum and divided in the midline
3. Lack of a light-reflecting layer in the retina of the eye (i.e., the eyes do not glow in the dark), combined with the presence of a specialized central area (**fovea**) that increases visual acuity
4. Development of the placenta and fetal membranes
5. Blood supply to the middle ear from the same (promontory) branch of the carotid artery
6. Position of the olfactory bulbs of the brain

However, tarsiers resemble prosimians in the form of the uterus, features of the brain, and numerous other traits. They also have the oldest-known fossil record of any primate, and their teeth, at least, appear to have remained unchanged for 45 million years. In short, tarsiers seem to be unique creatures that have managed to survive an extraordinarily long time. In this book, tarsiers are retained as an infraorder of prosimians (see Fig. i-6), partly for the sake of tradition and partly because of the many physical and behavioral features that make them very much at home among the prosimians.

The Spectral Tarsier (Tarsius Spectrum)

Tarsiers are curious-looking little creatures (Fig. 3-14, color plate) that weigh only 108 to 142 g (about 4 to 5 oz). They have the biggest eyes of any primate but lack the light-reflecting tapetum of other nocturnal primates, and the retina contains a fovea, which is also unusual for a nocturnal primate. Although their eyes are said to be truly bigger than their stomachs, muscular control of the enormous eyeballs is poorly developed. Consequently, tarsiers move their eyes by moving

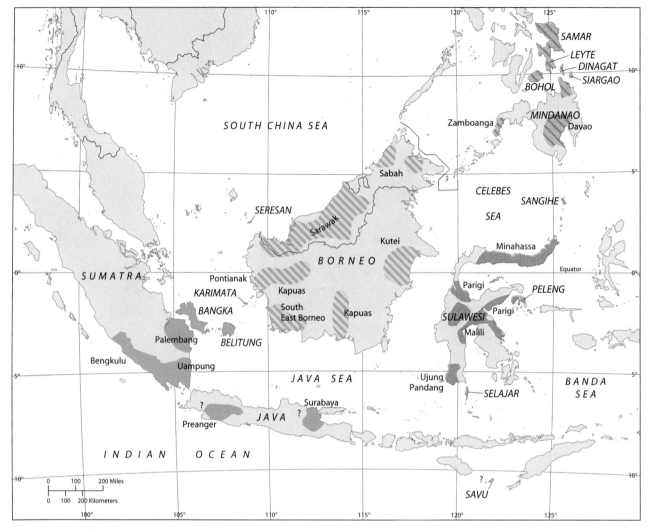

Figure 3-15 This map illustrates the approximate distribution of extant tarsiers.

their whole head, which can be swiveled around 180° so that they are looking directly backward. As befits their nocturnal lifestyle, tarsiers have large mobile ears that can be folded. They lack dental combs, but have a grooming claw on the second (as well as third) digits of the feet. The long tail, which is scaly with raised ridges of skin that resemble fingerprints underneath, is used in social contact although it is not prehensile. The fingers are relatively long with bulbous tips that act as suction pads when tarsiers move across vertical surfaces.

Five, or possibly six, species of tarsiers live in the tropical rain forests of Southeast Asia (Fig. 3-15). Of these, the two best known are *Tarsius bancanus* from the islands of Sumatra, Java, and Borneo and *Tarsius spectrum* from Sulawesi (formerly Celebes). They are superb vertical clingers and leapers with extremely long and powerful hind limbs that have elongated ankles (or tarsal bones, hence, the name of the primate) and fused leg bones. These specializations facilitate leaps of up to 6 meters (18 feet)—not bad for an animal that can fit into a teacup!

Tarsiers are the only fully carnivorous primate. They eat live prey exclusively, including insects, bats, birds, and snakes. Carsten Niemitz, an expert on tarsiers from the Free University of Berlin, has suggested that tarsiers and owls occupy similar niches in different parts of the world: Both are nocturnal predators with large eyes that lack a tapetum, and both are able to rotate the head to look directly behind. In addition, tarsiers and owls locate similar species of prey primarily through hearing, and then noiselessly ambush them near the ground.

Niemitz believes that the spectral tarsier, which in many ways resembles *G. senegalensis* in Africa, is the most primitive of the living tarsiers. Both the galago and spectral tarsier (unlike *T. bancanus*) sleep in nests and mothers move their young by carrying them in their mouths. Mothers of both species also park their babies in trees when they go off foraging during the night (see In the Field). Interestingly, subadult spectral tarsiers may guard or baby-sit infants when the mothers are absent, according to Sharon Gursky of Queens College in New York. Spectral tarsiers do not engage in the dispersed polygyny of lesser bushbabies. Instead a male and female adult form a strong pair-bond and share a territory of several hectares with up to four of their offspring (less often, spectral tarsiers live in small multimale, multifemale groups). This territory is defended with an array of vocalizations; by depositing scent from facial, chest, and anogenital glands; and by urine marking on the part of both sexes. Males and females exchange melodic calls (known as **duetting**), groom, playfight, intertwine tails, and otherwise engage in a good deal of touching.

As noted earlier, predators are one of the ecological factors that constrain, limit, or shape the behavior of populations. The locomotor patterns of fast leapers like galagos and tarsiers and slowly creeping lorises lend themselves to two different strategies for dealing with predators: (a) a rapid getaway or (b) avoiding detection by camouflage or freezing (i.e., by having cryptic coloration or locomotion). A slow creeper that is detected by an arboreal predator curls up, lunges, or bites the attacker. Although we humans might be tempted to think the fast leapers have the better strategy, Niemitz's observation of the interaction between a tarsier and a slow loris suggests otherwise:

> One slow loris was released in the jungle next to a Bornean tarsier. It caused the tarsier to leap away one or two jumps when it came too near. The tarsier, however, obviously forgot about the incident, as could easily be seen when the loris followed the tarsier. The loris, which was very hungry, actually tried to catch the tarsier, apparently in order to eat it. While the latter was noisily absorbed in eating the beetle, the slow loris crept down the thin tree trunk so stealthily that the author was forced to intercede at the very last second by vigorously kicking the tree trunk, causing the tarsier to fall to the ground. Besides their biological role as food competitor, living partly on insect food and possibly on small birds, the slow loris has also shown that it is able, at certain circumstances, to be a nocturnal predator to the tarsiers. (Niemitz, 1984:67)

REVIEW EXERCISES

1. How does the tarsier move, and what associated anatomical features exist in its hind limbs? What about the fingers?
2. How is the diet of tarsiers unusual for primates?
3. What nonprimate animals do tarsiers in some way resemble behaviorally and anatomically? This is an example of _____ evolution.
4. Describe the social system of *Tarsius spectrum*.
5. What general features cut across the three major groups of prosimians?
6. How do the major groups differ from each other?
7. Summarize the correlates of diurnal living.

Cache and Carry Infant Care

Sharon Gursky, a primatologist at Queens College in New York, is one of the world's leading experts on tarsiers. Below, she transports us back to moments in the field when she and her assistants made startling discoveries about infant caretaking practices of spectral tarsiers. In her own words:

SHARON GURSKY GAZES UP INTO A TARSIER SLEEP TREE. (PHOTO COURTESY OF SHARON GURSKY.)

"It's almost six o'clock in the evening and we are crouching underneath a large palm leaf near a huge strangling fig tree. The unseemingly never-ending rain is drizzling and it is becoming quite chilly, especially for tropical lowland rain forest in Sulawesi, Indonesia, located less than 2° from the equator. For the last 18 months, Celsius, Franz, and I have been following a small (120 g, or about 4 oz) elusive nocturnal primate, the spectral tarsier (*Tarsius spectrum*). Unlike other nocturnal mammals, the eyes of spectral tarsiers do not possess a reflective tape called the *tapetum lucidum*. Consequently, spectral tarsiers are difficult to observe in the forest at night because they lack the yellow glowing eyes that cats and other nocturnal mammals possess.

A few days ago, female 288 stopped returning to the sleep tree she had been sharing with the rest of her family and began using another tree for her sleeping site. She was obviously very pregnant and we had been anxiously awaiting the birth of her infant. Tonight though, we are waiting for her to leave her sleep tree and begin her nightly activity. We finally see some movement in the fig tree above. The female leaps from inside the tree to a small neighboring tree. Celsius notices that she must have given birth and is carrying her infant in her mouth much like a cat or a dog. This is very unusual. The majority of primates carry their infants either on their chests or on their backs, while only a few prosimian species use oral transport. Because the literature has conflicting reports about how Bornean tarsiers (*Tarsius bancanus*) transport their infants, I did not know what to expect for the spectral tarsier.

Female 288 seems more skittish than normal, so Celsius and I hold back a little and do not follow her as closely as we usually do. She begins leaping from tree to tree with the infant in her mouth until she is approximately 60 meters from her sleep tree. The female stops traveling and begins to nurse her infant. Less than 15 minutes later, she picks the infant up in her mouth and places it in the fork of a small branch. This behavior is commonly referred to as infant parking, caching, or stashing. The infant grasps the branch and watches its mother leave, traveling higher in the canopy. Since all the literature I have read indicates that after the infant is parked, it remains immobile and the mother does not return to nurse it until dawn, we decide to follow the mother. She travels to a height of about 8 meters in the canopy—less than 4 meters from her parked infant—and begins foraging. We keep expecting the female to leave the tree and travel around her territory, but she remains foraging within the same tree where her infant is parked. A few minutes later, the female spots a large rhinoceros beetle, grabs it, and begins eating it, hard outer shell and all. Approximately 20 minutes after the female first parks her infant, she returns to nurse and groom it. This is quickly followed by another bout of traveling during which, once again, the female carries the infant in her mouth, and parks it in a new tree about 40 meters from the last foraging site. The female travels higher in the canopy from where the infant is parked and begins to forage.

We are surprised by female 288's behavior because it indicates that, contrary to the literature, one nocturnal primate, the spectral tarsier, provides substantial amounts of care to infants even though parents do not continually carry them throughout their nightly activity period (and with good reason, since newborns weigh one-fourth to one-third the adult's weight). Instead, spectral tarsiers intermittently cache (park) and carry their infants throughout the night. We have now observed this same pattern of caretaking for six infants, and our observations suggest that during the first 3 weeks of an infant's life, it remains in physical contact with the mother for about 60% of the time. Although this contrasts sharply with 3-week-old chimpanzees that are in contact with their mothers 100% of the time, the maternal care given to spectral tarsier infants is clearly more substantial than previously believed.

After infants are approximately 3 weeks old, a new behavioral pattern emerges. The mother begins foraging in trees at a distance from the parked infant, and begins leaving it alone for longer periods of time. However, the total amount of care given to infants does not change because at this time other family members (dad, older brothers and sisters) begin hanging out with the parked infant. Nonmaternal family members not only share food with parked infants, but also retrieve them when they fall, play tag with them, warn them of nearby predators, and groom them. So, during times when the mother is away, other family members care for infants. Overall, we are tremendously surprised to have discovered that, rather than being negligent parents, spectral tarsiers provide substantial amounts of care for their infants.

SUMMARY

The three major groups of prosimians are the lemurs, loroids (including galagos), and tarsiers. A general prosimian pattern has been enumerated into 12 features, a number of which converge on three complexes related to nocturnality, olfactory communication, and vertical clinging and leaping that are shared by many species across the three groups. Certain features distinguish these groups from each other. Lorises and galagos tend to be small, are nocturnal, and eat little plant food in favor of less variable but more nutritious insects, vertebrates, and gums. Tarsiers are even more carnivorous than lorises. Some lorises, galagos, and tarsiers use nests and carry their newborns in their mouths. The practice of baby parking (which need not correlate with nest use) is also common in these groups. As detailed in the next chapter, lemurs constitute a much more diverse group of prosimians, although certain species share some of the traits described in this chapter for lorises, galagos, and tarsiers.

FURTHER READING

BEARDER, S. (1987) Lorises, bushbabies, and tarsiers: Diverse societies in solitary foragers. In: B. B. Smuts, D. L. Cheney, R. M. Seyfarth, R. W. Wrangham and T. T. Struhsaker (eds) *Primate Societies*. Chicago: University of Chicago Press, pages 11–24.

BEARDER, S. (1995) Calls of the wild. *Natural History* **104**(8):48–57.

JOLLY, A. (1988) Madagascar's lemurs: On the edge of survival. *National Geographic* 174:132–161.

4 LEMURS

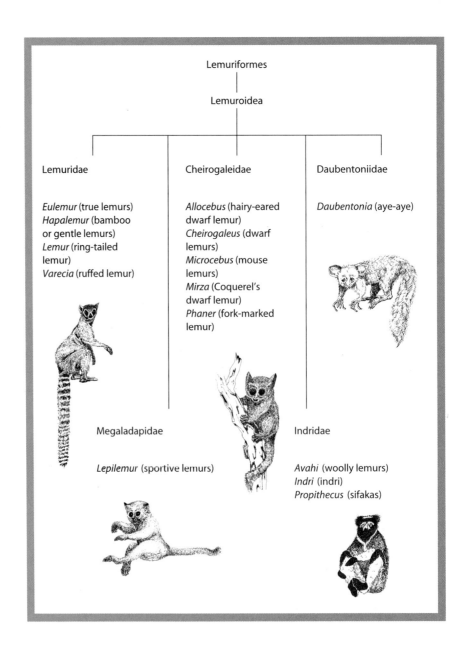

Lemuriformes

Lemuroidea

Lemuridae

Eulemur (true lemurs)
Hapalemur (bamboo
or gentle lemurs)
Lemur (ring-tailed
lemur)
Varecia (ruffed lemur)

Cheirogaleidae

Allocebus (hairy-eared
dwarf lemur)
Cheirogaleus (dwarf
lemurs)
Microcebus (mouse
lemurs)
Mirza (Coquerel's
dwarf lemur)
Phaner (fork-marked
lemur)

Daubentoniidae

Daubentonia (aye-aye)

Megaladapidae

Lepilemur (sportive lemurs)

Indridae

Avahi (woolly lemurs)
Indri (indri)
Propithecus (sifakas)

INTRODUCTION

Some primatologists regard lemurs as the living primates that are generally nearest the primitive mammalian condition. They evolved without competition from monkeys or apes on the island of Madagascar, which is somewhat smaller than Texas and is located about 250 miles off of the east coast of Africa. Presumably, the ancestors of modern lemurs rafted there on floating islands of vegetation over 40 mya during the Eocene epoch (Fig. 4-1). If Madagascar was anything like it is today, these pioneer lemurs found themselves on an island with an assortment of habitats ranging from desert to plains to rain forests. By the time humans first arrived about 2000 years ago, lemurs had radiated into at least 49 species. Unfortunately, around 20 of them are extinct today, partly because of humans (see Box 4). Even so, living lemurs are quite varied, ranging in average size from the 31-g (about 1-oz) pygmy mouse lemur to the 6- to 7-kg (13- to 16-lb) indri. Lemurs may be terrestrial or arboreal, social or solitary, and vegetarian, frugivorous, gummivorous, or insectivorous. Despite the fact that their name derives from the Latin *lemures*, which means "ghosts," some of the larger lemurs have been able to shake their nocturnal past and become diurnal. A lack of competitors and major predators no doubt contributed to this.

In this chapter, portraits are presented for five species that represent the five families of living lemurs shown in Figure i-6. Detailed portraits are possible for the large diurnal ring-tailed lemur and the gray mouse lemur because these species have been thoroughly studied. The remaining portraits for the sportive lemur, indri, and aye-aye are less detailed.

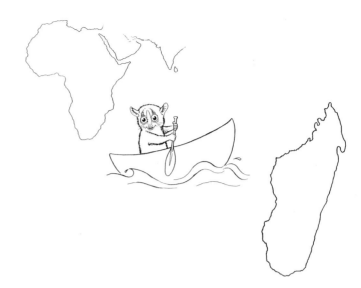

Figure 4-1   An ancestral lemur makes its way to Madagascar.

Box 4

Madagascar's Extinctions: What the Subfossils Tell

Although Madagascar boasts a sumptuous array of modern flora and fauna, **subfossils** (bones that have not had enough time to fossilize completely) indicate that about 1.5 times the number of living species of lemurs were resident a mere few thousand years ago. In fact, all but one family of living lemurs appear to have extinct relatives represented by subfossils. Reconstructions of some of the better known extinct Malagasy subfossil lemurs are illustrated here. The two specimens of *Archaeolemur* shown at the bottom of the tree were approximately the size of living baboons (15 to 25 kg, or 33 to 55 lb) and seem to have spent a good deal of time on the ground in open country. *Palaeopropithecus* (upper right), on the other hand, was arboreal although it weighed over twice as much as the largest of the *Archaeolemur* (i.e., up to 59 kg, or 130 lb). Its movement has been described as slothlike, with a tendency to progress suspended from trees with hooklike hands and feet. The remaining extinct lemur shown in the figure is *Megaladapis*, the best known and perhaps most unusual of the subfossils that became extinct only a few centuries ago. Despite a weight of up to 77 kg (170 lb), this lemur was arboreal with long grasping extremi-

Madagascar subfossils include the extinct lemurs *Archaeolemur* (two shown here on the ground), *Megaladapis* (upper left), and *Palaeopropithecus* (upper right).

ties. Alan Walker describes its locomotion as closest to that of the marsupial koala from Australia, which progresses up trees in a series of hops and makes short leaps between branches. Another lemur expert, Ian Tattersall of the American Museum of Natural History, notes that the shape of the skull transformed the entire head into a long extension of the neck and that this, together with a mobile snout, would have allowed *Megaladapis* to feed on leaves from a large surrounding area while sitting in a tree.

As pointed out by William Jungers of the State Uni-

LEMURIDAE

Eulemur ("true lemur") is the most diverse genus in the lemurid family (see Fig. i-6), with five species that include crowned, black, brown, mongoose, and red-bellied lemurs. As their names imply, these lemurs vary a good deal in coat color, both between species and between the sexes. Some of these species are active during parts of the night and the day; that is, they are **cathemeral**. The three species of bamboo lemurs (*Hapalemur*) are so-named because of their diets, while the one species of ring-tailed lemur (*Lemur catta*) gets its name (and its fame) from—you guessed it! The ruffed lemur (*Varecia variegata*) has long shaggy hair on its ears, which accounts for its name. "If anything in paleontology is certain," writes Ian Tattersall (1993:112), "it is that all anthropoids are descended from a diurnal common ancestor. So if we wish to find analogies to our remote Eocene ancestors,

REVIEW EXERCISES

1. What and where is Madagascar? Which two groups of primates presently live there?
2. Explain the derivation of the name for lemurs.
3. What is a subfossil?
4. Identify and describe *Megaladapis*. Is it alive today?
5. What is the significance of the Ranomafana National Park Project?

versity of New York at Stony Brook, one of the most consistent and interesting differences between the living and subfossil lemurs is that the extinct species were usually considerably larger than their living relatives. Other giant animals such as certain birds and tortoises also died out on Madagascar. But why have so many species of **megafauna** become extinct? Two intertwined causes seem to have been a changing climate and the arrival of humans. The selective extinction of large slow lemurs suggests that humans may have hunted them. Human settlers also cleared away large areas of forests in conjunction with farming. This would have been disastrous for resident lemurs, especially since forests and wetlands were already shrinking in response to an increasingly dry climate. Ross MacPhee of the American Museum of Natural History suggested still another possibility, that the first lemurs to come into contact with people may not have had resistance to human diseases. Indeed, any combination of these features could account for the unfortunate loss of so many spectacular species. Worse yet, many of the species that managed to survive on Madagascar are presently threatened with extinction.

But there is hope. In an effort to help conserve Malagasy tropical forests and their fauna, anthropologist Patricia Wright of the State University of New York at Stony Brook helped to organize the Ranomafana National Park Project (RNPP) in southeastern Madagascar. This multifaceted, international effort began in 1987 and is already seeing results with the identification of one previously unknown lemur (the golden bamboo lemur), another species thought to be extinct (the greater bamboo lemur), and a third species that was not thought to occur within this region of Madagascar (the aye-aye). Madagascar is the tenth poorest country in the world, and has already lost 90% of its native vegetation to farming, mostly as a result of slash-and-burn agriculture, according to Alison Richard of Yale University and Robert Sussman of Washington University. For this reason, Wright emphasizes that a key to the future success of the project is for conservation efforts to take place within the context of local economic development and education, and to be of practical benefit to the people who live in the area. Several research projects have been completed at Ranomafana and others are in progress. The RNPP is extremely ambitious and hopefully will be successful and of use as a model for primate conservation in other parts of the world.

we must turn to the primates of Madagascar." Although we do not share a diurnal common ancestor with lemurs (i.e., diurnality arose independently in the common ancestor of tarsiers and anthropoids and the ancestors of some of the lemurs), Tattersall's suggestion that diurnal lemurs might provide some insight into the evolutionary significance of diurnal living is worth pursuing. We therefore begin our survey with a diurnal species that lives among the largest of social groups of any lemur, and which is also the most terrestrial prosimian in the world.

The Ring-Tailed Lemur *(Lemur catta)*

Ring-tailed lemurs are sleek cat-sized animals (averaging about 2.7 kg, or 6 lb; Table 4-1) that have a rather startling appearance because they appear to be wearing masks, and have very long, bushy, ringed tails (Fig. 4-2). These animals have longer hind limbs than forelimbs and are at home both in the trees, where they do a fair amount of vertical clinging and leaping as well as scampering quadrupedally along branches, and on the ground. They are therefore semiterrestrial quadrupeds. Because they are diurnal, their eyes are not as large as those of nocturnal primates. However, they retain a light-reflecting *tapetum*, which suggests that they have become diurnal relatively recently, probably because of the lack of monkey and rodent competition that has always graced Madagascar. Their nocturnal past is also reflected in a pointed wet rhinarium that functions in olfaction, vibrissae, and glands that secrete scents on the wrists and insides of the upper arms.

Lemur catta is unusual because of its combination of nocturnal and diurnal adaptations. Nowhere is this better illustrated than in its spectacular tail, which is used for both visual and olfactory displays. Although the stripes and a high S-shaped carriage of the tail act as visual signals, this appendage is used by males in even more dramatic olfactory communications that are called **stink-fights** by Alison Jolly of Princeton University, who has done much of the classic research on this species. In these conflicts, the tail is rubbed with secretions from the wrist glands and then flicked at an opponent, an act of aggression that Jolly describes for the dominant male in one of the groups that she first observed:

> Often he gouged a sapling with his wrist spurs, or used them to comb a glandular scent through his black-and-white-striped tail. He then shivered the tip like a feather duster, an act of insult he usually aimed at the third-ranked male. That bullied lemur, lacking courage to stink-fight, would leap away with a spat call, a series of pleading squeaks pressed past lips drawn back in a grimace of fright. (Jolly, 1988:145)

Ring-tailed lemurs live in southwestern Madagascar (Fig. 4-3), where Sussman and his students have studied nine groups of individuals in the Beza Mahafaly Special Reserve for many years. All of the

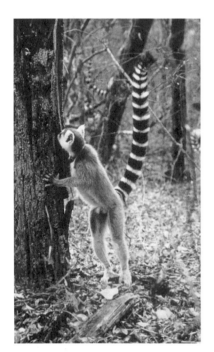

Figure 4-2 Ring-tailed lemurs are cat-sized animals that appear to be wearing masks, and have long, bushy, ringed tails.

TABLE 4-1	Summary Statistics for Lemurs				

	Lemur catta (ring-tailed lemur)	Lepilemur leucopus (white-footed sportive lemur)	M. murinus (gray mouse lemur)	I. indri (indri)	D. madagascariensis (aye-aye)
Adult mean body weights					
Female	2.68 kg (5.9 lb)	580 g (20.3 oz)	63 g (2.2 oz)	7.14 kg (15.7 lb)	2.52 kg (5.6 lb)
Male	2.71 kg (6.0 lb)	544 g (19.0 oz)	59 g (2.1 oz)	5.83 kg (12.9 lb)	2.69 kg (5.9 lb)
Adult brain weight (g)	26		1.8	35	45
Maximum known life span (yr)	27		16		
Nocturnal (N) or diurnal (D)	D	N	N	D	N
Tapetum	+	+	+	0	+
Olfactory communication[a]	sf, sm	um, sm?	sm, uw, um, fm	sm, cd	um, sm?
Location (habitat)[b]	sc, d, gf	d, gf	sf, d, gf, b	mf	pf, sf, rf
Locomotion[c]	sq	vcl	aq	vcl	aq
Diet[d]	f, l, fl, b, g, h	l, f	f, i, fl, g, n, v	l, f, s, fl, d	i, s, n, fu
Solitary foragers	0	+	+	0	+?
Nests	0	+	+	0	+
Mean no. of offspring	1	1	2–3	1	1
Mouth carry	0		+	0	0
Season breeder	+	+	+	+	0
Gestation (days)	134–145		59–62	120–150	158–172
Social organization[e]	mm	dp	dp	mo	dp
Female dominance	+	0?	+	+	
Territorial	0	+	0	+	
Ratio of adult females to males	5:4		1–4:1	1:1	
Infanticide/allomothering	+/+				
Average group size	5–27		1–15	2–5	1–2?
Density	1.5–3.5/ha	200–450/km²	60–800/km²	0.09–0.16/ha	
Home range	6–23 ha	0.2–0.5 ha	0.2–3.5 ha	17.7–18.0 ha	F: 31.7–39.5 ha M: 126–214 ha
Conservation status[f]	H	H	H	H	H

Note: Information in this table has been selected from many sources and should be viewed as tentative (i.e., subject to change in light of new information). Sources include Rowe, 1996 (including conservation statuses); Harvey et al., 1987 (brain weight); Bearder, 1987; Tattersall, 1982; Martin, 1973; Mittermeier et al., 1994; Niemitz, 1984; Feistner and Sterling, 1994; and other references mentioned in the text. Abbreviations: +, Yes; 0, no. Blank entries indicate that no information is available. Dietary items are listed in approximate order of decreasing importance to the species.

[a]Olfactory communication: cd, communal defecation; fm, fecal marks; sf, males stink-fight; sm, scent marks; um, urine marks; uw, urine washes.

[b]Habitat: b, brush; d, desert; gf, gallery forest; mf, montane forest; pf, primary forest; rf, rain forest; sc, scrub; sf, secondary forest.

[c]Locomotion: aq, arboreal quadruped; sq, semiterrestrial quadruped; vcl, vertical clinger and leaper.

[d]Diet: b, bark; d, dirt; f, fruit; fl, flowers; fu, fungus; g, gums (sap); h, herbs; i, insects; l, leaves; n, nectar; s, seeds; v, small vertebrates.

[e]Social organization: dp, dispersed polygyny; mo, monogamy; mm, multimale, multifemale groups.

[f]Conservation status: H, highly endangered.

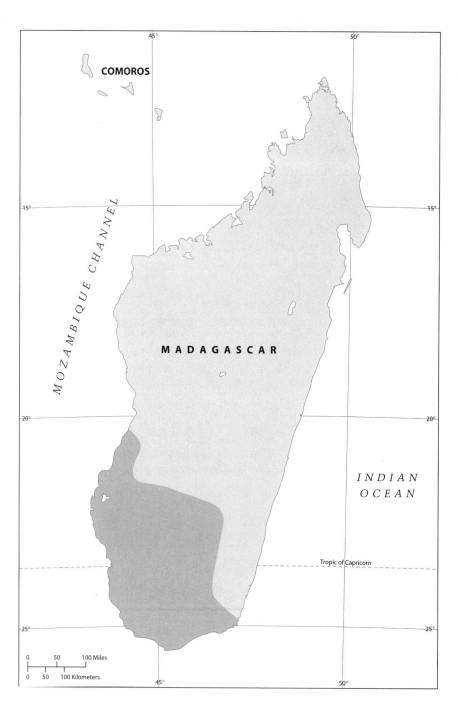

Figure 4-3 Ring-tailed lemurs live in southwestern Madagascar.

adults in the groups are collared and tagged with numbers so that they can be easily identified. The habitat at the reserve consists of varied riverine forests and is very seasonal, with a specific hot-wet season and a cool-dry season. This situation is ideal for studying ring-tailed lemurs in multiple ecological contexts.

According to Sussman and Michelle Sauther of the University of Colorado (Fig. 4-4), groups average 13 to 15 individuals (ranging from

5 to 27) with about the same number of adult males and females. Females remain in their natal groups, while males migrate away when they are 3 to 5 years old and change groups about every 3.5 years thereafter. Groups, or **troops** as they are sometimes called, are therefore centered around a core of mature females and their offspring. Dominance hierarchies have been observed especially for females, based on winners and losers in aggressive episodes and approach-retreat behavior. A single top-ranking female seems to head the troop, which is unusual for primates. A consistent hierarchy is less clear for males because they move around so much, although there is one non-natal male in each group that dominates all of the other males for access to food and receptive females.

Ring-tailed lemurs forage opportunistically on a wide range of leaves, buds, flowers, and fruits at all levels of the forest, including the ground. They eat their food at its source by plucking it directly with their mouths. Not surprisingly, groups with small home ranges expand their ranges to obtain wider access to food during the dry season. Thus, home ranges of different groups may overlap and effectively be time-shared for sleeping, resting, or feeding. In fact, because groups do not defend and maintain *exclusive* use of their home ranges, Sauther and Sussman reject the usual claim that ring-tailed lemurs are territorial in the strict sense of the word.

This is not to say that ring-tailed lemurs are docile. Nothing could be further from the truth. In a fine piece of socioecological detective work (that builds on earlier work on ring-tailed lemurs by Jolly, and on sifakas by Alison Richard of Yale University), Sauther has convincingly argued that the high level of aggression, particularly in females, seems to be related to an unusual mating system that is constrained in part by the dietary requirements of lactating females. Ring-tailed lemurs are one of the few primate species in which adult females consistently evoke submissive behavior (retreating, running away) from adult males. They do so by approaching, cuffing, lunging toward, biting, and chasing the males which, unlike those in male-dominated species, are of approximately the same body size as females. This aggression frequently occurs over food, as illustrated by another of Jolly's early observations:

Figure 4-4 Michelle Sauther is shown with a ring-tailed lemur.

An amber-eyed harridan I called Aunt Agatha . . . one day . . . astounded me by bouncing up to the troop's dominant male . . . to cuff him on the nose and appropriate a ripe tamarind he had just begun to eat. I waited for him to retaliate; instead, he meekly looked for other fruit. (Jolly, 1988:145)

True female dominance like that seen in *Lemur catta* and other lemurs is so rare that it begs for an explanation. Sauther interprets this behavior in terms of reproductive physiology and socioecology. She points out that females are normally in estrus for less than 1 day each year, as indicated by increased size and flushing of the vulvae and vaginal orifice. The mating season is confined to May, during which time adult females come into estrus on different days. Females aggressively rebuff males that approach them sexually during the rest of the year. Although males fight ferociously over estrous females (and lose weight doing so), the order in which females choose their several mates reflects the dominance hierarchy of the nonnatal males. As a result of this, and the fact that males engage in postejaculatory guarding of females to ward off other suitors for as long as possible, the dominant male stands a better chance of actually being the one to impregnate the females in his group (a finding that is supported by preliminary DNA fingerprinting analyses).

The birth season lasts for a few weeks around October, during which most females usually give birth to single offspring after a gestation of about 141 days. Lactating females have higher protein and calcium requirements than those of other females, and the birth season is exquisitely timed so that mothers are nursing and weaning their young during the wet season (December to April) when the leaf buds and young leaves that provide these nutrients are readily available. Since all reproducing females undergo similar stresses and must compete with each other for a limited food supply, female dominance helps to decrease further nutritional competition from males. Under these rather harsh conditions, one might wonder why males bother to stay in social groups at all. However, given the socioecological constraints that characterize ring-tailed lemurs, group membership is a reasonable, if trying, male (unconscious) strategy for passing genes along to future generations.

Dietary constraints in the setting of changing seasons are not the only pressures that mold the social behavior of *Lemur catta*. Another important factor is predator pressure. Ring-tailed lemurs are wary of cats, dogs, large snakes, and birds of prey including the harrier hawk and the buzzard. Curiously, these last two elicit distinctive sequences of vocalizations from ring-tailed lemurs. Instead of the usual clicks and yaps that are directed at other potential predators, the two avian predators cause *Lemur catta* to emit chirps, moans, and (at a more intense level) shrieks. Because much of their travel is done out in the open on the ground, ring-tailed lemurs are thought to be especially vulnerable to predation by harrier hawks and buzzards. Furthermore, Sauther observes that peaks in encounters with these birds of prey occur at a time of year when infant lemurs are especially vulnerable to attack, and also when hawk and buzzard chicks still require feeding. Interestingly, it has been suggested that the shrieks of ring-tailed lemurs indicate representational signaling that alerts troop members to aerial predators. (Such signaling has been observed in only a few other nonhuman primates, most notably Old World vervet monkeys.) These findings are important for a number of reasons. It is generally thought that the reason why some lemurs have become diurnal is because Madagascar lacks the monkey and rodent competitors that occupy diurnal niches in other parts of the world. As illustrated, however, *Lemur catta* has encountered some predators in its newfound diurnal niche, and this may be one reason why it has formed multimale social groups that are absent in nocturnal lemurs. It may also be a reason why highly dominant females allow males to live in their groups. That is, once aerial predators are sighted, they are harassed and mobbed by lemurs, and this is one of the few aggressive activities in which males may actually play a greater role than females.

As we have seen, the extensive female dominance toward males that is seldom seen outside of lemurs may be understood in terms of reproductive physiology that is adapted to dramatically fluctuating seasons and food supplies. *Lemur catta* also shows an interesting combination of nocturnal retentions (related to olfaction) and diurnal adaptations (including visual signals and social responses to diurnal predators). And don't forget, this last observation is best illustrated with the part of the anatomy for which the ring-tailed lemur was named!

MEGALADAPIDAE

The one genus in the megaladapid family, *Lepilemur*, contains seven recognized species of sportive lemurs, and still counting since the taxonomy of this group is currently under dispute. Sportive lemurs are widely distributed vertical clingers and leapers that live in Madagascar's forests. They prefer to eat leaves, and all species are nocturnal. They are also territorial and, in fact, are called sportive lemurs because, when threatened, they assume aggressive postures reminiscent of boxers (Fig. 4-5). *Lepilemur* is thought to be related to the extinct giant lemur *Megaladapis* (see Box 4), although all species weigh less than a kilogram. Adults lose the upper incisors, resulting in an unusual dental formula of 0.1.3.3./2.1.3.3. The white-footed sportive lemur has been selected for a brief portrait because more is known about this species than the other sportive lemurs.

The White-Footed Sportive Lemur *(Lepilemur leucopus)*

The white-footed sportive lemur is one of the smaller species of *Lepilemur*, with adults of both sexes weighing a little over a pound (see

Figure 4-5 The sportive lemur can assume an aggressive posture befitting a boxer.

Figure 4-6 This white-footed sportive lemur clings to a branch.

Table 4-1). This species has round ears and a long brown tail (Fig. 4-6). The fur on its back is grayish, with touches of brown around the head and shoulders, while the belly and underneath part of the tail are white. White-footed sportive lemurs live in the southern part of Madagascar (Fig. 4-7) in gallery forests or desert regions, where they subsist primarily on leaves, with the addition of flowers during the dry season. Leanne T. Nash of Arizona State University reports that

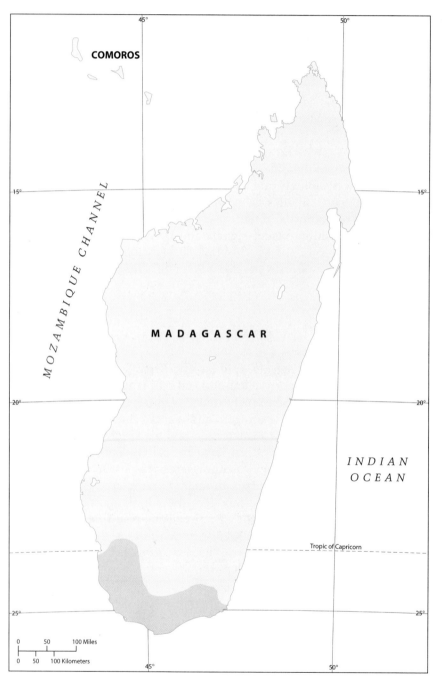

Figure 4-7 *Lepilemur leucopus* lives in the southern part of Madagascar.

this species copes with the poor quality of its diet through inactivity, especially during the cooler season. Like all sportive lemurs, *Lepilemur leucopus* is a nocturnal vertical clinger and leaper.

Lepilemur leucopus has been reported as both solitary and occurring in pairs. These observations, and the fact that male territories overlap those of females, are consistent with a dispersed polygynous social system. This species uses urine to mark branches and watches for intruders at the borders of its territories. Both sexes defend their territories with vocalizations, and scars on the bodies of all adults suggest that they engage in a good deal of actual fighting. Males and females have been reported to sleep both separately and together in tree hollows or thick vines. Breeding is seasonal and results in the birth of one offspring. Outside of the breeding season, the vulvas of females become sealed and the testes of males regress. Mothers are reported to park their babies in nests, holes, or vegetation. When mothers are active during the night, they frequently relocate their small offspring, possibly as a precaution against potential predators.

Night raptors prey on white-footed sportive lemurs. Other threats to their survival include habitat destruction from fires or overgrazing by livestock. They are also hunted for food by humans.

CHEIROGALEIDAE

There are five genera of small nocturnal dwarf lemurs in the cheirogaleid family (see Fig. i-6). These nocturnal quadrupeds have long bodies and short legs, and the average body weights for the eight recognized species are under 600 g, with some species being truly tiny. Dwarf lemurs sleep in small nests or tree hollows, and some species like the fat-tailed dwarf lemur (*Cheirogaleus medius*) actually hibernate for up to 6 months during the dry season. The cheirogaleids are thought to be the lemurs that most closely resemble African galagos. Because the genus *Microcebus* may be the most abundant and widespread of all the Malagasy primates, one of its better-known species, the gray mouse lemur, has been selected for an in-depth portrait.

The Gray Mouse Lemur *(Microcebus murinus)*

Mouse lemurs (*Microcebus*) include the smallest living primates in the world (Fig. 4-8). The classic fieldwork on the gray mouse lemur (*Microcebus murinus*) was done some 30 years ago by Robert Martin, and important new findings have recently been provided by a number of other workers. This species is interesting because it retains numerous primitive features that it shares with one of the more primitive lorises from Africa, *Galago demidovii* (Fig. 4-9). Martin suggests that both species remain morphologically and behaviorally close to the condition of the ancestor that gave rise to both lorises and lemurs. In other words, the earliest ancestors of lemurs that rafted across the

REVIEW EXERCISES

1. Describe the combination of nocturnal and diurnal adaptations found in *Lemur catta*. In this respect, how is the tail significant?
2. What is a stink-fight and which primate does it?
3. Describe the social groups of ring-tailed lemurs.
4. Which sex rules the roost in ring-tailed lemurs? What about other lemurs?
5. Adult female ring-tailed lemurs are in estrus for approximately —————————————— each year.
6. How do ring-tailed lemurs respond to birds of prey? What significance has this been given?
7. What sets sportive lemurs apart from other lemurs?
8. Discuss territoriality for *Lepilemur leucopus*.

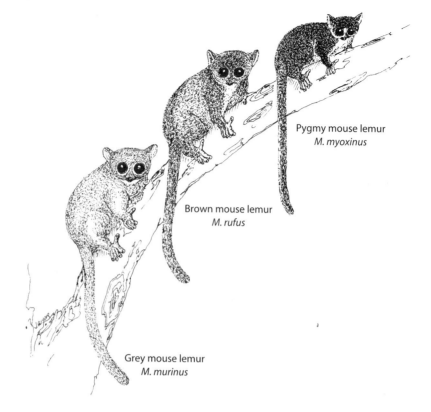

Figure 4-8 Three species of mouse lemurs are currently recognized: the gray mouse lemur (*Microcebus murinus*), the brown mouse lemur (*Microcebus rufus*), and the pygmy mouse lemur (*Microcebus myoxinus*).

Pygmy mouse lemur
M. myoxinus

Brown mouse lemur
M. rufus

Grey mouse lemur
M. murinus

Mozambique channel to colonize Madagascar may have resembled *M. murinus*. Gray mouse lemurs are unusual because adult females are considerably larger than adult males, and because they endure the severest predator pressure known for any living primate. These peculiarities beg for explanation and, as discussed later, they begin to make sense only when interpreted within socioecological contexts.

A

B

Figure 4-9 One of the galagos from continental Africa, *Galago demidovii* (**A**) bears a strong resemblance to the mouse lemur *Microcebus murinus* (**B**).

Gray mouse lemurs weigh a minuscule 5 to 6 g (about one-fifth of an ounce) at birth, according to Martine Perret of the National Museum of Natural History, Paris. Adults average about 60 g (roughly 2 oz), but are still so small that they can easily fit into the palm of a human hand. Another mouse lemur expert, Peter Kappeler of the German Primate Center in Göttingen, measured 60 adults and found that the average body weight of males is only 83% of that for females. This is an extraordinary reversal from the usual pattern in which males are bigger in sexually dimorphic primates. Not only are male gray mouse lemurs smaller than females, but also their canines are no larger than those of females. Clearly, the enlarged physical traits that characterize the males of most sexually dimorphic primates are lacking in this species.

Besides being tiny, gray mouse lemurs look rather like the animals for which they are named, having pointed little noses and tails that are sometimes longer than their furry bodies. They are strictly nocturnal, which is reflected in their large mobile ears, vibrissae, and big eyes that have a light-reflecting tapetum lucidum. In fact, Martin was able to locate his subjects after sunset because their tapetums readily reflected the light from his headlamp. Considering the difficulties of pursuing small, active animals in dark forests, Martin was able to learn quite a lot about the social organization as well as other details about his subjects. However, unlike diurnal species such as ring-tailed lemurs, he did not recognize most of his subjects as individuals. Olfactory communication, which plays an important role in regulating sexual behavior and social structure, is manifested in fecal markings, punctuated urination, and urine washing. However, the gray mouse lemur lacks the specialized scent glands that are present in many other prosimians (e.g., the glands inside the wrists of *Lemur catta*).

Gray mouse lemurs are found in dry as well as humid forests around the periphery of western and southern Madagascar (Fig. 4-10). Although tiny, this species is a bit larger than its eastern brown cousin (*Microcebus rufus*), and approximately twice the size of the pygmy mouse lemur (*Microcebus myoxinus*) which, at an average adult weight of about 30 g (approximately 1 oz), holds the record for being the world's smallest living primate. Gray mouse lemurs scurry quadrupedally along fine branches that are usually 1 to 5 meters above ground (Fig. 4-11). However, they frequently come to the ground to cross pathways or to catch beetles among the fallen leaves. Here, they move in froglike hops in which they thrust the hind legs and also use the forelimbs (i.e., they do not hop bipedally). These little omnivores eat insects, fruits, seeds, and an occasional small animal. They also use their dental combs to scrape trees for sap.

Years ago, Martin observed that *M. murinus* dwells along pathways at the edges of its habitat rather than in the interior. Greg Corbin of Yale University and Jutta Schmid of the German Primate Center at Göttingen confirmed this observation by placing radio collars on a number of individuals and tracking their nocturnal wanderings. They discovered that the larvae of insects that resemble cicadas deposit a sugary substance on plants at the edges of mouse lemur habitats during the dry season, and *M. murinus* adjusts its ranging patterns to take advantage of this food source. Seasonal changes in body weight and fat storage occur in this species and are related to length of day as well as food supply. During the rainy season when food is abundant, fat reserves are built up in the base of the tail. Although gray mouse lemurs do not actually hibernate like their relative the

Figure 4-10 Mouse lemurs (*Microcebus*) are probably the most abundant and widespread of the Malagasy primates.

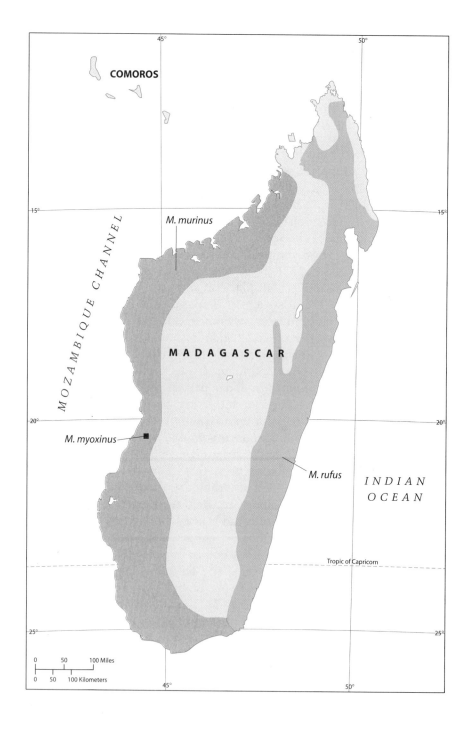

Figure 4-11 *Microcebus* assumes a variety of postures in trees and on the ground.

fat-tailed dwarf lemur (*Cheirogaleus medius*), their activity rates slow down during the dry season and they may not leave their nests for days. It is during this period of torpor that the fat reserves are consumed.

Not surprisingly, range size and population density vary with food supply. As summarized by Perret, when resources are evenly distributed as they are in northwestern Madagascar, there is little competition for food, population densities are low (60 to 250/km²), and extensive overlap occurs between individual home ranges that are relatively large (1.6 to 3.5 ha). Under these circumstances, the sex ratio is about 1:1. On the other hand, patchier distribution of resources in the drier forests of the south is associated with higher competition, higher densities of local populations (300 to 800/km²), and smaller home ranges (0.2 to 0.5 ha) that do not overlap. In this case, the female-male sex ratio is about 4:1 because surplus males have been excluded.

During the day, mouse lemurs sleep in nests in many kinds of trees, either in hollows or in spherical nests constructed from leaves. The diameters of these nests are rather small, averaging about 6.35 to 8.90 cm (2.5 to 3.5 inches). Because of tree felling by humans, a large number of nests have been collected and their inhabitants counted before being released. From this, it is known that as many as 15 females may sleep in the same nest, whereas males are often alone or share their nests with only one other male. During the winter quiescent period, however, nests include both sexes, but these groups break up in early spring. Relationships between females may be relatively stable, and it is likely that nestmates with highly synchronized reproductive cycles bear and rear their offspring in groups. Thus, although individuals are observed foraging alone during their active night hours, *M. murinus* is now known to be more social than previously thought. Similarly, the fact that males at times tolerate each other as nestmates is an indication that they are not strictly territorial.

The gray mouse lemur engages in dispersed polygyny, with relatively large ranges of central males overlapping smaller ranges of numerous females that the males presumably visit and mate with. As noted earlier, peripheral *M. murinus* males may be tolerated at the fringes of each population or excluded, depending on the distribution of resources, and sex ratios vary accordingly. In other words, the availability of resources in some way affects the expression of dominance in free-ranging males. But what are the mechanisms for expressing dominance? The answer to this is very interesting and helps to explain the lack of weaponry noted earlier. According to Perret, dominance relationships entirely disappear during the quiet winter months. However, as day length increases (and neurologically stimulates hormone production), mutual tolerance is replaced with chasing, fighting, and urine marking that reach a peak just before the mating season:

> Aggressive behaviour is relatively simple in mouse lemurs, consisting mainly of chasing and fighting. Aggression is essentially unidirectional, but threats, submissive postures or displacement behaviours are not easily recognized. Moreover, in mouse lemurs general activity is seasonal, and social behaviours vary accordingly. During the quiescent period (short day length), animals show great mutual tolerance. (Perret, 1992:8)

By the time mating season begins, linear dominance relationships between males have clearly been established and will be maintained until the next quiescent period. With gray mouse lemur males, the key to dominance seems to be in the urine rather than in relatively big bodies and teeth. As summarized by Kappeler and Perret, the urine of females contains chemical signals that generally stimulate males at the kickoff of mating season. On the other hand, a change in the length of daylight in combination with odors from the urine of females induces dominant males to produce substances that stimulate a reduction in body size, testes size, and hormone levels in subordinate males. It thus appears that the sense of smell interacts with light-regulated reproductive function in *M. murinus* so that dominant males do most of the marking and mating within a group—no weapons needed!

The mating season for gray mouse lemurs occurs in mid-September, and all adult females enter into a brief period of estrus within a period of 3 to 4 weeks. During estrus, a female's vulva, which is normally sealed, becomes swollen and opens. Each female is receptive and mates for only a few hours, and her vulva closes about a week after this. A female's entire estrous cycle therefore occurs within less than 2 weeks. On the day of receptivity, females show significantly increased locomotor and scent-marking activity, and utter advertisement trill calls, according to C. D. Buesching and colleagues at the German Primate Center in Göttingen. As is the case with female ring-tailed lemurs, a temporary vaginal plug may form after mating. During the mating season, there is intense competition among males for mates. Interestingly, male testes increase in size about a month before mating season and begin to decrease about a month after the season has ended.

Most adult females conceive during the mating season and most give birth to two or three infants. In keeping with the tendency to have multiple births, females have two pairs of nipples. Some females may produce two litters per year. Gestation is approximately 60 days, and infants are weaned about 1½ months after they are born, which happens to be during the rainy season when food is relatively plentiful. Thus, like ring-tailed lemurs, a strictly seasonal breeding pattern seems to be adaptive for infant mouse lemurs. As with *Lemur catta*, it also appears that female mouse lemurs have priority over males for access to food. Infants are born relatively helpless, are carried by mouth by their mother, and are left in the nest when she goes out to forage. Mouse lemurs grow up very quickly, however, and newborn females are likely to bear their own young the following year. Energetic needs associated with the unusually high rate of reproduction in this species may partially explain why females have larger bodies than males.

The high reproductive rate of *M. murinus* is adaptive because this species is subjected to the highest predation rate known for any primate. According to Steven Goodman of the Field Museum of Natural History and his colleagues, about 25% of the population is believed to be removed by predators every year. Although snakes and carnivores prey on gray mouse lemurs, most of this predation is due to two species of owls. Such a high rate of predation would bring the species to the brink of extinction, were it not for the high annual birthrate. However, predators are not all that threatens *M. murinus*. Given that an analysis of historical documents and satellite images (by Glen Green of Washington University and Sussman) reveals that 66% of the area covered by Madagascar's forests at the turn of the century was denuded by 1985, one can only wonder what will become of this important species that offers a number of hints about the possible nature of the common ancestor of lemurs and lorises.

INDRIDAE

Two genera in the indrid family (*Indri* and *Propithecus*) contain the large diurnal indris and sifakas, while the third genus (*Avahi*) consists of the small nocturnal woolly lemurs. All indrids are vertical clinger

REVIEW EXERCISES

1. Which species is the smallest and most widespread of all the lemurs?
2. What might the ancestors of lemurs that first rafted to Madagascar have looked like?
3. Name some physical and behavioral features that are associated with being nocturnal.
4. What is unusual about the sexual dimorphism of *Microcebus murinus*, and how is it explained?
5. How do male gray mouse lemurs compete for dominance?
6. What is the relationship between distribution of resources, range size, and population density for *Microcebus murinus*? How is sex ratio affected by these variables?
7. Why is the gray mouse lemur not strictly solitary? (Hint: What are its sleeping arrangements?)
8. What is thought to be the adaptive advantage of seasonal breeding patterns for lemurs?
9. In what way is the high annual rate of reproduction in *Microcebus murinus* adaptive?

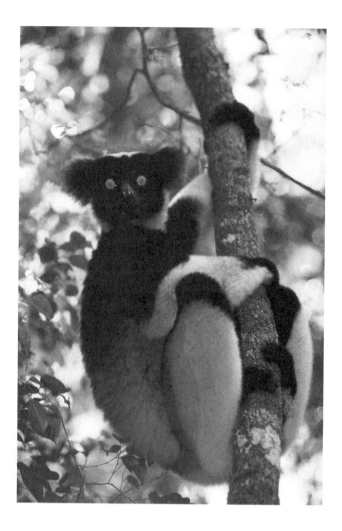

Figure 4-12 *Indri indri* is the largest of the extant lemurs.

and leapers, and are distinguished by an unusual dental formula of 2.1.2.3./2.0.2.3. The indri, which is the largest living lemur, and one of the most interesting socially, is portrayed here.

The Indri *(Indri indri)*

In many ways, the indri (or babakoto) differs markedly from the tiny mouse lemur. Only one species is recognized for this genus, and it is the largest of all extant lemurs, averaging 6 to 7 kg (approximately 13 to 16 lb), with captive animals being even heavier. Instead of resembling a rodent, this species appears almost teddy bear-like with an attractive black and white coat, round eyes, and stub of a tail (Fig. 4-12). The indri is fully diurnal. Its eyes lack the light-reflecting tapetum, and are not as large as those of nocturnal species. Like most prosimians, however, tactile vibrissae are present and olfactory communication continues to be important to the indri, which participates in communal defecation as well as anogenital scent marking of territories by both sexes. Males also mark territories with their cheeks.

Indris live in a restricted area of rain forest on the east coast of Madagascar, but were once more widely distributed as indicated by subfossils (Fig. 4-13). As discussed earlier, a relationship exists between body weight and the type of food a species eats, with only smaller primates like mouse lemurs able to subsist on a large portion of high-quality insect food, for which they forage alone. Large-bodied primates, on the other hand, tend to rely heavily on plentiful but

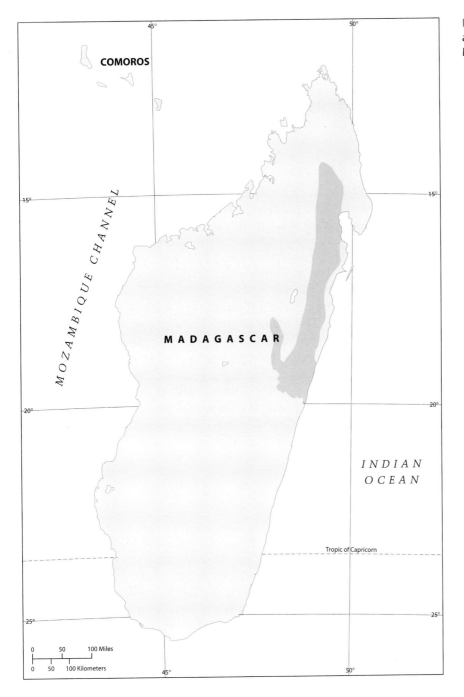

lower-quality leaves, and they forage in stable social groups. The latter is the case with the arboreal vertically clinging indri, which relies heavily on plants, fruits (using its dental comb as a scoop), and flowers. Curiously, for reasons that are not understood, it also eats dirt. Fruit is picked with the mouth, then transferred to and eaten from the hand. To eat leaves, the indri simply pulls branches down to its mouth and snacks. All of this is done anywhere from 2 to 40 meters above the ground.

J. I. Pollock of University College London, who studied indris in their natural habitats, observed that they consume at least 62 species of plants and that, from these, 5 to 12 species are eaten daily, including up to one or two ripe fruits. Thus, the overall diet of indri is extremely varied and plentiful. In fact, unlike ring-tailed lemurs, the seasonal changes associated with the indri's habitat do not seem to constrain the availability of food. However, Pollock notes that groups are likely to travel farther to reach fruiting trees, the distribution of which therefore affects ranging behavior. Probably as a partial result of this set of behaviors, the indri has developed a social organization that stands in startling contrast to those that have already been described for lemurs.

Indris live in small groups consisting of a **pair-bonded** adult male and female (i.e., mates) and their one to three immature offspring. The group lives, moves about, and changes its sleeping trees nightly within a territory of approximately 17 to 18 ha. Indris spend a good deal of time feeding, and less in social activities such as grooming. Different territories overlap at the edges, and group members defend their space vocally with loud modulated howls. In fact, some regard indris as the noisiest lemur. Urs Thalmann of the University of Zurich-Irchel and his colleagues have found that the song structure of indris is more complicated than previously reported, and that males and females sometimes duet. Although indris are seasonal breeders, a pair produces only one offspring every 2 or 3 years. In feeding situations, females dominate males despite the apparent freedom from the seasonal stresses due to competition for food that plague other lemurs. This may indicate that female dominance preceded the evolution of a variety of social organizations in the ancestors of lemurs.

DAUBENTONIIDAE

There is only one living species in this family, the aye-aye. As described in the following portrait, the aye-aye is extremely specialized in a number of different features. It appears to be descended from an extinct species, *Daubentonia robusta*, that was larger bodied than its living descendants. As was the case for the families of sportive lemurs and indrids, this one is distinguished by a unique dental formula, 1.0.1.3./1.0.0.3.

The Aye-aye (*Daubentonia madagascariensis*)

No review of lemurs would be complete without discussion of the largest of the nocturnal primates (around 2.6 kg [5.7 lb], or about the size of a large cat), the mysterious and exquisitely specialized aye-aye. Villagers are ambivalent about aye-ayes. Some are frightened of them because of their appearance, which is characterized by little pointed faces, widespread glowing eyes that are enclosed in dark circles, and extremely large membranous ears (Fig. 4-14). Other villagers think that aye-ayes bring good luck, according to one of the leading experts on this species, Eleanor Sterling of the American Museum of Natural History (see In the Field). The coat of *Daubentonia* appears scraggly because it contains many long dark coarse hairs interspersed with softer ones, and the tail is long and bushy. Except for the big toe, or **hallux**, the digits have claws instead of nails. Furthermore, the third digits of the hands are thin and wiry, like a skewer (Fig. 4-15). As if this were not enough, in addition to the unusual dental formula, the constantly growing large chisel-like front teeth project in a manner that gives the mouth a rather macabre appearance (see Fig. 4-15). Being nocturnal, this species has a keen sense of hearing, which is indicated by enormous ears, and of smell (both sexes urine mark).

Until recently, it was thought that aye-ayes were restricted to a

Figure 4-14 The aye-aye is the largest of the nocturnal primates.

Figure 4-15 The aye-aye has extremely specialized manual digits (note the elongated middle digit) and chisel-like front teeth related to its feeding habits.

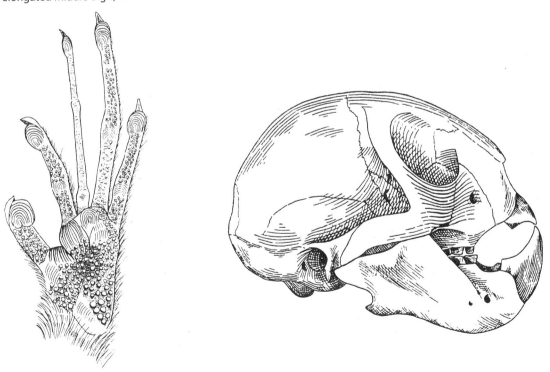

small segment of Madagascar's east coast. However, Sterling recently showed that *Daubentonia* may be one of the most widely (but perhaps not densely) distributed primates in Madagascar (Fig. 4-16). Aye-ayes are arboreal quadrupeds that engage in walking, climbing, vertical clinging, and jumping from branch to branch in a wide range of trees and shrubs. They are also comfortable descending head first, or leaping on four feet along the ground. As they move, the tail is carried extended in a curve that is concave downward.

Figure 4-16 The distribution of the aye-aye recently was shown to be wider than previously believed. The squares represent recent sightings.

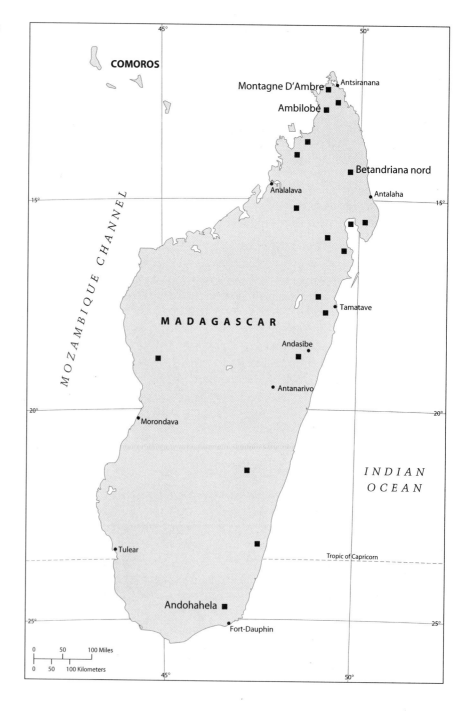

The aye-aye's unusual anatomical features have fascinating functions that are related to its particular ecological niche. It uses its keen senses along with its peculiar fingers and teeth in its nightly pursuit of favorite foods including wood-boring insect larvae. The aye-aye locates them by tapping along wood with its third finger and listening (termed **tap-scanning** by Carl Erickson of Duke University, and also known as **percussive foraging**). Once larvae are located, a hole is made in the wood with the front teeth, and the larvae are skewered with the finger, extracted, and eaten. Although this specialized anatomy is also used to open and eat green coconuts, mangoes, and other fruit, the aye-aye has been described as occupying a woodpecker niche in Madagascar because of its percussive pursuit of insects that live in wood. The aye-aye also uses its middle finger in a little-remarked-upon Q-tip manner, as described by Osman Hill:

> The aye-aye will, on occasion, suspend itself by its hind-feet, using its hands to feed or to perform its toilet. In this latter operation it uses its long third finger to comb its fur, scratch itself and cleanse its face, corners of its eyes, ears and nose, keeping meanwhile its other fingers flexed. (Hill, 1953:696)

Brain size may also be related to the aye-aye's highly specialized niche. The brain of *Daubentonia* is surprisingly large for a prosimian of its body size. For example, its brain is significantly bigger than that of *Indri*, whose total body weighs over twice that of the aye-aye. *Daubentonia* uses its hands to process fruit in a more skilled manipulative manner than is typical for prosimians, and this together with its enhanced sense of hearing may require more neural machinery. In a passing remark in one of his classic monographs, Hill (1953:696) noted that some workers believed "the aye-aye uses only its left hand in feeding." Sterling, on the other hand, believes that aye-ayes are probably ambidextrous. These observations are thought provoking from the point of view of primate brain evolution (see Neural Note 4).

Although aye-ayes were discovered over 200 years ago, very little has been learned about their social organization. However, an important volume of *Folia Primatologica* that appeared in 1994 ("The aye-aye: Madagascar's most puzzling primate," edited by Anna Feistner of Jersey Wildlife Preservation Trust, and Sterling) has done much to rectify this situation. It is now clear that aye-ayes nest in leaf nests in much the same way that mouse lemurs do. Aye-ayes have been observed foraging alone, in small groups of adults, and in mother-offspring pairs, so like mouse lemurs they may be somewhat social.

The reproductive behavior of aye-ayes is also better understood than it was a few years ago. Males fight for access to estrous females and, surprisingly, Sterling has shown that aye-ayes do not appear to be seasonal breeders as was previously believed. A copulatory position that is sometimes used involves the female hanging suspended from a branch and the male mounting her so that she is carrying both of their weights (Fig. 4-17, see In the Field). Unlike mouse lemurs, females give birth to only one offspring at a time, after a gestation that aver-

REVIEW EXERCISES

1. Which species represents the largest lemurs?
2. How does the indri differ in appearance from ring-tailed and mouse lemurs?
3. Describe social organization in the indri.
4. Of all the nocturnal primates, which species is largest? What is the size of its brain?
5. Describe three unusual anatomical specializations of *Daubentonia*. How are they used?
6. What, besides skewering larvae, do aye-ayes use their impressive middle manual digits for?
7. Are any nonhuman primates right-handed to the extent that people are?

Figure 4-17 Aye-ayes sometimes copulate suspended upside-down from a branch.

MacNeilage's Model of Left-Handed Reaching and Brain Evolution

The right hand is controlled largely by the left side of the brain, whereas the left hand is dominated by the right hemisphere. Unlike other primates, over 92% of humans are right-handed for most tasks. Thus, right-handedness is a manifestation of brain lateralization or cerebral dominance of the left hemisphere for manual activities. Since the last century, thousands of case histories of patients with language deficits and a multitude of scientific experiments have documented that speech is another activity that is controlled by the left hemisphere. In fact, this is true for over 94% of humans, including 95% of the right-handers and 85% of the left-handers. The fact that speech and handedness are dominated by the same (left) hemisphere in most people is not surprising because the parts of the brain that control the right hand and speech are located very close to each other (i.e., forming sort of a neurological package) (see figure). As you are well aware, nonhuman primates do not communicate in the same way we do. Unlike

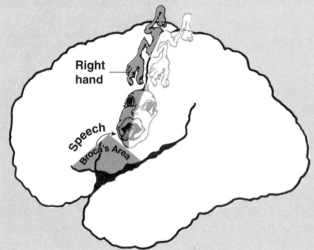

The part of the frontal lobe that controls movements of the right hand is close to Broca's area, which has an important role in speech.

ages 167 days according to Kenneth Glander of Duke University Primate Center. Aye-ayes have only one pair of nipples that, curiously, are located low on the abdomen rather than on the chest.

Add to their unusual appearance the fact that aye-ayes have a reputation for occasionally killing chickens and being fearless of humans, and it is not difficult to understand the unfortunate genesis of the Madagascar folklore that aye-ayes are harbingers of grave misfortune. Needless to say, this superstition hampers conservation efforts for this species because it encourages the killing of aye-ayes.

SUMMARY

The Malagasy subfossil lemurs, such as *Megaladapis*, recently became extinct. Hopefully, current efforts to promote conservation of the re-

humans, they also do not use the same hand for virtually all of their one-handed tasks. For these reasons, brain lateralization has traditionally been viewed as the result of relatively recent evolution in early humans.

This picture has changed dramatically within the last 10 years, partly because of surprising research on prosimians. In 1987, Peter MacNeilage of the University of Texas at Austin and his colleagues decided to take a hard look at the earlier literature on handedness in nonhuman primates. As expected, most studies indicated that populations of nonhuman primates fail to show the dramatic preference for general use of one hand that characterizes *Homo sapiens*. On closer inspection, however, it became clear that hand preference sometimes exists in nonhuman primates, but only for certain tasks. In particular, various prosimian species (possibly including the aye-aye) prefer to reach for food with the left hand, whereas monkeys and apes often use the right hand for other types of tasks that involve manipulation of objects.

MacNeilage theorizes that prosimians retain an ancestral pattern of clinging to a support (e.g., branch) with the right hand while reaching for food with the left, and that this goes along with the right hemispheric specialization for visuospatial skills that characterizes most primates in-cluding humans. Thus, the right hemisphere became specialized early on for the hand-eye coordination needed to see food and to use the left hand to reach for and carry it to the mouth. It was not until later in evolution when some primates became ground living that the right hand gave up its role as a "hanger-on." Freed of its grasping responsibilities, the right hand (having nothing better to do) took on new challenges under the tutelage of the left hemisphere (which had been subtly developing the ability to process socially meaningful vocalizations while the other hemisphere was concerning itself with visually guided feeding). Through this process, the right hand of terrestrial primates eventually became adept at skilled manipulation (i.e., more complex than reaching for and grasping food), while the activities of the left hand became more generalized and supportive of the right hand's undertakings. By the time *Homo* appeared, the left hemisphere had become dominant for virtually all manual activities including reaching for food, as well as another behavior that sets humans apart from all other primates—language. In sum, MacNeilage's investigation of hand preferences in prosimians suggests that, rather than being new to *Homo*, brain lateralization has enjoyed an extremely long evolutionary history.

maining lemurs and other biota at Ranomafana may provide a model for conservation in other parts of the world. Representatives of the five families of lemurs include the ring-tailed lemur, white-footed sportive lemur, gray mouse lemur, indri, and aye-aye. Some species of lemurs share with some of the other prosimians the practices of using nests and carrying their newborns in their mouths (see Tables 3-1 and 4-1). All lemurs except the aye-aye are seasonal breeders and female dominance, which is rare among other primates, is common among them. Within a socioecological framework, these features (as well as others) have been recognized as reproductive strategies in the setting of erratic food supplies.

Because the ancestor of all anthropoids is believed to have been diurnal, the ring-tailed lemur and indri are of particular interest. Interestingly, these species continue to share certain features with nocturnal lemurs, such as reliance on olfactory communication, sea-

Rendezvous in the Dark

*Eleanor Sterling, an anthropologist, is one of the world's leading experts on aye-ayes (*Daubentonia madagascariensis*). She spent 2 years studying a population of aye-ayes on an uninhabited island off the northeast coast of Madagascar. Much of the recent breakthrough in our understanding of these extraordinary primates is due to Sterling's experiences in the field. In her own words:*

ELEANOR STERLING HAS CONTRIBUTED MUCH TO RECENT INSIGHTS ABOUT THE ONCE-MYSTERIOUS AYE-AYE.

A constant refrain during my undergraduate physical anthropology classes went, "All primates are distinguished by (some feature), except the aye-aye." For instance, primates have nails on their digits, except the aye-aye; primates have mammary glands at the chest level, except the aye-aye (at the abdominal level); no primate species has continuously growing incisors, except the aye-aye. My desire to learn more about this puzzling primate eventually led me to the only place in the world where it is found in the wild—Madagascar. When I first started my field research on the aye-aye, comparatively little was known about aye-aye ecology and social behavior. The short-term studies that had been undertaken on the aye-aye yielded a few tantalizing facts about its diet and nest-building behavior, but its nocturnal, cryptic habits precluded in-depth observations. Managers of a recently instigated captive breeding program sought information about aye-aye social—and particularly mating—behavior to effectively manage the captive population of this endangered species. Consequently, I was anxious to collect any data relevant to their social behavior. Night after night two Malagasy guides and I followed individual

sonal breeding patterns, and a tendency for females to dominate males. However, these diurnal lemurs have given up sleeping in leaf nests or tree hollows during the day and the relatively solitary pursuit of insects at night, in favor of a more active diurnal rhythm that not only is visually stimulating, but also facilitates a gregarious social life. Diurnal living is further correlated with a more-varied diet (including a large component of leaves), bigger body, larger brain, and birth of a single infant that is able to cling to fur, thereby releasing its mother from the task of transporting it by mouth (although a number of these features are also seen in the large nocturnal aye-aye).

As we have seen, prosimians manifest a wide variety of anatomical specializations and socioecological adaptations. Their social systems range from nearly solitary, to dispersed polygyny, to monogamous pair-bonds, to large multimale, multifemale social groups. According to Tattersall, this extraordinary variety of biological and behavioral adaptations suggests that

> from the beginning, well before the increase in brain size that we associate
> with the higher primates today, primates showed a behavioral flexibility

radio-collared aye-ayes, wondering (among other things) when and how they reproduce. One night a young male aye-aye acted strangely. He traveled quickly over long distances, then stopped in one place for a while, but not to feed as expected. He seemed to pace along tree branches, then headed off on a long trek again. In the early morning he arrived at a nest inhabited by another aye-aye. He tried to enter the nest, but was repelled by the inhabitant. He ended the night by sleeping out in the open above the nest. I turned the receiver through the channels, trying to see if the animal within the nest was one we had tagged with a radio collar. I hit the frequency of our one collared female and suddenly heard the characteristic beeping of her collar. We had never before seen an aye-aye sleeping outside, unprotected by a nest, nor had we seen two aye-ayes sleeping in such close proximity.

I spent the following day trying to work out what was happening, retracing our steps to look for signs of feeding on something unusual that could explain the male's long walkabout, but found nothing. We arrived in the later afternoon at the place where we had left the female and young male to make sure we were there before they woke up. We settled in below the tree, taking time to prepare ourselves for the long night.

That turned out to be the last bit of rest we had for the next three nights. During this exhausting time, we learned that aye-aye females "call in" males with vocalizations and scent marking and several males compete for access to the female. When a male finally wins, he and the female hang upside down from a tree limb; she grasps the limb and he grasps her ankles and about her chest. They remain in this position for over an hour, during which time several other males are climbing nearby trees and attempting to dislodge the male by pulling his tail. All the animals call excitedly throughout this event. At the end of the copulation, the female sprints to a new area 500 to 1000 meters away and calls in males to start the whole adventure over again.

Unfortunately, the female's collar malfunctioned before she would have come to term, so we do not know the results of all this frenetic activity. However, the information we gathered helped the managers of the captive population to decide when and how to house the aye-ayes to stimulate reproduction. The captive population has produced 12 surviving aye-ayes to date. With the future of aye-ayes in the wild uncertain, this successful breeding program provides a safety net for the wild population.

and adaptability that belie the inference most people would draw from the description of these early forms as "primitive." It seems to me that this is the essential evolutionary heritage of our order.... (Tattersall, 1993:114–115)

FURTHER READING

ALEXANDER, J. P. (1992) Alas, poor *Notharctus*. *Natural History* **101**(8):55–59.

CULOTTA, E. (1995) Many suspects to blame in Madagascar extinctions. *Science* **268**:1568–1569.

JOLLY, A. (1988) Madagascar's lemurs: On the edge of survival. *National Geographic* **174**:132–161.

KAPPELER, P. M. AND J. U. GANZHORN (1993) *Lemur Social Systems and Their Ecological Basis*. New York: Plenum Press.

TATTERSALL, I. (1993) Madagascar's lemurs. *Scientific American* (January) **268**:110–117.

5 MONKEYS OF THE NEW WORLD: THE CALLITRICHINES

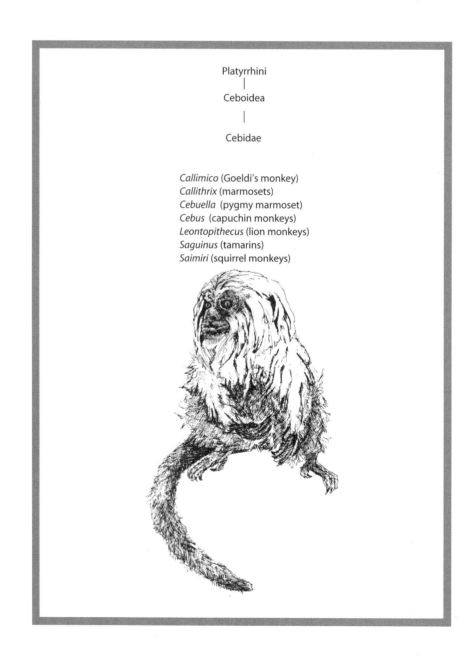

Platyrrhini
|
Ceboidea
|
Cebidae

Callimico (Goeldi's monkey)
Callithrix (marmosets)
Cebuella (pygmy marmoset)
Cebus (capuchin monkeys)
Leontopithecus (lion monkeys)
Saguinus (tamarins)
Saimiri (squirrel monkeys)

INTRODUCTION

The ancestors of New World monkeys arrived in South America by about 30 mya during the Oligocene epoch, although it is still not clear whether they originated in North America or Africa. Because the relevant fossil record is so skimpy, nobody knows for sure just how long before that time the first primates arrived in the New World, exactly how they got there (perhaps on rafts of floating vegetation), or if they resembled prosimians or monkeys. What is known, however, is that for millions of years after their arrival, South America remained a relatively isolated island continent. It finally joined with Central America after establishment of the Panamanian land bridge 5 to 6 mya. At that time, a variety of animals including various species of monkeys moved north into Central America and southern Mexico (Fig. 5-1).

Because of their isolation from primates in other parts of the world, today's New World monkeys are the products of very long evolutionary histories that were separate from those of primates living in other parts of the world. This provides us with a chance to study parallel evolution between New World and Old World monkeys as well as the environmental features that may have channeled the similarities in their adaptations (see Box 5). Recall that the term *monkey* indicates a grade of primate that is between prosimians and apes (see Fig. 1-14). Researchers vary in their opinions about the extent of the similarities between **neotropical** monkeys from the New World and the Old World monkeys from Africa and Asia. Some workers see no fundamental differences between New and Old World monkeys, while others think there is only a limited resemblance between them.

Extant New World monkeys are divided into 16 genera that contain approximately 50 recognized species, and we are still counting

Figure 5-1 Living New World monkeys are found in parts of South America, Central America, and southern Mexico.

(Table 5-1, Fig. 5-2). Because the evolutionary history of neotropical primates is so poorly understood (see Chapter 1), workers sometimes differ in their opinions about which genera are most closely related to each other. Consequently, the taxonomic classifications of these researchers also differ. The classification used in this text (see Table 5-1) follows Alfred Rosenberger of the Smithsonian Institution, and is based on a combination of anatomical and molecular evidence summarized by H. Schneider of the Federal University of Para in Brazil and Rosenberger. The superfamily Ceboidea is divided into two families, the Cebidae with seven genera and the Atelidae with nine genera. This classification departs from earlier ones that included the marmosets, tamarins, and Goeldi's monkey in their own family (the Callitrichidae) because recent molecular and anatomical studies now

TABLE 5-1 *Ceboidea*

Cebidae	Atelidae
Callithrix (marmosets)	*Aotus* (owl monkeys)
Cebuella (pygmy marmoset)	*Callicebus* (titi monkeys)
Leontopithecus (lion tamarins)	
Saguinus (tamarins)	*Cacajao* (uakaris)
Callimico (Goeldi's monkey)	*Chiropotes* (bearded sakis)
	Pithecia (sakls)
Saimiri (squirrel monkeys)	
Cebus (capuchin monkeys)	*Alouatta* (howler monkeys)
	Ateles (spider monkeys)
	Brachyteles (woolly spider monkeys)
	Lagothrix (woolly monkeys)

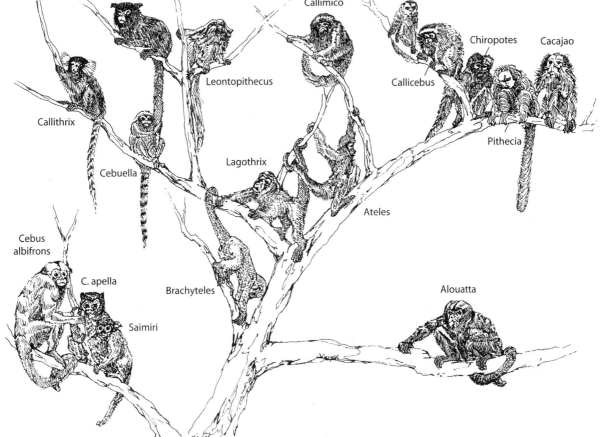

Figure 5-2 New World monkeys are highly variable in their appearance.

Box 5

Convergent and Parallel Evolution

It is well known that organisms in different parts of the world may appear surprisingly alike, even if they are not particularly closely related. For example, a number of marsupials that evolved on the island continent of Australia have clear counterparts in placental mammals from other parts of the world. Australia has its own "wolves," "anteaters," "cats," "flying squirrels," "ground hogs," "mice," and "moles" (see figure). These coincidences are examples of **convergent evolution** because they are the result of independent adaptations to similar habitats in animals that are extremely remotely related.

Parallel evolution is similar to convergent evolution, but occurs in organisms that are somewhat more closely related. Although parallelisms appear independently after lineages split from a common ancestor, their occurrences are believed to have been channeled not only by similar selective pressures, but also by characteristics that were retained from the common ancestor. Despite the fact that New and Old World monkeys underwent long evolutions in isolation from each other, a visitor to a zoo immediately recognizes the monkiness in members from both groups precisely because of parallel evolution. John Robinson of the Wildlife Conservation Society in New York and Charles Janson of the State University of New York at Stony Brook have identified examples of parallel evolution in the configurations of body weights, diets, and ranges in a number of New World and Old World monkeys, including the following pairs: squirrel monkey (*Saimiri*)/talapoin monkey (*Miopithecus*), capuchin (*Cebus*)/mustached guenon (*Cercopithecus cephus*), spider monkey (*Ateles*)/gray-cheeked mangabey (*Cercocebus albigena*), and woolly spider monkey (*Brachyteles*)/red colobus monkey (*Colobus badius*). (A number of these monkeys are discussed in the following chapter.) For example, both species in the last pair eat highly folivorous diets of mature and immature leaves, and have day ranges that are approximately the same size. Groups of woolly spider monkeys and red colobus monkeys are similar in size, frequently include a number of adult males and females, and may be characterized by female migration. Further, estrous females of both species sometimes copulate with a number of males in quick succession. There are also numerous differences between these New World and Old World pairs. For example, groups of woolly spider monkeys are reported to divide into smaller subgroups, contrary to red colobus monkeys, and they also have larger home ranges. The fact that certain socioecological factors are subject to parallel evolution is consistent with discussions in earlier chapters about the relationships between body size, type of diet, morphology of teeth, and ranging patterns.

Convergent evolution occurred numerous times between placental and marsupial mammals.

CONVERGENT EVOLUTION OF PLACENTAL AND MARSUPIAL MAMMALS

PLACENTALS

MARSUPIALS

Tasmanian
wolf
(Thylacinus)

Wolf
(Canis)

Native cat
(Dasyurus)

Ocelot
(Felis)

Flying
squirrel
(Glaucomys)

Flying
phalanger
(Petaurus)

Ground
hog
(Marmota)

Wombat
(Phascolomys)

Anteater
(Myrmecophaga)

Anteater
(Myrmecobius)

Mole
(Talpa)

Mole
(Notoryctes)

Mouse
(Mus)

Mouse
(Dasycercus)

group these species in the Cebidae along with *Saimiri* and *Cebus*. Although some workers would also shift *Aotus* to this family, the evidence is inconclusive and, for the present, *Aotus* is retained with the other monkeys in the Atelidae. Genera that are grouped together in Table 5-1 are believed to be closely related. For example, the marmosets, tamarins, and Goeldi's monkey form a subfamily, the callitrichines.

As was the case for lemurs in Madagascar, the long isolated evolution of New World monkeys led to a wide variety of species with different locomotor patterns, social systems, and feeding adaptations. Rosenberger points out that South America contains as much as 60% of the world's expanse of continuous closed tropical forests. He suggests that the ecological diversity of neotropical primates resulted partly from radiation into all levels of the three-dimensional arboreal habitat, and also from interactions between innumerable coexisting nonprimate species. As discussed in subsequent chapters, living Old World monkeys are not as diverse as the New World monkeys, perhaps because a sequence of successful adaptive radiations replaced each other through time. Rosenberger (1992:216) speculates that in the New World, on the other hand, "the fossils we are sampling may in fact reflect all that there ever was—a single adaptive radiation that has continued without much replacement since its origins."

THE NEW WORLD MONKEY PATTERN

Despite the variability of New World monkeys, one can delineate a general pattern in much the same way that was done for prosimians. It must be noted, once again, that no one feature distinguishes neotropical primates from other primates, but that a set of particular traits makes up a general pattern.

BONY FEATURES (Fig. 5-3)

1. Dental formula of 2.1.3.2. for *Callithrix, Cebuella, Leontopithecus,* and *Saguinus*; dental formula of 2.1.3.3. for other ceboids
2. Clawlike nails on all but the big toe of marmosets and tamarins; curved nails on the digits of other ceboids
3. Curved finger bones in some species

OTHER FEATURES (Figs. 5-3 and 5-4)

4. Oval nostrils that are widely separated
5. Prehensile tails in some species
6. All species arboreal
7. Primitive frontal lobe sulcal pattern

What functional explanations account for the New World monkey pattern? First of all, it is difficult to interpret feature 4, from which

REVIEW EXERCISES

1. What is known about the place and time of origin of New World monkeys?
2. Distinguish between convergent and parallel evolution. Which applies to neotropical and Old World monkeys?
3. Name the two families of platyrrhine monkeys. Approximately how many genera and species of neotropical primates are there?
4. What does "platyrrhine" mean?
5. What pattern of features is shared by most New World monkeys?
6. What is the likely adaptive significance of the New World monkey pattern?
7. In what ways are the brains of capuchin monkeys unusual for platyrrhine monkeys?

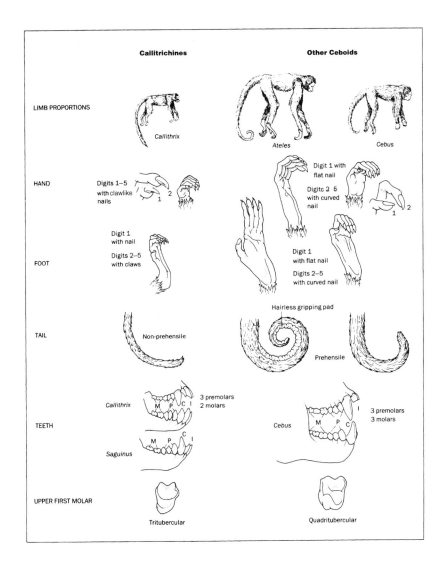

Figure 5-3 A number of features distinguish marmosets and tamarins from the other platyrrhine monkeys. Interestingly, *Callimico* resembles the other ceboids rather than the other callitrichines in its dental formula and form of the upper first molar.

the infraorder Platyrrhini (see Fig. i-6) takes its name (*platy* means "flat"; *rhine*, "nose"). Feature 7 may simply be the result of the way in which relatively small brains are scaled, in which case it too is difficult to interpret (see Neural Note 5). However, the remaining features of teeth, hands, and tails are probably all related to adaptations for moving and feeding in a variety of arboreal habitats.

CEBIDAE

The Callitrichines: Marmosets, Tamarins, and Goeldi's Monkey

The callitrichines consist of closely related species of small, long-tailed, diurnal New World monkeys including marmosets (*Callithrix*, *Cebuella*), tamarins (*Leontopithecus*, *Saguinus*), and Goeldi's monkey

(*Callimico*) (see Fig. 5-2). Although it was once thought that they retained the primitive morphology and behavior of ancestral New World primates, this is no longer the case. Rather, it is now recognized that these monkeys manifest a mixture of derived and primitive features and that, if anything, the group is generally more specialized than the other platyrrhines. For example, a derived trait that distinguishes this group is clawlike nails on all digits except the big toe (see Fig. 5-3). All genera except *Callimico* are also characterized by one less molar than the other ceboids (i.e., a dental formula of 2.1.3.2.), a three-cusped upper first molar, and a high frequency of twinning. Until fairly recently, the callitrichines had also been viewed as squirrel-like, monogamous, and territorial. As discussed later, this perception is changing based on recent field studies.

The number of species in each callitrichine genus and their geographical distributions are variable. Some workers recognize up to 12 species of *Callithrix*, which are relatively widely dispersed in South America. The one species of pygmy marmoset (*Cebuella*), on the other hand, is confined to the tropical forests of the upper Amazon region, and the four species of *Leontopithecus* are located in a few restricted areas in eastern Brazil. There are currently 11 recognized species of *Saguinus*, which are widely distributed over parts of Central and South America. Goeldi's monkey (*Callimico*) again has only

Figure 5-4 New World monkeys (platyrrhines) have flat noses with widely separated oval nostrils. The noses of Old World monkeys, apes, and humans (catarrhines) are turned down at the tips rather than flat, and the bottom ends of their nostrils tend to be closer together. *Platyrrhine* means "flat nosed," and *catarrhine* means "downward-turned nose."

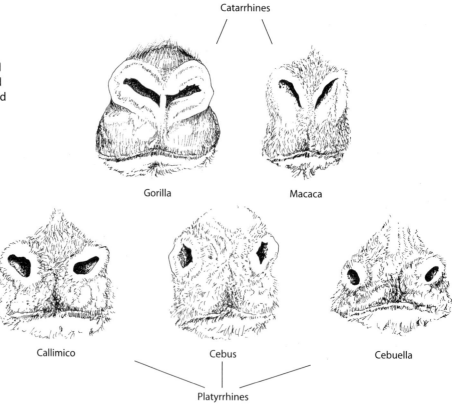

The Strange Brain of the Capuchin

Although the brains of higher primates share numerous features, the pattern of the grooves that separate the convolutions (sulcal pattern) differs in some respects between New and Old World monkeys because of the groups' long, independent evolutions. For example, the sulcal pattern is markedly different in the frontal lobes. New World monkeys usually have a single, long curved sulcus at the edge of their frontal lobes (named the rectus), whereas this sulcus is straighter and its back end is surrounded by an additional sulcus (the arcuate) in Old World monkeys. Farther back on the brain's surface, the sylvian fissure and superior temporal sulcus are usually parallel in New World monkeys, while these two features converge in many Old World monkeys. No one knows what, if any, the functional significances are of these particular differences (unlike sulcal patterns that

reflect prehensile tails in New World monkeys, discussed in Neural Note 6). Curiously, only one New World monkey fails to exhibit these two features in its brain—the capuchin (*Cebus*) or organ-grinder monkey (see figure). Instead, the brains of capuchins appear like those of the Old World cercopithecines because they consistently have both an arcuate sulcus and a converging sylvian fissure and superior temporal sulcus. If all New World monkeys are descended from one founding population, then *Cebus* evolved these features in parallel with the Old World monkeys. If not, it would be logical to assume that *Cebus* may be descended from a line that diverged more recently from Old World primates than did that which gave rise to the other New World monkeys. The latter alternative, however, is supported by neither anatomical nor molecular evidence.

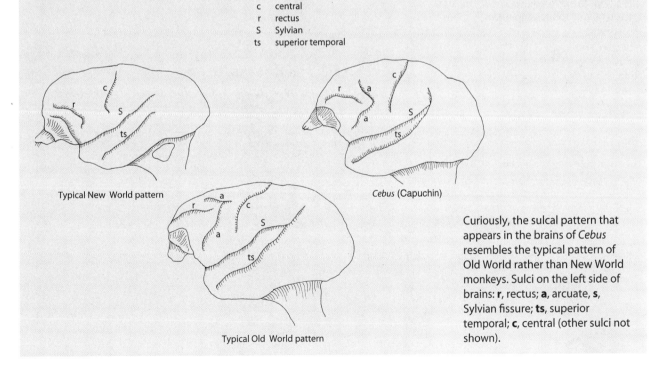

Sulci: a   arcuate
    c   central
    r   rectus
    S   Sylvian
    ts   superior temporal

Typical New World pattern

Cebus (Capuchin)

Typical Old World pattern

Curiously, the sulcal pattern that appears in the brains of *Cebus* resembles the typical pattern of Old World rather than New World monkeys. Sulci on the left side of brains: **r**, rectus; **a**, arcuate, **s**, Sylvian fissure; **ts**, superior temporal; **c**, central (other sulci not shown).

TABLE 5-2 Statistics for Representative Callitrichines

	Callithrix jacchus (common marmoset)	Cebuella pygmaea (pygmy marmoset)	L. rosalia (golden lion tamarin)	S. fuscicollis (saddleback tamarin)	Callimico goeldii (Goeldi's monkey)
Mean adult body weights					
F	236 g	126 g	578 g	403 g	
	(8.3 oz)	(4.4 oz)	(20.2 oz)	(14.1 oz)	
M	256 g	130 g	607 g	387 g	278 g
	(9.0 oz)	(4.6 oz)	(21.2 oz)	(13.5 oz)	(9.7 oz)
Unknown (wild)					500 g (17.5 oz)
Adult brain weight (g)	7.9	4.2	12.9	9.3	10.8
Maximum known life span (yr)	12.0	11.7	14.2		17.9
Olfactory communication[a]	sm, um	sm, um	sm	sm, um	sm, um
Location (habitat)[b]	Br	uA	Br	uA	uA
	df, sc, sw, tp	re, ff	lf	rf, lf, ff	rf, bf
Locomotion[c]	qr, cl	qr, vcl	qr, l, c, s	qr, vcl	qr, vcl
Diet[d]	f, g, i	g, f, n, i	f, n, fl, g, i	f, n, g, i, sp	f, i, sp
Mean no. of offspring	2	2	2	2	1
Seasonal breeders	+	Peaks	+	+	+
Gestation (days)	148	131	129	148	154
Interbirth interval (mo)	5.1	5–7	6–12	6–12	6–12
Mating systems[e]	cp, pg	mo, cp?	mo, cp, pg	cp, mo, pg	pg, mo?, cp?
Territorial	0		+	+	
Ratio of adult females to males	Var	1:1–2	1:1, 2:1, 1:2	1:2	2:1
Alloparenting	+	+	+	+	+
Group size	3–15	2–9	2–16	2–10	2–8
Density		34–247/km²	0.05–10.00/km²	7–20/km²	
Home range	0.5–6.5 ha	0.1–0.5 ha	20–40 ha	30–120 ha	30–60 ha
Conservation status[f]	L	L	CE	L	H

Note: Information in this table has been selected from numerous sources and should be viewed as tentative, i.e., subject to change in light of new information. Sources include but are not confined to Ford and Davis, 1992; Rowe, 1996 (including conservation statuses); Goldizen, 1987; Harvey et al., 1987; Soini, 1993; and Garber, 1993. Abbreviations: +, yes; 0, no; var, variable. Blank entries indicate that no information is presently available. Dietary items are listed in appproximate order of decreasing importance to the species.

[a]Olfactory communication: sm, scent marks; um, urine marks.

[b]Location (habitat): Br, Brazil; uA, upper Amazon region; bf, bamboo forest; df, dry forest; ff, floodplain forest; lf, lowland forest; re, river edge; rf, rain forest; sc, scrub; sw, swamp; tp, tree plantations.

[c]Locomotion: c, climbing; cl, clinging and leaping; l, leaping; qr, quadrupedal branch runners; s, suspension; vcl, vertical clinging and leaping.

[d]Diet: f, fruit; fl, flowers; g, gums (sap); i, insects; n, nectar; sp, seed pods.

[e]Mating systems: cp, cooperative polyandry; mo, monogamy; pg, polygyny.

[f]Conservation status: CE, critically endangered; H, highly endangered; L, lower risk.

one species, which is found in mixed habitats in parts of western South America.

The callitrichines are all small, weighing under 1 kg (about 2 lb). The tamarins nevertheless tend to be larger bodied than marmosets, and *Leontopithecus* (the lion tamarin) is among the largest of all the callitrichines (Table 5-2). The marmosets are also distinguished from tamarins and *Callimico* by their anterior lower dentition (Fig. 5-5). In marmosets, the lower canines are roughly the same height and shape as the incisors. These lower teeth also have thick layers of enamel on their outside surfaces. Marmosets use the resulting short tusk primarily for gouging holes in trees to obtain saps and gums. The other callitrichines have lower front teeth with relatively long canines that are more typical of anthropoids. Although, except for *Callimico*, they too feed on gums, they do not bore holes in trees in order to do so.

Rosenberger describes callitrichines as "masters of the marginal habitat" that colonize the boundaries between different vegetation zones, fallen trees, and remains of forests. Compared to atelines, most callitrichines prefer low levels in the trees, secondary forests, and edge habitats. A mix of habitat types is found within the home ranges of callitrichines. These ranges frequently contain numerous species of New World monkeys that keep track of each other (often through vocal communication) while avoiding direct competition for similar foods, that is, by favoring different levels of the canopy, or foraging at

Anterior view

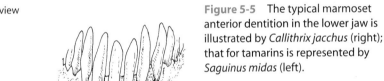

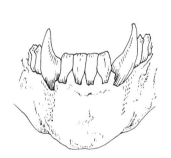

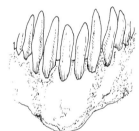

Figure 5-5 The typical marmoset anterior dentition in the lower jaw is illustrated by *Callithrix jacchus* (right); that for tamarins is represented by *Saguinus midas* (left).

Oblique side view

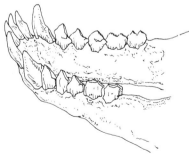

Saguinus midas   *Callithrix jacchus*

different times of the day. For example, two species of *Saguinus* studied at Manu National Park in Peru, emperor and saddleback tamarins, eat the same plant foods and use vocal contact to jointly patrol the borders of their shared territory.

Callitrichines are essentially quadrupedal branch runners and springers with hind limbs that are longer than forelimbs. They are capable of leaping, and their sharp clawlike nails permit vertical clinging onto tree trunks and large branches. Some are able to turn their heads 180°, much like tarsiers. In general, marmosets retain quick stop-and-go jerky locomotor patterns, tamarins move slightly more smoothly, and Goeldi's monkeys are even more fluid in their coordination.

Callitrichines were once thought to occupy a niche similar to that of arboreal squirrels. However, Robert Sussman and the late Warren Kinzey have shown that, unlike tamarins, squirrels tend to travel directly down vertical trunks and also avoid leaping on thin flexible branches. Furthermore, despite their clawlike nails, callitrichines use their hands to grasp food, which is not the tendency for squirrels. In sum, Sussman and Kinzey (1984:427) believe that "squirrels use their claws mainly as *locomotor* adaptations, aiding the animals as they move vertically up and down large oblique and vertical supports within the forest, whereas *Saguinus* and *Cebuella*, as well as other callitrichids, use their claws mainly for *postural* activities related to specific feeding behaviors."

The diets of callitrichines rely on the same types of food, but in different proportions. Insects are an important component, and are particularly abundant in the understory or lower levels of vines and shrubs in the forests. Fruit, flowers, and nectar are eaten by tamarins and, to a lesser extent, marmosets. These plant foods (like insects) are patchily distributed throughout the callitrichines' ranges. All callitrichines except *Callimico* also eat plant **exudates** including sap, gum, and resin, although this food provides a larger component of the diet of marmosets. Insects are high in phosphorus and low in calcium. Since the reverse is true for exudates, it is believed that exudates and insects help provide a balanced diet for animals that eat both. As noted earlier, marmosets use their front lower teeth to gouge holes in trees (Fig. 5-6) and vines, which causes gum to flow. Although tamarins lack the short gouging front teeth of marmosets, they too eat some exudates, but only from sources that occur naturally. According to Paul Garber of the University of Illinois, Rosenberger, and Marilyn Norconk of Kent State University,

> overall, the feeding ecology of callitrichines is distinguished from other platyrrhines by the ability of these primates to exploit a range of resources that are associated with tree trunks in the forest understory. This includes plant gums, bark refuging insects, small vertebrates concealed in knotholes, prey hidden in bromeliads [tropical plants] that grow along the main axis of the tree, as well as use of vertical trunks to scan for insects and small vertebrates located on the ground. Given the highly faunivorous diet of all callitrichines, the evolution of claw-like nails is best understood

Figure 5-6 This tree is riddled with holes that were gouged by *Callithrix jacchus*.

as a foraging adaptation enabling these small primates to exploit high protein and carbohydrate resources restricted to particular microhabitats in the forest understory. (Garber and Norconk, 1996:92)

THE SOCIAL LIFE OF MARMOSETS AND TAMARINS Only a few of the 20 currently recognized species of marmosets and tamarins have been thoroughly studied, and these studies reveal a wide range of variability in group size, ranging patterns, territorial behavior, and to some extent, even mating patterns. Nevertheless, the social life of callitrichines may be characterized with certain broad, interesting generalizations. Marmosets and tamarins usually live in multimale, multifemale groups of 5 to 15 animals, of which several are likely to be adults. Callitrichines sleep in roost trees (which may be returned to on consecutive nights), in dense foliage, or in holes in trunks of trees that are located within home ranges of variable size depending on the species. For example, a home range as small as 0.1 ha has been reported for the pygmy marmoset, while some species of *Saguinus* have home ranges of up to 120 ha. (The area that a species moves within during any given day, or day range, is generally considerably smaller than their entire home range.) The home ranges of *Callithrix* are usually smaller than those of *Saguinus*.

Home range sizes depend on the availability and distribution of resources such as roosting sites, exudate trees, and understory in which insects may be sought. The extent to which a given species is territorial is also constrained by the availability of resources. For example, all species of *Callithrix* appear to have overlapping ranges that they do not defend, although at least one species reportedly defends and scent marks its main exudate tree. This species is not considered to be territorial, however, because it fails to defend its entire home range. The picture for the many species of *Saguinus* is much different. Home ranges of different groups within a species frequently overlap, and territorial behavior such as vocalizing and physical aggression may or may not occur at their overlapping boundaries. Different species of *Saguinus* also share and jointly defend home ranges, while other species exhibit no sign of territoriality. Unlike marmosets, tamarins are clearly variable both within and between species when it comes to defending home ranges.

Until fairly recently, callitrichines were regarded as living in either monogamous or extended-family units. This view, which was based on studies of captive breeding pairs and artificially assembled zoo populations, has been replaced by a much more interesting and complex picture of mating and rearing systems derived from studies of marmosets and tamarins in the wild. As summarized by Garber, Rosenberger, and Norconk, an extremely broad range of mating and grouping patterns including polyandry, polygyny, and less often monogyny has been found at times in all species that have been carefully studied. Contrary to earlier views, the authors (1996:93) note "there is no evidence from the wild that a single male and a single female maintain an exclusive mating relationship over an extended period of time."

Although mating within groups appears to be promiscuous, only

REVIEW EXERCISES

1. Are callitrichines now believed to retain the primitive morphology of ancestral New World monkeys? How does current thinking on this differ with previous suppositions?

2. Are callitrichines as squirrel-like, monogamous, and territorial as traditionally believed?

3. What part of the body distinguishes marmosets from tamarins, and how (and for what) is it used?

4. Why have marmosets and tamarins been described as "masters of the marginal habitat"?

5. What do callitrichines use their clawlike nails for?

6. Describe the diet of callitrichines. What food balances their high insect intake?

1. What size and composition typify the groups that marmosets and tamarins live in?
2. Where do they sleep?
3. What variables constrain the sizes of callitrichine home ranges?
4. Are all species of callitrichines territorial?
5. How has knowledge of callitrichine mating and rearing systems changed in recent years?
6. Which parent usually provides the main infant care after callitrichine twins are born?
7. Explain the concept of co-operative polyandry as it applies to callitrichines.
8. What explanations have been offered to explain the unusual social system of callitrichines?
9. What animals prey on marmosets and tamarins, and how do callitrichines defend themselves?
10. Which callitrichines are highly endangered?

one adult female in each group usually gives birth. Groups with two breeding females, however, have been reported for *Leontopithecus rosalia*, *Saguinus fuscicollis*, and *Callithrix jacchus*. Studies of captive callitrichines reveal intense competition and aggression between adult females, as well as suppressed hormonal changes mediated through olfactory cues that apparently interfere with ovulation in subordinate (but not dominant) females of most species (*Leontopithecus* is an exception). Sexually mature females (at the age of 1½ to 2 years) do not menstruate, but have estrous cycles that average a little more than 2 weeks in length. Gestation takes 4 to 5 months and, as noted earlier, females usually give birth to twins (*Callimico*, however, usually bears a single infant). There is not a strict birth season, although there may be two birth peaks annually. According to Garber, the potential for females to produce two litters each year results in intrinsic rates of population increase that are greater than those found in other anthropoids. The combined weights of newborn twins is extremely large compared to the mother, and they require a large amount of food because of their high metabolic rates (which are typical for small animals).

In response to the extremely high investment required to gestate and raise offspring, callitrichines have evolved a highly unusual rearing system in which adult males become the principal caregivers that carry and provide solid food for infants. Other group members including nonbreeding adult females also participate in infant care. This type of infant rearing in which extra helpers care for the young is best described as a communal rearing system. Because only one of the mating females in a group usually gives birth, Sussman and Garber describe the breeding system of marmosets and tamarins as "functionally polyandrous." That is, despite a promiscuous mating system, the pattern of births is like those seen in systems in which females mate with more than one male, while males mate exclusively with one female (i.e., the reverse of a unimale group, or cooperative polyandry).

A good deal of variability in the details of social organization is observed within and between the species of callitrichines that have been studied in the field. However, the above generalizations do float to the top, and, as always, one wonders about the functional significance of such unusual behaviors. These patterns are ripe for sociobiological hypotheses, such as the suggestion that two or more males are willing to invest heavily in caring for infants because, having each mated with the mother, any of them could be the father of the twins. (No conscious thought is implied here, rather the behavior of the parties involved, for whatever reasons, maximizes perpetation of their particular genes.) However, as Sussman and Garber discuss, this explanation does not help explain the caregiving behavior of adult females that are not even related to the mother, let alone her offspring. Captive studies suggest that there is a relationship between previous baby-sitting experience on the part of mothers and the likelihood that their own offspring will survive. In any event, field studies of two species of tamarins in Peru by Garber (see In the Field) suggest that the presence of at least two adult males in a group is critically associated with infant survival and that female promiscuity may be a means to ensure the continued presence of a number of male helpers. In sum, marmosets and tamarins show fascinating role reversals in which (1) females rather than males compete aggressively with each other for reproductive partners and their caregiving efforts, (2) females rather than males are the leaders of "harems" of males (at least with respect to breeding if not mating), and (3) males rather than mothers assume most of the child care for infants. Certainly, callitrichines can no longer be described as monogamous primates in which individuals reproduce sexually with only one partner. Or, as Garber, Rosenberger, and Norconk (1996:93) conclude, "Any notion that tamarins, marmosets, or Goeldi's monkeys live in simple social or mating systems is completely in error."

THREATS Predators provide an ecological constraint on social organization as discussed earlier, and the small callitrichines certainly have their share. Cats such as ocelots and jaguarundis, and a weasel-like creature called the tayra may prey on marmosets and tamarins. Dogs and large snakes are also a threat to callitrichines. Perhaps their most serious predators (aside from humans) are birds of prey, which typically evoke either high-pitched alarm calls followed by fleeing, or the alternative defense of silence and freezing. (Callitrichines sometimes react similarly to airplanes.) In fact, callitrichines give the impression of being constantly watchful and frightened, with at least one adult (often a male caregiver) keeping an eye out while others feed. Tamarins are known to vocalize and mob potential predators when threatened. The coats of marmosets and tamarins frequently blend in with their surroundings (are cryptic), and they choose relatively protected roosting sites in which to sleep.

According to Russell Mittermeier of Conservation International, approximately one-third of all living primate species live in the New World and, of these, a number are endangered, some to near extinction. As elsewhere in the world, the declining numbers are largely due to tropical forest destruction. Among the callitrichines, at least three species of *Callithrix* and one of *Saguinus* are endangered, as are all four species of lion tamarin and Goeldi's monkey. Besides deforestation, the pet trade has contributed to this unhappy situation.

Because in-depth field studies have only been carried out on a limited number of callitrichine species beginning in the 1980s, the above discussion relies on observations of captive as well as wild populations, and is therefore somewhat general. Below, portraits are presented for several representative species of callitrichines.

THE COMMON MARMOSET (*CALLITHRIX JACCHUS*) Except for the tiny pygmy marmoset, which is portrayed next, the up to 12 species of *Callithrix* (depending on who's counting) represent the smallest of the New World monkeys (see Table 5-2). Marmosets are generally considered to be more specialized than tamarins. The common marmoset (*Callithrix jacchus*) is gray and black and has a striking appearance because of its large white ear tufts and striped tail (Fig. 5-7). As expected, this species has the short tusk that is typical of marmosets. It lives in

Figure 5-7 The common marmoset (*Callithrix jacchus*) has a striking appearance because of its large white ear tufts and striped tail.

Tamarin Secrets

Primatologist Paul Garber is a leading expert on the feeding ecology and social behavior of callitrichines. He describes his subjects as "tiny, ornately colored, and beguilingly coifed." When Garber began studying tamarins in the forests of Panama in 1978, little was known about their behavior in the wild. One of his first discoveries was that the unusual clawlike nails that typify callitrichines are used to cling to large tree trunks in order to feed on insects and saps. Now, with two decades of fieldwork under his belt, Garber observes that "one look into the darting eyes of these squirrel-sized creatures is more than sufficient to reveal a curiosity and awareness consistent with their complex patterns of social interactions and feeding strategies." In his own words:

PRIMATOLOGIST PAUL GARBER HOLDS A MOUSTACHED TAMARIN. (PHOTO COURTESY OF PAUL GARBER.)

As I watched tamarins search among the giant trees for insects and plant foods to share with their young, cooperatively carry infants across gaps in the rain forest canopy, and even jointly hold and consume fruits that were too large and heavy for one animal alone to pluck from the stem, I became less and less aware of their small size and seemingly primitive appearance, and began to understand the degree to which much of our own behavior is rooted in our primate past. It was not until I went to the field to study moustached tamarins in the Amazon Basin in 1981, however, that I began to fully understand the complex and intricate nature of tamarin social and mating behavior. My study in Peru involved trapping, marking, and releasing all individuals from a local tamarin population. It became clear as we tranquilized these tiny elflike creatures and examined them, that groups commonly contained not one adult male and one

scrubs, swamps, forests, and tree plantations of Brazil (Fig. 5-8, see Table 5-2), where it uses its tusk to gouge numerous holes in trees (see Fig. 5-6) in pursuit of one of the largest amounts of gum eaten by any callitrichine (70% to 80% of its plant food, or about 15% of its total diet, which also includes fruits and insects). Not surprisingly, although common marmosets' home ranges are relatively small compared to those of tamarins (see Table 5-2), their core areas contain a high density of exudate trees. Marmosets sleep in vines, on branches, or in tree holes.

As is the case for most callitrichines, social organization is variable, ranging from multimale, multifemale groups to mixed groups that have only one adult male or female. Estrous females solicit mating with stares and tongue flicks. The matings last less than 25 seconds, during which the female opens her mouth and looks over her

adult female, but two or three adult males and one to three adult females. Clearly this did not fit our expectations of monogamy. We put color-coded identification collars on all individuals, and then released each group back into their home range. Over the next several months, we followed the tamarins and observed their behavior. I still vividly remember the first time we documented two adult males from the same group mating with the dominant adult female. These copulations occurred minutes apart and in full view of the adult males. A mating system in which several males copulate with a sovereign breeding female, or polyandry, is extremely rare among mammals. Polyandry has not been reported in any other group of primates (it has now been reported in other tamarin and marmoset species), except for an extremely small number of human cultures.

What I next observed was even more remarkable. In most animals in which polyandry has been documented, the males generally are believed to be related, commonly brothers. This is called fraternal polyandry, and explained through kin selection and inclusive fitness. However, over the years we have observed migrations of adult males into and out of our study groups. Because these animals were marked, it became apparent that a group with three adult males might contain a male originally from the red group, another from the blue group, and a third male that originated in the green group. Thus, many of our groups contained unrelated males. To our amazement, even these unrelated males were extremely cooperative. All cared for the groups' infants, and each was observed to carry and provision the young with juicy fruits and succulent insects. Given that not all of these males could have been the infants' father (it is possible, however, that different fathers each sired one of the infants—tamarin twins are dizygotic and the two ova are independently fertilized), individual selection rather than kin selection may help to explain such high levels of cooperation among unrelated males. I now believe that tamarins are characterized by a highly flexible breeding system in which groups may be polyandrous, polygynous, and on rare occasions, serially monogamous, and in which males compete for access to reproductive partners by cooperation instead of aggression. That is, male caregiving is a form of "courtship behavior" and a breeding female preferentially mates with those males who cooperate in caring for her young. As I continue to return to the field and observe the behavior and ecology of these amazing primates, I can't even begin to imagine what other secrets they will teach me about their world.

shoulder. Females are around 2 years old when they first give birth. Usually only one dominant female in a group produces young (but exceptions have been reported), and the litter most often consists of twins. Because their combined size is heavy relative to the mother that nurses them, twins are a large energy drain from the point of view of producing enough milk, and also garnering enough strength to carry them about during the first 2 to 3 months of their lives. As described by Leslie Digby of Duke University, *Callithrix jacchus*, like other marmosets and tamarins, engages in cooperative rearing of the young by adult males and juveniles of both sexes. Help for the mother comes primarily in the form of carrying her offspring.

Digby and Claudio Barreto of the Federal University of Rio Grande in Brazil have quantified the activity patterns in groups of common marmosets with and without young, with interesting results. During

Figure 5-8 The common marmoset lives in Brazil.

times when there were no young, adults spent approximately 43% of their time foraging or feeding and 30% of the time resting. During the first 2 months after a birth, however, individuals that carried infants decreased their foraging and feeding time to 12% and increased their resting time to 59%. The carriers also decreased their day ranges compared to other group members. Digby and Barreto concluded that these changes in activity are because helpers save their strength to sustain enough energy to carry the infants. This conclusion should be of no surprise to parents of young children.

Odors are important for this species, in which even infants mark by scent. The suppression of reproduction in nondominant females is thought to be mediated by odors. Other forms of communication consist of a variety of vocalizations including squeals, twitters, and mobbing calls. Although predation on common marmosets has rarely been observed, it is safe to assume that raptors and other animals that prey on small primates (e.g., snakes) are potential threats.

THE PYGMY MARMOSET (*CEBUELLA PYGMAEA*) The diminutive pygmy marmoset is the most specialized of the tamarins and marmosets. Its body weight averages about 130 g (less than 5 oz) in the wild (see Table 5-2), which makes this species the smallest of all the New and Old World monkeys, but not the smallest primate (an honor reserved for the mouse lemur). These tiny primates are rather handsome with longish manes of hair on their faces and heads that cover their ears, and an interesting tawny coat color that appears frosted with lighter hairs (coloration that is known as *agouti*, Fig. 5-9). This somewhat patchy coat color helps them to blend in with their environment, and the upper surface of the tail is barred with black horizontal stripes.

The present distribution of pygmy marmosets includes tropical forests in the upper Amazon region of South America, encompassing parts of Brazil, Colombia, Ecuador, Peru, and Bolivia (Fig. 5-10). Much of what is known about this species in the wild is from Pekka Soini of the Department of Agriculture in Peru, who has conducted long-term field studies in the Manu National Park in Peru. According to Soini, pygmy marmosets are habitat specialists that prefer river edge and floodplain forests, although they are sometimes found elsewhere. These primates are extremely fearful in the wild, perhaps because of their tiny size. When moving from tree to tree, pygmy marmosets prefer to keep low, possibly as a means of avoiding birds of prey. They can be extremely quiet (e.g., in the presence of humans) or very noisy. Depending on circumstances, vocalizations may include hisses, trills, whistles, screams, ultrasonic cries (that humans cannot detect), and cricket-like chirps. One of the most charming portraits of this species was offered years ago by the late Philip Hershkovitz as he described the June afternoon in 1936 when he first met the pygmy marmoset:

> I was resting on a fallen log when I heard an insistent chirp, but feeling too listless to raise my head and investigate I shrugged off the sound as that of the ubiquitous crickets. When the chirping became a chorus, however, I looked up suddenly and found myself staring into the face of the tiniest monkey I had ever seen alive. The pygmy was clinging to the terminal twig of a bush no more than an arm's length from my face. Its little eyes were ablaze, tusks bared, and mane bristling, while it hissed, grimaced, and swung menacingly as if to gain momentum for springing upon me. As I watched entranced, the entire shrub came alive with pygmy marmosets. My eyes shifted from the first sprite, no doubt the leader, to the trembling twigs of the bush where the pygmies, perhaps a score in all, were running and leaping back and forth, chirping, hissing, posturing, and grimacing. I was tempted to seize one with my hands, but as I straightened out slowly to do so, the imps simply vanished, leaving only a rear

Figure 5-9 Pygmy marmosets (*Cebuella pygmaea*) are the smallest of the anthropoids.

guard crisscrossing the branches of the trees to confuse pursuit and taunt the enemy with its chirpings. (Hershkovitz, 1977:472)

As is apparent from Hershkovitz's observations, pygmy marmosets run and leap along branches as is typical for callitrichines. They cling to vertical supports when feeding on gums, and can turn their heads 180° to scan for predators while they're at it. *Cebuella* also engages in protective types of locomotion similar to that of squirrels, that is, frequently dodging behind trunks or branches, and advancing in spurts of

Figure 5-10 Pygmy marmosets are found in the upper Amazon region of South America.

dashes alternating with frozen immobility. They may also move exceeding slowly, much as sloths do.

Pygmy marmosets spend almost half of their waking hours feeding or hunting for food, most often exudates (followed by fruit, nectar, and insects). In fact, home ranges seem to vary in size depending on the density and distribution of the various species of exudate trees, dense thickets, and sources of nectar (which is important when gums become scarce during the dry season). These ranges are often quite small (0.1 to 0.5 ha). Exudate trees may be riddled with hundreds, or even thousands, of feeding holes. Pygmy marmosets move up and down trees head first, and adopt any of a variety of positions when feeding or excavating holes. M. Flores Ramirez of the Puerto Rico Nuclear Center and colleagues note, however, that hole digging is usually accomplished in an upright position in which the protruding lower dentition is used to gouge a hole by moving the head up and down (kind of a vertical sawing motion). Additional holes appear to be added to trees over time, from the bottom up.

During a typical day, some individuals may repeatedly visit and feed at holes for durations that last anywhere from several seconds to half an hour, while other troop members might hunt for insects (sometimes pursuing them to the ground), forage for fruit, or rest. Soini reports that the group shifts from mostly foraging activities to huddling, grooming, and playing after feeding first thing in the morning, and again later in the day. Thus, at most times the troop is likely to be spread out within its small range and therefore its members are difficult for an observer to count. Pygmy marmosets sleep in holes in trees, which they enter while it is still light out (before 6 P.M.).

The social life of pygmy marmosets is quite interesting. Soini observed 2 to 9 individuals in 80 different troops, with an average of about 5 individuals. Although the extent to which pygmy marmosets are territorial is unclear, both sexes engage in scent and urine marking of objects in their environment including exudate trees. Adult threat displays are practiced by high-ranking males and females. As colorfully described for males by Hershkovitz (1977:473), a threat display "is a frontal confrontation with back arched, fur erect, gait slow and stiff-legged. This is followed by a half-circle turn and presentation of genitalia from the rear with tail raised nearly perpendicularly, head turned back with eyes glaring at his opponent." Curiously, females also rebuff unwanted sexual advances of males by presenting their genitalia, in which case the males lick but do not mount the female. According to Kinzey, the genital presentation of *Cebuella* is unique among callitrichines.

The mating system of pygmy marmosets may be either functionally monogamous or polyandrous. According to Soini, if two adult males are present, the dominant one attempts, with greater or lesser success, to restrict access to the single dominant female by intimidation and mate guarding. As is true for most callitrichines, only the dominant female usually gives birth. Twins are the norm, and help from other group members including adult males is extremely impor-

tant. Infants are carried on the backs of helpers for the first 1 to 2 weeks of life. After that, parents begin to leave infants alone in protected places for periods of time while group members forage. Although pygmy marmosets are potentially vulnerable to predators because of their tiny size, their diminutive stature has one thing in its favor: They are too small to be hunted by people for food and are therefore not as endangered as some of the other callitrichines. The pet trade, however, does take its toll.

THE GOLDEN LION TAMARIN (*LEONTOPITHECUS ROSALIA*) We turn now to the rarest, most endangered, and one of the largest of the callitrichines, the golden lion tamarin (averaging about 600 g or somewhat over a pound, see Table 5-2). *Leontopithecus* contains four species that live in separate small refuges in southeastern Brazil. The teeth of these species are relatively large, and their ears and genitals reveal some unique details. A well-developed laryngeal air sac is consistently present in males. Lion tamarins have elongated arms and hands, and their fingers are webbed together by a membrane near the palm. According to Hershkovitz, this membrane binds the elongated fingers (especially the middle one) together into a strong tool that is used to probe for insects in holes or under loose bark. (Hershkovitz also notes that the aye-aye is the only other primate with a hand so obviously specialized for insect probing.)

The four species of lion tamarins are easily distinguished by their coat patterns and colors, which span from pale golden or orange, to reddish, to black. Rather than the color of their tresses, however, it is their hairdos that give lion tamarins a spectacular style (from which they take their name). The face is relatively bare, except for short hair on the forehead that may form a widow's peak, and a small goatee. Longer hairs on the crown of the head appear to be parted down the middle and are whorled back on both sides in a kind of pompadour that joins hairs on the cheeks and neck. The whole effect is one of a glorious lion-like mane that surrounds the face and covers the ears. These features and its striking golden-reddish color make *L. rosalia* one of the most beautiful of all the monkeys (Fig. 5-11, color plate).

Golden lion tamarins, which live in patches of primary and secondary forest in Rio de Janeiro, Brazil (Fig. 5-12), have only recently been studied in the wild. According to Brian Stafford of the City University of New York and his colleagues, *L. rosalia* is an arboreal quadruped that leaps a good deal and climbs. Additionally, these tamarins engage in a small amount of suspensory behavior and, curiously, a kind of lopsided gallop in which the forelimbs and hind limbs are placed off center and to the same side on the substrate. Much of what is known about this species comes from an ongoing long-term study at the Poco das Antas Biological Reserve in Rio de Janeiro by James Dietz of the University of Maryland at College Park and colleagues. This team reports that golden lion tamarins spend a surprising amount of time in undisturbed swamp forests, probably because of a relatively greater availability there of cryptic insect and small vertebrate prey. Their diet also includes a wide variety of fruits, nectar, flowers, and gums. Golden lion tarmarins usually sleep in tree holes or vines.

Figure 5-12 Golden lion tamarins are found in remnant patches of forests in Rio de Janeiro, Brazil.

Details about the social lives of wild lion tamarins are beginning to emerge. According to Dietz, groups at Poco das Antas contain 2 to 11 individuals of mixed ages and sex, and dominance hierarchies exist in both sexes. Further, Anthony Rylands of the Federal University of Minas Gerais in Brazil reports that males dominate females. Monogamy is frequent at Poco das Antas as 50% of the groups contain only one adult male and female. The potential for polyandry is also high (40% of the groups consist of two adult males and one adult female), while that for polygyny is less so (10% contain one adult male and two females). Unlike most other callitrichines, dominant females do not appear to suppress ovulation of subordinate females through olfactory stimuli. At Poco das Antas, individuals of both sexes leave their natal groups by the age of 4.

Golden lion tamarins fiercely defend their home ranges. They have been reported, however, to travel with other species of primates. Adults of both sexes use urine and secretions from sternal and circumgenital glands to mark their scent. During aggressive interactions, *L. rosalia* arches its back and thrashes its tail, and its hair stands on end. Vocalizations are also used to maintain spacing between groups, and to reinforce the pair-bond between mates that duet.

Females give birth, frequently to twins, during the rainy season when food is relatively abundant. Unlike most callitrichines, mothers do not allow other members to carry their infants until they are at least 2 weeks old. After that, all group members help to carry infants, and may even share food with them. Andrew Baker of the Philadelphia Zoological Gardens notes that males tend to carry male infants more often, but that females show no sex preference for the infants they carry. Previous parental experience seems to be important for infant survival, at least in captivity, and there are even reports of inexperienced adults killing and eating their offspring.

As summarized by Kinzey, golden lion tamarins are preyed on by a number of animals. For example, Dietz observed a boa constrictor attack and eat an adult female, and an ocelot kill another adult. He has also seen golden lion tamarins chased by capuchin monkeys and birds of prey. Other more serious threats have brought this species to the brink of extinction. Intensive logging has contributed to extensive destruction of their habitats in coastal rain forests. The illegal pet trade also threatens lion tamarins, but less so than in the past. In recent decades, heroic efforts have been made to conserve this species (nicely summarized by Devra Kleiman of the National Zoological Park in Washington, D.C., and colleagues). These include the creation of the Poco das Antas Biological Reserve and development of a breeding facility in Brazil. Breeding and management programs have also been implemented for captive individuals in the world's zoos, and efforts have been made to protect habitats and to reintroduce individuals into the wild. As with any conservation program, ongoing research and public education are important components. Sadly, however, the number of estimated remaining golden lion tamarins in the wild has dwindled to a few hundred.

THE SADDLEBACK TAMARIN (*SAGUINUS FUSCICOLLIS*) Saddleback tamarins are the smallest, most widespread, and most studied of the 11 species of *Saguinus*. *S. fuscicollis* is also the most diverse species, containing 12 subspecies that differ from each other primarily on the basis of coat colors and patterns. According to Hershkovitz, these subspecies evolved in virtually identical habitats that were isolated from each other by rivers. Random genetic drift, inbreeding, and sexual selection therefore may have contributed to the proliferation of so many subspecies.

Saddleback tamarins are distinguished from other callitrichines by a number of features including their saddleback coat pattern, which is the result of having different colors for the shoulders (mantle), back (saddle), and rump regions (Fig. 5-13, part A, color plate). The 11 species of *Saguinus* fall into three groups based on the appearance of their faces: hairy faced, bare faced, and mottled faced. Saddleback tamarins are in the first group. They also have black exposed ears, and a band of short grayish to white hairs surrounds the mouth (sort of a decorative muzzle). Some of the subspecies appear to have dramatic white eyebrows. Although all saddleback tamarins have three color zones along their backs, the dominant hues of different subspecies vary from shades of black to orange to white.

Saddleback tamarins live in humid forests in the upper Amazon region of Brazil, Colombia, Ecuador, Peru, and Bolivia (Fig. 5-14). In

Figure 5-13 **B**. Two saddleback tamarins cling to a tree. (Photo courtesy of Paul Garber.) (**A** is a color plate.)

Figure 5-14 Saddleback tamarins are widely dispersed in the upper Amazon region of South America.

addition to fruit and gums, insects are an important source of food throughout the year. According to Garber (1993:278), tamarin insect foraging falls into several patterns based on "hunting techniques, substrates exploited, vertical stratification, and modes of positional behavior." The most distinctive pattern is seen in *S. fuscicollis* (and *Saguinus nigricollis*), which forages at all levels of the canopy and uses stealth to capture large cryptic or hidden insects such as grasshoppers, katydids, and crickets. Saddleback tamarins utilize a high proportion of vertical clinging postures, and Garber (1993:293) thinks that they are able to selectively forage on tree trunks and other large vertical substrates that would otherwise be "too wide to be spanned by their tiny hands and feet" because of their clawlike nails.

Saddleback tamarins don't just live on insects, however. John Terborgh of Duke University reports that the territories of *S. fuscicollis* at the Cocha Cashu Biological Station in the Manu National Park in Peru are extremely variable and very extensive for such small-bodied primates (ranging from 30 to 120 ha). Terborgh attributes this need for space to food requirements associated with being near the bottom of the primate pecking order. Saddleback tamarins thus avoid competing with other primates at preferred food sources such as large concentrations of flowers or fruiting trees. Instead, they gather small amounts of food (including nectar during the dry season) from less-desirable sources as they travel over relatively long distances. Not only are the territories of *S. fuscicollis* large, but also their individual boundaries remain fixed over long periods of time. Terborgh's analysis suggests that both the relative rarity of this species compared to other primates at Cocha Cashu and its large fixed territories are related to living at the edge, that is, to the necessity of occupying and defending ranges that include enough of the different food sources that sustain this little primate in the face of rather stiff competition from its larger cousins.

Wild saddleback tamarins tend to live in small groups that are remarkably variable in their social organization, with different groups practicing cooperative polyandry, monogamy, and sometimes even having two simultaneously reproductive females. As with most callitrichines, adult males are extremely important caregivers that provide the majority of carrying and food sharing for infants. However, Anne Goldizen of the University of Michigan and Terborgh also document the importance of the other group members in helping to care for offspring (usually twins) in the wild, noting that they have never observed a simple pair of adults produce young. In other words, because of the extensive energy required to raise twins, saddleback tamarins must have helpers. In keeping with this, monogamous pairs that produce offspring are always accompanied by juveniles, subadults, or grown offspring that help with infant care, and mothers rarely carry their offspring after their second day of life. If no such helpers are available, groups typically accept an additional male as a second helper and breeder. When polyandrous mating situations occur, they appear to be amazingly free of strife. Goldizen and Terborgh's research adds a dimension of social flexibility to this chapter's generalizations about cooperative polyandry in callitrichines.

The saddleback tamarins at Cocha Cashu are extremely territorial. They patrol and defend the borders of their territories against other groups of saddleback tamarins with vocalizations and, if necessary, physical aggression. Scent marking is concentrated in feeding and resting areas, where individuals usually sleep near each other in vines or in tree holes. At Cocha Cashu, the territories of *S. fuscicollis* overlap with the ranges of at least nine other species of primate including the emperor tamarin (*Saguinus imperator*), pygmy marmoset, and a number of larger platyrrhines such as howler monkeys, capuchins, and spider monkeys (see Chapter 6). Tamarins are preyed on by snakes, raptors, ocelots, and humans. The ranges of saddleback tamarins throughout the Amazon usually overlap with those of a larger species of *Saguinus*, such as *S. mystax*, that provides protection against arboreal predators in the upper canopy (e.g., through warning calls), while *S. fuscicollis* keeps an eye out for terrestrial predators in the lower canopy.

GOELDI'S MONKEY (*CALLIMICO GOELDII*) As noted earlier in this chapter, the genus *Callimico* has been difficult to classify over the years. Most evolutionary biologists now group it with the cebids, although a few still consider it to be an atelid, and a third group prefers to include it in its own family (the Callimiconidae). In this text, Goeldi's monkey is retained in the callitrichine subfamily of the cebids (see Table 5-1), while recognizing that it shares some features with marmosets and tamarins and others with atelids. For example, *Callimico* has clawlike nails on all digits except the hallux, but shares the atelid form of the molars and dental formula of 2.1.3.3. rather than the cebid formula of 2.1.3.2. Also like the atelids, *Callimico* usually gives birth to single infants instead of twins. Unlike the other callitrichines, *Callimico* does not eat gums, which led Susan Ford of Southern Illinois University to suggest that it represents an earlier stage in evolution when callitrichines had developed vertical clinging postures using their clawlike nails, but had not yet become adapted for feeding on exudates.

Goeldi's monkeys are generally black to dark brown, although the hair on the head and tail is sometimes highlighted with red, white, or silvery brown (Fig. 5-15). *Callimico* also has a unique hairdo that is layered on the crown and pompadoured on the forehead (a human might find this rather appealing in a punky sort of way). Goeldi's monkey is among the larger callitrichines (see Table 5-1), but smaller than any atelid.

The distribution of Goeldi's monkey coincides with those of *Cebuella* and *S. fuscicollis* in the upper Amazon region (Fig. 5-16) and overlaps with the ranges of numerous other primates as well. *Callimico*, however, is extremely rare and therefore has not been studied as extensively in the wild as some of the other platyrrhines have.

Nevertheless, research in Bolivia by A. G. and G. Pook of the New York Zoological Society suggests that *Callimico* is a habitat specialist that prefers low-lying bamboo forest and spends most of its time traveling quietly by vertical clinging and leaping within 3 meters of the ground.

As Hershkovitz (1977:904) poetically described Goeldi's monkeys, "their dark, shadowy figures, comparatively silent movements, infrequent vocalizations, cryptic habits, and less aggressive, more retiring nature—as compared with sympatric tamarins and squirrel monkeys—render them unobtrusive." Goeldi's monkeys engage in a good deal of scent and urine marking and, like some species of tamarins, are likely to mob their enemies (including innocent primatologists). Insects and fruits are important foods and, during the difficult dry season, *Callimico* may resort to eating the sticky gum that adheres to seed pods (although the Pooks are careful to note that they have not observed the type of feeding on exudates seen in marmosets and tamarins).

Goeldi's monkeys sleep in dense vegetation rather than holes in

Figure 5-15 Long-term field studies are needed to learn more about Goeldi's monkey (*Callimico goeldii*).

Figure 5-16 Goeldi's monkeys live in parts of western South America.

trees, and at least some groups may not retire as early in the afternoon as marmosets and tamarins. (According to Hershkovitz, *Callimico* may be active at twilight, or **crepuscular**, although this is contrary to the Pooks' observation that their subjects settled down for the night before dusk.) The social systems of wild Goeldi's monkeys are not yet well understood, but it appears that they live in small social groups that frequently consist of multiple adult males and females. As noted earlier, births usually result in one offspring rather than twins. Similar to golden lion tamarins, the mother is the primary caregiver for the first 2 to 3 weeks, after which the father assumes responsibility (except, of course, for nursing). Hershkovitz (1977:909) notes that *Callimico* is "one of the most docile, trusting, and inquisitive of platyrrhines."

Because *Callimico* is so sparsely distributed within its range, very little information exists about its population size or the extent to which it is endangered. The mere fact that this unique species is rarely observed, however, is cause for concern about its future.

SUMMARY

The ancestors of New World monkeys arrived in South America by about 30 mya during the Oligocene, although it is still not clear whether they originated in North America or Africa. Because of their isolation from primates in other parts of the world, today's neotropical monkeys are the products of very long evolutionary histories that were separate from those of primates living in other parts of the world. New World monkeys are divided into two families of platyrrhines (the Cebidae and Atelidae). The general New World monkey pattern includes seven features, most of which are related to adaptations for moving and feeding in a variety of arboreal habitats. Marmosets and tamarins (the callitrichines) are relatively small habitat specialists that exhibit variable social organization depending on the availability of suitable food (insects, fruits, and gums), places to sleep, and protective covering from predators. Most callitrichines live in small groups in which only one female usually gives birth at a time, most often to twins. Except for nursing, the major care for the infants is often provided by adult males (frequently two) that have mated with the mother, and other helpers. This system of cooperative polyandry apparently enhances survivorship of the young. Callitrichine social systems are variable, however, and depending on circumstances may also include mating patterns based on monogamy or polygyny. Portraits have been presented for the common marmoset, pygmy marmosets (the smallest monkey anywhere), golden lion tamarins (rare, endangered, and relatively large), saddleback tamarins (the smallest and widest-ranging tamarin), and Goeldi's monkey (which in some ways resembles atelids more than callitrichines).

REVIEW EXERCISES

1. Which species contains the smallest monkeys in the world?
2. Describe pygmy marmosets and name their genus and species.
3. In what ways are golden lion tamarins unique? Name their genus and species.
4. What is the common name for *Saguinus fuscicollis*? What accounts for its name?
5. What position does *S. fuscicollis* have in the primate pecking order at Cocha Cashu, and how and why does this affect its range size?
6. In what ways does *Callimico* appear like marmosets and tamarins? Like other platyrrhines?
7. How does the diet of Goeldi's monkey differ from those of other callitrichines?

FURTHER READING

GARBER, P. A., ROSENBERGER, A. L. AND M. A. NORCONK (1996) Marmoset misconceptions. In M. A. Norconk, A. L. Rosenberger and P. A. Garber (eds) *Adaptive Radiations of Neotropical Primates*. New York: Plenum Press, pages 87–95.

KINZEY, W. G. (1997) Synopsis of New World primates (16 genera). In W. G. Kinzey (ed) *New World Primates: Ecology, Evolution, and Behavior*. New York: Aldine de Gruyter, pages 169–305.

6

OTHER NEOTROPICAL MONKEYS: CEBIDAE AND ATELIDAE

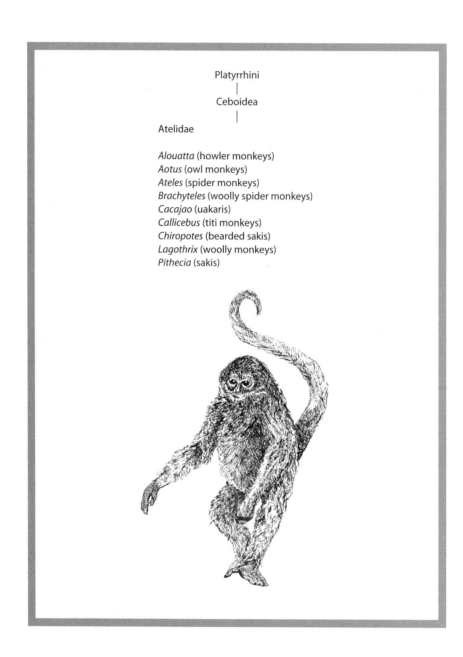

Platyrrhini
|
Ceboidea
|
Atelidae

Alouatta (howler monkeys)
Aotus (owl monkeys)
Ateles (spider monkeys)
Brachyteles (woolly spider monkeys)
Cacajao (uakaris)
Callicebus (titi monkeys)
Chiropotes (bearded sakis)
Lagothrix (woolly monkeys)
Pithecia (sakis)

INTRODUCTION

We say good-bye to the callitrichines (see Chapter 5), and now turn to the remaining New World monkeys, which include two genera of the Cebidae (*Saimiri* and *Cebus*) and all nine genera of the Atelidae (see Table 5-1). Because these monkeys represent a greater number and variety of species than the callitrichines, their classification can be tricky. The particular classification that has been adopted for this text is Alfred Rosenberger's, which is nicely illustrated by panels that represent the two families and four subfamilies of living platyrrhines (Fig. 6-1). These panels also root the living genera to the fossil record, and provide information about dietary preferences and locomotion. Three of the four subfamilies that comprise Rosenberger's panels are the focus of the present chapter: Spider (*Ateles*), woolly (*Lagothrix*), woolly spider (*Brachyteles*), and howler (*Alouatta*) monkeys are members of the ateline subfamily. Of all the monkeys in the world, these large-bodied species are particularly fun to watch because of their acrobatic use of prehensile tails. As is true for callitrichines, atelines are viewed as a separate, unified group based on both molecular evidence and shared evolutionary adaptations linked to moving and feeding within their particular arboreal habitats. The pitheciines represent another subfamily, which includes the sakis (*Pithecia*, *Chiropotes*), uakaris (*Cacajao*), titi monkeys (*Callicebus*), and owl monkeys (*Aotus*). Although the sakis and uakaris remain somewhat mysterious owing to their hard-to-reach, seasonally flooded habitats in Amazonia, primatologists again agree that they form a unified group based on both molecular evidence and shared adaptations for eating tough fruits and seeds.

There is less agreement about how to classify squirrel (*Saimiri*), capuchin (*Cebus*), titi (*Callicebus*), and owl (*Aotus*) monkeys, however, and various workers interpret different combinations of these mon-

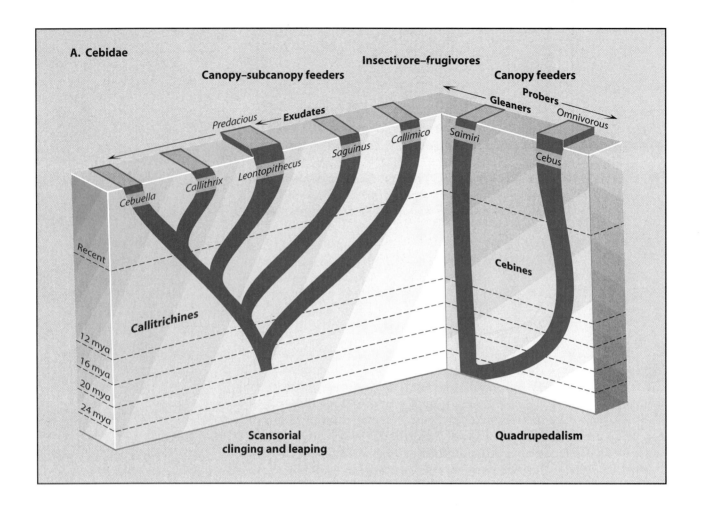

keys as belonging to either the cebid or atelid families. In this text, we follow Rosenberger, who includes capuchin and squirrel monkeys in the cebine subfamily of the cebids, and places the titi and owl monkeys in with the pitheciine subfamily of the atelids (see Fig. 6-1). The confusion regarding these monkeys is not surprising when one remembers that the evolutionary history of New World monkeys is poorly understood, partly because of a fragmentary fossil record. Although there is no one right classification for platyrrhines, we have adopted Rosenberger's because it incorporates information about locomotor patterns and feeding adaptations. It should be emphasized, however, that the four problematical genera remain just that, because to varying degrees (as discussed below) they resemble callitrichines in some respects and pitheciines in others.

CEBIDAE

The Cebines: *Saimiri* and *Cebus*

Although adult capuchins (*Cebus*) are four to six times heavier than adult squirrel monkeys (*Saimiri*), many workers classify *Saimiri* and

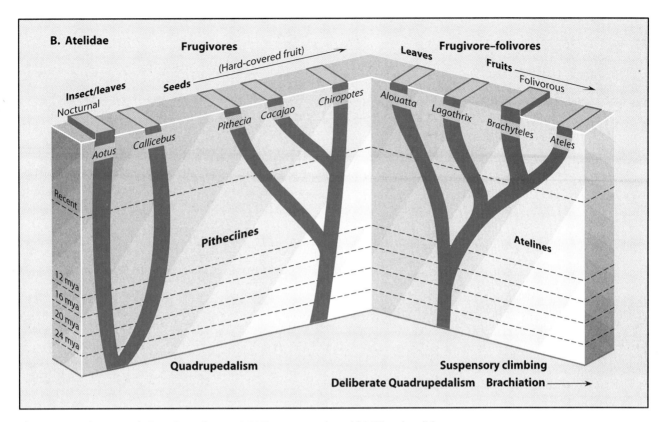

Figure 6-1 These panels from Rosenberger (1992) represent the cebid (**A**) and atelid (**B**) families of New World monkeys. Living genera are listed at the top of each panel, and the numbers at the left refer to millions of years before the present. Note the different locomotor patterns and feeding adaptations that typify the four subfamilies.

Cebus in the same subfamily because of suggestive molecular data (summarized by Schneider and Rosenberger), and because they resemble each other in their frugivorous-insectivorous diet, aspects of social organization, and certain features of the skull and teeth (see Fig. 6-1). For example, Linda Marie Fedigan of the University of Alberta in Canada and colleagues note that *Cebus* and *Saimiri* both have short faces, narrow noses, broad premolars, tiny third molars, similar canines that are highly sexually dimorphic, and rounded braincases that contain relatively large brains. Like callitrichines, these monkeys eat a good deal of insect food and urine wash (some workers group them with *Aotus* and *Callicebus*, which also engage in these behaviors).

An alternative view suggested by cladistic studies of certain details of teeth, muscles, and bones by Susan Ford, among others, is that capuchins have been separated from all other platyrrhines for a very long time (perhaps 20 mya) and therefore should not be placed in the same subfamily as squirrel monkeys (Fig. 6-2). Ford's systematic scheme is worth considering because *Cebus* differs remarkably from the other ceboids in numerous respects. For example, capuchins reach puberty later and have much longer lives than other New World mon-

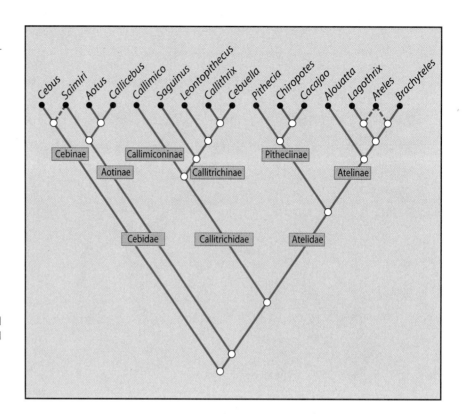

Figure 6-2 In Ford's (1992) systematic arrangement of New World monkeys, *Cebus* and *Saimiri* are placed in separate subfamilies, and *Cebus* is interpreted as the end-product of an ancient radiation that evolved in isolation from the other platyrrhines.

keys. As discussed in the last chapter (see Neural Note 5), the frontal lobes of *Cebus* also differ from those of the other New World monkeys. Although *Cebus* is classified with *Saimiri* in the cebine subfamily in this text, this decision is not an easy one and could change with the addition of more information from future molecular and morphological studies.

THE COMMON SQUIRREL MONKEY (*SAIMIRI SCIUREUS*) The taxonomy of squirrel monkeys is currently in a state of confusion, with some workers recognizing as many as seven species. Minimally, there are at least two species, the common squirrel monkey (*Saimiri sciureus*) from South America, and the red-backed squirrel monkey (*Saimiri oerstedii*) from Central America. Although field studies have been carried out on squirrel monkeys since the early 1980s, only a few long-term studies have focused on individually recognized troop members, beginning in 1986 with research on *S. oerstedii* in Costa Rica by Sue Boinski (see In the Field) of the University of Florida, and followed shortly thereafter with comparable studies by Boinski and others on *S. sciureus* in Manu National Park, Peru.

S. sciureus is roughly squirrel sized and, partly because of its small size, has commonly been kept as a pet and used in laboratory research. This pert little monkey has a distinctive appearance with its dense coat of short yellow or greenish-yellow hair, large ears, pink face, black muzzle, and white markings around the eyes that make it look like it is wearing goggles (Fig. 6-3). (Although squirrel monkeys are sometimes divided into two groups based on the shape of these

Figure 6-3 Squirrel monkeys look like they are wearing goggles. (Photo courtesy of Sue Boinski.)

goggles, i.e., one form with "Roman" rounded arches and the other with "Gothic" pointed arches, Warren Kinzey warned that these designations do not correlate with current taxonomy.) Adult males are generally larger than adult females (Table 6-1) and there is also marked sexual dimorphism in upper canine size. As noted for *Cebus*, the third molars are reduced and the face is foreshortened in *Saimiri*. Interestingly, while the tails of adult squirrel monkeys are not prehensile, those of infants are to some degree.

Common squirrel monkeys live in a variety of habitats throughout the Amazon basin (Fig. 6-4). They run and walk along tree branches with great agility. When resting, *Saimiri* monkeys typically sprawl

Figure 6-4 Common squirrel monkeys (*Saimiri sciureus*) are found throughout the Amazon basin of South America.

TABLE 6-1 Statistics for Representative New World Monkeys

	S. sciureus (common squirrel monkey)	Cebus apella (tufted capuchin)	Aotus trivirgatus (owl monkey)	Cacajao calvus calvus (white uakari)	Alouatta palliata (mantled howler)	Brachyteles arachnoides (woolly spider monkey)
Mean adult body weights						
F	0.68 kg	2.39 kg	0.95 kg	2.88 kg	5.35 kg	9.45 kg
	(1.5 lb)	(5.3 lb)	(2.1 lb)	(6.3 lb)	(11.8 lb)	(20.8 lb)
M	0.85 kg	3.10 kg	0.92 kg	3.45 kg	7.15 kg	12.13 kg
	(1.9 lb)	(6.8 lb)	(2.0 lb)	(7.6 lb)	(15.8 lb)	(26.7 lb)
Adult brain weight (g)	24.4	71	18.2	73.3	55.1	120.1
Maximum life span (yr)	21	48	20	20	25	30
Olfactory communication[a]	uw	uw	uw, sm	uw, sm	sm, uw	uw, um
Location (habitat)[b]	Ab	ncSA	uA	Br	M-E	Br
	rf, rif, m, sw	rf	rf, df	ff	rf, df, m	rf, sf
Locomotion[c]	qwr, l	qr, l, c	qr, l	qwr, l, c, d	qr, cl, s, b, d	br, s, qwr, c, l
Diet[d]	i, a, f, n	f, s, p, n, i, a	f, fl, l, i, n	s, f, fl, a	l, fl, f	l, f, s, fl, n
Mean no. of offspring	1	1	1	1	1	1
Seasonal breeders	+	+	Peak?	+	0	Peak
Gestation (days)	170	155	120	183?	186	233
Interbirth interval (mo)	13.6	22	10		22.5	36.4
Mating systems[e]	mm	mm	mo	mm	mm	mm
Territorial	0	0	+?		0	0
Ratio of adult females to males	2–4:1	1–2:1	1:1	1:1	1–4:1	1–2:1
Alloparenting	+	+	+		+	+
Infanticide	0	+	0	0	+	0
Group size	25–54	8–14	2–5	20–50	2–45	8–45
Density	50–175/km²	5–40/km²	14–40/km²		16–90/km²	3–17/km²
Home range	33–250 ha	125–260 ha	5–10 ha	500–550 ha	3–60 ha	70–168 ha
Conservation status[f]	L	L	L	H	L	H

Note: Information in this table has been selected from numerous sources and should be viewed as tentative, i.e., subject to change in light of new information. Sources include but are not confined to Ford and Davis, 1992; Rowe, 1996 (including conservation statuses); Kinzey, 1997; Robinson and Janson, 1987; Robinson, Wright, and Kinzey, 1987; Harvey et al., 1987; Crockett and Eisenberg, 1987; Glander, 1996; and Strier, 1996. Abbreviations: +, yes; 0, no. Blank entries indicate that no information is presently available. Dietary items are listed in order of decreasing importance to the species.

[a]Olfactory communication: sm, scent marks; um, urine marks; uw, urine washes.

[b]Location (habitat): Ab, Amazon basin; Br, Brazil; M-E, Mexico to Equador; uA, upper Amazon region; ncSA, northern and central South America; df, dry forest; ff, floodplain forest; m, mangroves; rf, rain forest; rif, riverine forest; sf, seasonal forest; sw, swamp.

[c]Locomotion: b, bridging; br, brachiation; c, climbing; cl, clinging and leaping; d, dropping; l, leaping; qr, quadrupedalism; qwr, quadrupedal walking and running; s, suspension.

[d]Diet: a, animal prey; f, fruit; fl, flowers; g, gums (sap); i, insects; l, leaves; n, nectar; p, pith; s, seeds.

[e]Mating systems: mm, multimale, multifemale (promiscuous); mo, monogamy.

[f]Conservation status: H, highly endangered; L, lower risk.

with their abdomen down on a branch and the limbs extended. To sleep, squirrel monkeys huddle squatting with their head down and their tail tucked between the legs and draped over a shoulder. In Peru, squirrel monkeys take advantage of capuchin monkeys' knowledge of the locations of large fruiting trees, following them to feed. They supplement the fruit component of their diet with cryptic, frequently immobile insects such as caterpillars that they snatch or glean from the surfaces of leafs, and eat only small amounts of leaves.

Groups of *S. sciureus* average around 25 to 54 individuals but vary enormously, sometimes coalescing into several hundred animals, while at other times dividing into relatively small subgroups based on age, sex, or both. Social organization varies a good deal between species of squirrel monkeys. In common squirrel monkeys, males maintain dominance with a conspicuous penile display and, contrary to *S. oerstedii*, dominance hierarchies and alliances are discernible among females. In Peruvian populations at least, *S. sciureus* females are dominant to males. Also unlike red-backed squirrel monkeys, males usually migrate away from their natal groups before reaching sexual maturity.

Matings occur during a restricted season and are potentially polygamous in the multimale, multifemale groups that are typical during the mating season. Interestingly, males become fattened before the mating season, during which they compete for mates by displaying and wounding each other. However, the extent to which males compete is apparently variable. For example, Boinski reports that it is female choice rather than male competition that distinguishes the mating season of *S. oerstedii* in Costa Rica. There, at least, squirrel monkeys might be viewed as rather egalitarian. After gestations of about 5½ months, females give birth to one infant. Unlike marmosets, tamarins, and the owl monkey, male squirrel monkeys take no part in the direct care of infants.

Because they are relatively small, *Saimiri* monkeys are vulnerable to, and produce alarm calls upon seeing, cats, snakes, and birds of prey. Squirrel monkeys may be especially susceptible to predators because of their small bodies, which may partly explain the large congregations of hundreds of squirrel monkeys that have been reported. It may also explain the greater emphasis on contact calls (called "security-blanket" vocalizations by Fedigan and Boinski) that has been reported for, at least, *S. oerstedii*. Humans are also a threat to squirrel monkeys, which are considered to be pests because they raid crops. Although *S. sciureus* is not regarded as endangered, *S. oerstedii* is.

Finally, much has been made of the fact that squirrel monkeys have brains that are as big relative to body size as those of humans. However, this observation is not nearly as impressive as it seems when one considers that smaller species of mammals generally have brains that are relatively bigger than those of larger species, owing to simple scaling factors associated with different body sizes. This can even be seen during the development of one species, say *Homo sapiens*, in which infants have much bigger heads (brains) relative to body

The Moment I Became a Scientist!

Sue Boinski is a primatologist at the University of Florida who is best known for her field studies of the social, foraging, and vocal behavior of squirrel monkeys (Saimiri) in Central and South America. She also studies capuchin monkeys (Cebus). Her research on both of these cebines is characterized by intense attention to their natural histories, focusing on the costs and benefits of foraging decisions, predation risks, and current food availability. Boinski remembers well the moment she became a scientist. In her own words:

SUE BOINSKI IS ONE OF THE WORLD'S LEADING EXPERTS ON SQUIRREL MONKEYS. (PHOTO COURTESY OF SUE BOINSKI.)

My original intention as a graduate student was to study the behavioral ecology of lemurs in Madagascar. However, in my second month as a graduate student, I met a zoology professor who had just received a big National Science Foundation grant to study butterflies at an isolated lowland wet forest site in Costa Rica, Corcovado. This site also happened to have four species of monkeys: howlers, spiders, capuchins, and squirrel monkeys. The professor

size than do adults. (What *is* impressive is that humans have brains as relatively large as those of squirrel monkeys despite the fact that their bodies are so much bigger.)

THE TUFTED CAPUCHIN (*CEBUS APELLA*) Capuchin monkeys are so-named because the hair on their heads appears to form dark caps that are reminiscent of the hoods worn by Capuchin monks. Because they are clever, capuchins have been used to solicit coins to the tunes of organ-grinder music, which is why they are also called organ-grinder monkeys. *Cebus* includes three allopatric "nontufted" species (*C. albifrons*, *C. capucinus*, and *C. olivaceus*) and one "tufted" species (*C. apella*) that overlaps with the nontufted species except *C. capucinus*. A fifth species (*C. kaapori*, or the Ka'apor capuchin) has only recently been recognized. *Cebeus apella* (the tufted, brown, or black-capped capuchin) is distinguished by its flat-top hairdo with two tufts that appear somewhat like horns (Fig. 6-5), and the fact that, at about 2 to 3

Figure 6-5 The tufted capuchin (*Cebus apella*) appears to be wearing a small, somewhat flat fur hat.

asked me if I would like to go down to Corcovado the next summer to study monkeys and, of course, I said, "Sure." I spent the intervening months planning a study for my master's thesis. Once there, I did a study on the interactions and dominance relations among the four species of monkeys when they met at fruit sources. At the time, I found the squirrel monkeys to be the least interesting of the four species—there were too many in a troop, they moved too fast, and they all looked alike. However, although they were the second most commonly studied primate in captivity, squirrel monkeys had not been studied in the field in other than short-term, qualitative, 10-week studies of poorly habituated animals. I therefore calculated that it would be worthwhile to focus on these beasts for my Ph.D. dissertation. Now, 19 years later, I find them absolutely fascinating, and every new gobbet of data I trip across scratches the bottom of my brain, provoking new questions about the wonderful world of *Saimiri*.

However, it was not always this way. Shortly after beginning my dissertation research and habituating the beasts so that I could closely observe them, a problem arose. The squirrel monkeys were not behaving the way they were supposed to! Although the extensive literature on the behavior of captive squirrel monkeys repeatedly emphasized the high levels of aggression and strict dominance hierarchies in squirrel monkeys, I just couldn't see any. The monkeys never fought or directly competed over anything, including food. In fact, there was so little competition among them that I had no data whatsoever that would permit me to construct a dominance ranking for the animals. They were completely egalitarian in every respect (this changed during mating season, however). I was convinced that I was a failure as a primatologist because I hadn't found what I was supposed to, and even considered abandoning my study and returning to the States. Then I had a cathartic moment when I accepted that I just had to go with the data and not force them into any previous paradigm. If they didn't fit a particular hypothesis, I just had to reject it and construct new, more appropriate and valid hypotheses to test. That was the exact moment I became a scientist.

kg (5 to 7 lb), it is one of the larger-bodied capuchins. Like the other species of *Cebus*, its tail is prehensile but lacks the bare skin at the tip that typifies the (independently evolved) prehensile tails of the atelines. *Cebus apella* is the most widely studied of the five species of capuchins. According to Kinzey, its distribution in northern and central South America (Fig. 6-6) suggests that it is also the most widespread nonhuman species of neotropical primate.

As noted earlier and detailed by John Robinson and Charles Janson, capuchins share many features with squirrel monkeys, including habitat, quadrupedal locomotion with some leaping and climbing, and home ranges that contain patchy distributions of fruit and insects and that can be quite large (see Table 6-1). Both genera are sexually dimorphic. Furthermore, both capuchins and squirrel monkeys live in multimale, multifemale polygamous groups, have birth seasons, rely partly on olfactory communication, and defend resources such as fruiting trees rather than fixed territories.

Capuchin monkeys also differ from squirrel monkeys in a number of ways. They are more widely dispersed in the New World, and do not congregate in as large of social groups as squirrel monkeys do.

Figure 6-6 Located across northern and central South America, tufted capuchins may be the most widespread of any nonhuman neotropical primate.

While *Saimiri* seems to get along with other species of monkey, *Cebus* interacts hostilely with some of the atelines. Male-male competition for estrous females is not as marked in *Cebus* as it is in at least some populations of *Saimiri*, but dominance hierarchies are more obvious in the former (for both sexes). Capuchins are generally less variable in their social organization than squirrel monkeys, and tend to have stable matrilineal dominance hierarchies in which the highest-ranking female is dominant to all but the alpha male. Allogrooming between adults, food sharing, and allomaternal nursing are important for *Cebus* but not *Saimiri* (although the last behavior has been reported in one subspecies of *S. sciureus*).

The most remarkable distinction of *Cebus*, however, is its agility and tool-using abilities. As Elisabetta Visalberghi of the Institute of Comparative Psychology in Rome puts it (1993:139), "Capuchins try to fill the gap between themselves and their goal by means of external objects, in a way that is neither rigid nor stereotyped. Stones are employed to crack nuts, paper towels serve to sponge liquids, and sticks can be used as multipurpose tools to rake, to treat wounds, to kill a snake, or to threaten an enemy." Capuchins come by their dexterity naturally. Compared to other monkeys, they are noisy, destructive foragers that use their manipulative skills as they probe under bark and poke into holes for insects, or forage for tough, hard-to-open fruits and nuts. For example, a population of *Cebus apella* in Colombia has developed a complex sequence of steps for opening tough palm-fruits, as described over 20 years ago by Kosei Izawa of the Japan Monkey Center. Their method entails removing the fruit of the cumare (a cocoid palm), moving to a nearby tree, sitting on a branch and using the prehensile tail to hang on, and then striking the fruit with both hands against the tree (Fig. 6-7). The capuchin then peels away part of the cracked husk and repeats the procedure until the fruit itself cracks, at which point the monkey uses its teeth and fingers to create a large-enough hole through which to sip the fruit's juice. After the juice is gone, the hardened coco is removed from inside the husk with the

Figure 6-7 Capuchins are able to open tough palm-fruits by smashing them against a tree.

REVIEW EXERCISES

1. Which four genera of neotropical primates are difficult to classify and why? Why has this text opted for Rosenberger's classification?

2. Discuss two ways in which *Cebus* has been classified by different workers and the reasons for each interpretation.

3. In this text, which two genera comprise the cebine subfamily?

4. What factors might explain why squirrel monkeys sometimes congregate in groups of hundreds of individuals?

5. In what ways do squirrel and capuchin monkeys resemble each other? How do they differ?

6. Discuss the agility and manipulative skills of *Cebus*? Do these abilities necessarily indicate cognitive skills higher than those of other neotropical monkeys? Explain.

teeth and consumed. All of this is done with just the right number and intensity of strokes, which vary with the hardness of the husk.

But are the manipulative skills of capuchins indicative of higher cognitive (representational) capacities, as has been interpreted for similar activities in chimpanzees (see Chapter 13)? To explore this question, Visalberghi tested six *Cebus apella* animals with a Plexiglas tube baited with a peanut in the middle. The peanut could only be obtained by pushing it out with a stick. Sometimes a satisfactory stick was provided; at others times the provided tool had to be modified by breaking, untaping, using two pieces together, and so on. Although the monkeys eventually obtained the peanuts, their success did not seem to be because they understood what the task was about (e.g., length of tool), but rather (1993:140) "because they tried every possible way to solve it, using both their own bodies and any external objects they came across." Visalberghi also concluded that capuchins do not learn by imitation, nor do they recognize themselves in mirrors as do chimpanzees (see Chapter 13). In short, although capuchin monkeys are incredibly manipulative when it comes to objects in their environment (and therefore may be trained to help severely handicapped people), they never get beyond the "trial-and-error" stage of problem solving.

ATELIDAE

We now leave the family of little gum-feeding, insect-probing, and leaf-gleaning cebids (see Fig. 6-1A), and turn our attention to the nine genera of atelids (see Fig. 6-1B). As already noted, molecular and morphological data support the grouping together of the prehensile-tailed atelines, on the one hand, and the seed-eating uakaris and sakis within the pitheciines, on the other. Titi monkeys (*Callicebus*) are aligned with the pitheciines both morphologically and molecularly. The owl monkey (*Aotus*) is an unresolved problem, however, because its genetic data suggest it is a cebid while its morphology aligns it with *Callicebus* in the atelids. For now, this text opts for the latter interpretation by including *Aotus* together with *Callicebus* as a branch within the pitheciines (see Fig. 6-1B).

The Owl Monkey (*Aotus trivirgatus*)

The taxonomy of the owl monkeys (sometimes called night monkeys) is currently in a state of flux. Until fairly recently, many regarded all owl monkeys as belonging to one highly variable species (*Aotus trivirgatus*) that contained 10 subspecies. At least two natural groups of owl monkeys are now recognized, however, a gray-necked species (*A. trivirgatus*) found primarily north of the Amazon, and a red-necked species (*A. azarae*) that lives to the south. Based on an extensive analysis of cranial and dental measurements along with fur color, Ford believes that these two groups should be divided into five to seven recognized species of *Aotus*. Whether or not Ford's classification gains acceptance remains to be seen or, to quote Kinzey

Although titi monkeys (*Callicebus*) resemble owl monkeys in many ways, they are diurnal rather than nocturnal.

Box 6

Why Be Nocturnal?: Owl Versus Titi Monkeys

Titi monkeys (*Callicebus*) resemble owl monkeys in many ways, which is why some workers place them together in a branch of the pitheciines. Both genera contain little monkeys that have small ranges, mark with scent, twine tails, use vocalizations to maintain territories (titi monkeys more so), and live in monogamous family groups in which males intensively parent offspring. Unlike owl monkeys, however, titi monkeys are diurnal. For this reason, Patricia Wright decided to compare the diets, sleeping habits, and movement patterns of dusky titi (*Callicebus moloch*) and owl monkeys to gain insight into the owl monkey's nocturnal lifestyle.

Wright discovered that the two groups differ in several ways. For one thing, titi families sleep in a large number of different trees over the course of a year, while owl monkeys use only a handful. *Callicebus* forages quietly in the low canopy during the day, compared to the noisy adventures of *Aotus* high in the trees at night. Wright began to suspect that diurnal life for titi monkeys must be associated with dangers that are not of concern for owl monkeys at night. But what were these dangers? Predators were an obvious possibility, and Wright soon discovered that diurnal hawks and eagles prey on titi monkeys. Remaining relatively quiet and low in the canopy therefore may be an adaptation to these predators. Birds of prey are not a threat at night, however, and the usual nocturnal snake and cat predators are primarily terrestrial and therefore no match for lively, arboreal owl monkeys. As mentioned earlier, because they are nocturnal, owl monkeys also do not have to compete for food with other larger monkeys, as do titi monkeys. Interestingly, Wright found that *Aotus* had reverted partially to day living in an open forest habitat with a limited number of diurnal monkey species and birds of prey. Wright concluded that avoidance of predators and food competitors may have been stimuli for the owl monkey's return to nocturnal living.

(1997:187), "the taxonomic tale of *Aotus* has not yet been completed."

The owl monkey's claim to fame is that it is the only monkey in the world that is nocturnal. In keeping with its night life, *Aotus* (like prosimians) has a low basal metabolic rate, and the largest eyes relative to body size of any anthropoid. Several lines of evidence suggest that the ancestors of the owl monkey were diurnal, however, i.e., that becoming nocturnal was a secondary adaptation for *Aotus*. For example, owl monkeys lack the reflecting tapetum lucidum that makes the eyes of most nocturnal prosimians appear to glow in the dark. They also have olfactory bulbs that, although large for New World monkeys, are relatively small for a nocturnal primate. These features are believed to reflect an earlier evolutionary shift from nocturnal to diurnal living. Furthermore, in certain parts of their range where predator pressures are reduced, owl monkeys have even shifted back (again) to partial diurnal living (see Box 6). *Aotus* is associated with a deep fossil

Figure 6-8 Owl monkeys have large eyes and endearing little faces.

record (see Fig. 6-1B), and it appears that the ancestors of owl monkeys were already nocturnal by 12 mya. Since there are no prosimians in the New World, owl monkeys may be viewed as occupying a convergent ecological niche.

Owl monkeys are small, with adults of both sexes averaging a little less than 1 kg (about 2 lb) each (see Table 6-1). Their large eyes are highlighted above by light patches, which are accentuated by three black stripes that run toward the top of the head (Fig. 6-8). Although there is some variation, the middle stripe generally courses from the bridge of the nose, while the two outside stripes frame the lighter face. Coat color varies from reds to grays to browns, and appears frosted (agouti) due to lightened bands on individual hairs. The ears are small, and the digits of the hands and feet have tactile pads and nails (except the fourth digits of the feet, which have clawlike nails). These digits do not have as much independent mobility as those of many of the other atelids, however. Owl monkeys have relatively long legs, and long tails that are usually black toward their tips. Although their tails are not prehensile, *Aotus* pairs frequently sit next to each other with tails entwined.

Owl monkeys inhabit all kinds of forests from Panama to northern Argentina (Fig. 6-9). Because of their low metabolic rates, they are able to live at relatively cool temperatures, which partially explains their wide distribution. Lack of competition from other primates also permits them to occupy all levels of the canopy, through which they move quadrupedally and with a good bit of leaping and springing. Home ranges of owl monkeys are quite small (5 to 10 ha), as are the distances groups travel within their ranges on any given night (although they travel farther on brightly moonlit nights). Small ranges are probably related to lack of competition for food from other primates. Owl monkeys are basically frugivorous, enjoying a variety of fruits, flowers, and young leaves. However, they also eat a high percentage of insects, which are nutritious, noisy, and abundant at night. During seasonal scarcity of favored fruits, *Aotus* is able to harvest nectar and figs without competition from the larger diurnal monkeys. These feeding advantages may help explain why *Aotus* became secondarily adapted to nocturnality.

Owl monkeys are among the few primates that are truly monogamous. They live in small family groups of a mated pair and up to three offspring of different ages that have not yet established their own families (i.e., are under 3 years old). Families repeatedly return to a few daytime sleeping sites, which include holes in trees, brush, or tangles of vines, and family life is usually peaceful. According to Patricia Wright's observations of *Aotus* in the wild, however, fights are likely to occur between neighboring groups in fruiting trees at the borders of territories when the moon is full. Preceding these fights, adults of both sexes inflate their throat pouches, producing loud whoops. As they vocalize, their hair stands on end and they jump about stiff-legged. The actual fights last about 10 minutes and involve more aggressive whoops, chasing, and wrestling. During encounters, both sexes urine wash and scent mark (from glands beneath the tail), and afterward each group retreats into its own territory. Wright also notes that, in captivity at least, individuals fight with others of their own sex.

Socially, owl monkeys seem somewhat reserved. Unlike other monogamous primates (including titi monkeys), adult males and females do not advertise and strengthen their pair-bonds by duetting, and seldom groom each other. Owl monkeys become sexually mature by about 2.5 years of age. Despite lack of a clear birth season, births

Figure 6-9 Owl monkeys are widely distributed in the neotropics.

may peak during seasons when fruit is plentiful. Usually one offspring is born annually, after about 4 months of gestation. As is true for most callitrichines, shortly after birth, *Aotus* males become heavily involved with carrying, feeding, and playing with offspring. Mothers do not play much with their offspring and, except for nursing, are generally less-intensive parents than fathers.

Although nocturnal predators such as owls, snakes, and cats rarely eat monkeys, and despite the fact that *Aotus* is legally protected, humans prey on owl monkeys by exploiting them for food, for fur, and as pets. Because they are susceptible to malaria and have interesting visual systems, owl monkeys are frequently used in medical and neurological research.

The Seed Predators: Sakis, Bearded Sakis, and Uakaris

If the pitheciines constitute "the evolutionary secret of the New World monkey radiation," as described by Rosenberger, Norconk, and Garber (1996:329), it is the three genera of seed predators that form the most mysterious and bizarre evolutionary offshoot within this subfamily. These monkeys include four or five recognized species of sakis (*Pithecia*), two species of bearded sakis (*Chiropotes*), and two or three species of uakaris (*Cacajao*) (see Fig. 6-1). The lower front teeth of seed predators form pseudodental combs that are specialized for opening nuts and tough (hard-surfaced) fruits in order to obtain the large nutritious seeds that are hidden within. This and other shared dental adaptations allow these monkeys to eat unripe fruit that is not palatable for most primates. *Pithecia* is the smallest of the three genera, with both males and females usually weighing under 3 kg (less than 7 lb). It is also the most versatile in its ability to survive in a variety of habitats. Sakis have long shaggy coats of darkish hair with striking white markings that vary to different degrees between the sexes (sexual dichromatism). Bearded sakis are a little larger and, as their name implies, have full beards as well as full puffs of hair over the temples and forehead (Fig. 6-10). The body of *Cacajao* weighs generally under 9 lb and, unlike any other New World monkey, the uakari has a short, somewhat stumpy tail. *Cacajao* has a widespread nose, as do the other seed predators. To some observers, uakaris appear rather weird because of their long shaggy coats, bald skull-like heads, and bright red faces that get even more so when they become upset (Fig. 6-11).

These genera are found largely in the Amazon River basin. *Pithecia* inhabits nonflooded forests, while *Chiropotes* and *Cacajao* prefer the upper canopy of undisturbed forests (which are seasonally flooded in the case of uakaris). All three are basically quadrupedal, with some leaping, climbing, dropping to branches below, and even occasional bipedal hopping and walking. Uakaris are good leapers and swingers and, like orangutans (see Chapter 11), assume a variety of agile postures when feeding in trees. During seasons when preferred fruits and seeds are difficult to find, leaves and insects are added to the diets of

Figure 6-10 Bearded sakis (*Chiropotes*) have amazing hairdos.

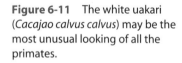

Figure 6-11 The white uakari (*Cacajao calvus calvus*) may be the most unusual looking of all the primates.

seed predators. As sometimes seen in other primates, some species also eat dirt, possibly for its mineral content.

To date, little is known about the social lives of seed predators. *Chiropotes* and *Cacajao* form multimale, multifemale groups of 15 or more individuals, and these groups split up to feed and join together to travel. Smaller multimale, multifemale groups are reported for *Pithecia*, which was once believed to be monogamous, an interpretation that is now seriously questioned. Births in these monkeys produce single infants, and there is little evidence of paternal care. Not much is known yet about possible territoriality for seed predators.

In terms of conservation, sakis have been hunted for food and for their tails (used as dusters or ornaments), although they are not viewed as particularly endangered. Bearded sakis have been subjected to the same pressures and are endangered, as are uakaris, which have been hunted for food, bait, and pets.

THE WHITE UAKARI (*CACAJAO CALVUS CALVUS*) Although uakaris have been studied to a limited extent in captivity and by Roy Fontaine and colleagues in seminatural conditions at Monkey Jungle in Goulds, Florida, they remain one of the least-studied primates in the wild. Part of the reason why uakaris have not been thoroughly researched is that they live in seasonally flooded forests in the Amazon that are extremely difficult to access. Jose Marcio Ayres of the Wildlife Conservation Society in Brazil overcame this obstacle by building a 33.6-km (21-mile) trail system that he navigated by dugout canoe to track the only white uakaris that live north of the Amazon. His study is one of the few recent reports on wild uakaris.

White uakaris are simply incredible looking, with long gleaming white coats that contrast sharply with nearly naked heads and bright red, extremely expressive faces (see Fig. 6-11). Adult males have large muscle masses on each side of the skull and forehead. The reddened face, which is due to a lack of pigment and extensive capillaries, is a sign of good health. White uakaris have the powerful jaws and pseudo-dental comb that typify seed predators, as well as the short tails that characterize all *Cacajao*.

The white uakaris studied by Ayres inhabit a 8060-km² (3100-square-mile) region in Brazil that is surrounded by waters from the Japurá and Amazon rivers and the Auati Channel (Fig. 6-12). At Ayres's site, uakaris overlap with red howlers, squirrel monkeys, and capuchins. During 6 months of each year, this region is completely flooded so that the tallest trees look like bushes floating on the water. Not surprisingly, white uakaris sleep high in the trees. Six or seven individuals may sleep spread out in one tree, with mothers and young offspring sleeping together. Early in the morning, the monkeys descend to lower trees, but they are still located high above water level. The large troops of white uakaris (about 50 individuals) are spread out most of the day as they travel, feed, and rest. (Ayres was able to keep track of the troop's general location by listening for the plopping sound of fruit remains hitting the water below.) By the end of the day, the uakaris have covered up to 4.8 km (3 miles), and loud vocalizations are heard as the monkeys begin moving upward to their sleeping sites.

Both mature and immature fruits are scarce during the low-water season. During this period, Ayres (1990:39) observed that white uakaris frequently descend to the ground and forage for new seedlings on the forest floor, where they "resembled white rabbits, with their quick leaping movements." Fortunately for the uakaris, there is not much competition for this food because many mammals seem to avoid regions that are prone to seasonal flooding. Interestingly, white uakaris travel farther (up to 6.4 km [4 miles] a day) when ripe fruits are abundant than during the low-water season (less than 2.4 km [1.5 miles]).

Receptive female and male white uakaris form pairs that leave the core of the troop during April and May. Young offspring of the female may accompany the pair. The male is likely to be harassed by bachelor troops of up to eight males that do not have mates. According to Ayres (1990:37), "these bachelors behave as badly as hooligans at a soccer match" and fights are frequent. It appears that females may copulate with only one male per pregnancy and that gestation lasts about 6 months. (However, some reports suggest that uakaris may mate promiscuously.)

Figure 6-12 The range of white uakaris in Brazil is completely surrounded by water, which explains why the area is flooded during half the year.

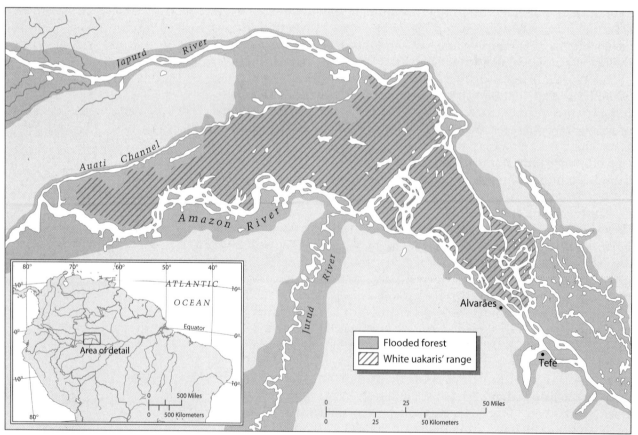

Some of the generalizations reported for other uakaris by Fontaine may also apply to white uakaris. For example, *Cacajao calvus calvus* engages in a high rate of allogrooming compared to other platyrrhines with which it is sympatric. Play is also important for uakaris in general, and interactions with other species of monkey may be friendly or hostile. Dominance relationships appear to be only weakly developed in semi-free-ranging uakaris, and these are expressed in noisy fighting, especially among females. Fontaine also reports that adult males urine wash with vigorous slapping, while both sexes engage in genitoanal scent marking and rub aromatics such as fruit juice into their fur. Interestingly, Fontaine (1981:488) also notes that "the male uakari's urine wash appears to involve the sternal glands as well, which may account for the pungent smell of large males."

Although uakaris are legally protected in Brazil and Peru, Frances D. Burton of the University of Toronto reports that they continue to be preyed on by humans who hunt them for food, for pets, and even for meat to be used as fishing bait! Consequently (and very sadly), this fascinating monkey is highly endangered.

THE ATELINES

Despite their famous prehensile tails (see Neural Note 6), most of the atelines have not been as thoroughly researched as many other neotropical monkeys have. This group comprises Rosenberger's fourth subfamily of New World monkeys (see Fig. 6-1), which includes (depending on who's counting) six to eight species of howler monkeys (*Alouatta*), two species of woolly monkeys (*Lagothrix*), one or two species of muriquis (or woolly spider monkeys, *Brachyteles*), and four to six species of spider monkeys (*Ateles*). Rosenberger's panel reveals a range of dietary preferences in atelines, spanning from a heavy emphasis on leaves in howlers to a predilection for fruit in spider monkeys. The panel also suggests that ateline shoulders, arms, hands, and prehensile tails are built for suspensory climbing and deliberate quadrupedalism in howler monkeys, to hand-over-hand, arm-swinging locomotion (brachiation) that is realized most fully in the muriquis and spider monkeys. The Old World gibbons are also wonderful brachiators but, unlike the atelines, these lesser apes (see Chapter 10) lack any kind of a tail. (Although some workers describe gibbons as true brachiators and the neotropical prehensile-tailed monkeys as semibrachiators, both are called brachiators in this text.) Relative to trunk length, brachiators in both worlds have long legs and especially long arms. The hands of platyrrhine brachiators are able to form hook grips, because the thumb is aligned next to elongated fingers in the case of howler and woolly monkeys, and is greatly reduced or even absent in spider monkeys.

Each of the four genera of atelines has interesting features. With their long slender limbs and prehensile tails, rapidly moving spider monkeys look like a flurry of hairy legs, which accounts for their

REVIEW EXERCISES

1. Which genera comprise the pitheciine subfamily?
2. Why is the classification of *Aotus* difficult?
3. Is the nocturnality of *Aotus* more likely to have (a) been retained from its earliest primate ancestors or (b) developed from diurnal ancestors as a secondary adaptation? What evidence supports your answer?
4. What advantages does *Aotus* gain by being nocturnal? How does comparison of *Aotus* and *Callicebus* shed light on this question?
5. Name the three genera of seed predators.
6. What anatomical features distinguish seed predators from the other neotropical monkeys and how are these traits related to their particular dietary and behavioral niche?
7. Why is so little known about uakaris and why are they often described as bizarre looking?
8. How do white uakaris cope with food shortages during the low-water season?

Therein Lies a (Prehensile) Tail

One of the greatest pleasures of visiting a zoo is to watch a group of prehensile-tailed monkeys (e.g., spider monkeys) as they navigate their arboreal habitats in a rapid hand-over-hand-over-tail acrobatic ballet. In these species, the tail functions not only as an extra arm, but also to suspend the entire body from supports, thus freeing the hands for grasping fruits or leaves. Prehensile tails are extremely muscular and sensitive to touch, with dermal ridges that are the equivalent of fingerprints on the underneath surfaces of their tips (see figure below). (However, the tail of the capuchin monkey, *Cebus*, in which another form of prehensile tail evolved convergently, lacks these ridges and does not have as strong of a grasp.)

Neurophysiologists have mapped the brains of both prehensile-tailed and non-prehensile-tailed monkeys. Not surprisingly, the brains of prehensile-tailed monkeys are relatively expanded in the primary sensory and motor areas that represent tails. In fact, the part of the brain that subserves the tail has enlarged to such a degree in the atelines (and to a lesser extent *Cebus*) that new sulci have formed at its edges on the surface of the cerebral cortex (see figure on right). Therefore, we can determine if a particular species of monkey has a prehensile tail by a quick glance at its endocast!

The undersurface of atelines' prehensile tails have dermal ridges that provide a firm grip.

Sensory and motor tail representations are enlarged in the brains of prehensile-tailed monkeys, and are associated with sulci that are not found in the brains of other monkeys. **A.** Left hemisphere of macaque. **B.** Left hemisphere of spider monkey. **SMI**, sensory cortex; **MSI**, motor cortex; **F**, face; **A**, forelimb; **H**, hind limb; **T**, tail (expanded on surface of brain in prehensile-tailed primates).

(A)

(B)

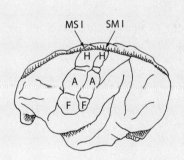

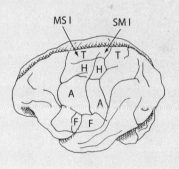

Left hemispheres of a macaque (A) and a spider monkey (B).

SM I	sensory cortex
MS I	motor cortex
F	face
A	forelimb
H	hindlimb
T	tail (expanded to surface of brain in prehensile-tailed primates)

Figure 6-13 Spider monkeys are so-called because of their long extremities and quick movements.

Figure 6-14 Humans find woolly monkeys quite attractive, which unfortunately makes them favorite targets of the pet trade.

common name (Fig. 6-13). *Ateles* lives throughout Central and South America, where it specializes in eating ripe fruits that are widely distributed in patchy habitats. Spider monkeys live in large multimale, multifemale groups that frequently divide into smaller parties and join together, which is similar to the fission-fusion societies of chimpanzees (see Chapter 13).

Woolly monkeys, which live in the Amazon basin, do not move as quickly or leap as much as spider monkeys. Their common name comes from their dense fur which, along with their rounded heads and potbellies, makes *Lagothrix* quite striking (Fig. 6-14). Curiously, these monkeys have a submissive gesture in which they make a sobbing noise while covering their eyes with a hand. According to Kinzey, the woolly monkey is probably the least-known platyrrhine genus (although its diet and, to a degree, social organization are believed to be similar to those of spider monkeys). Kinzey also notes that, of the neotropical monkeys, *Lagothrix* is the most vulnerable to hunting and habitat destruction, being hunted for the pet trade, their pelts, and meat.

A third genus of atelines includes the deliberately climbing howler monkeys (*Alouatta*) with their specialized anatomy for vocalizing (hence their name). These are the most widely distributed neotropical monkeys. The fourth genus (*Brachyteles*) is the most critically endangered New World monkey, and the largest of all of the neotropical primates. These monkeys are known as woolly spider monkeys because they combine the round head and dense fur of the woolly monkey with a body build like that of the spider monkey. Here, one species of howler monkey, and the woolly spider monkey are described.

The Mantled Howler (*Alouatta palliata*)

Alouatta palliata is called the mantled howler because its blackish or brownish coat has a golden mantle of long hairs along each side of the back. The face is bare and dark, with fairly large eyes that are emphasized by thick, short-cropped hair from the head that overarches each eye and forms a widow's peak in the middle (Fig. 6-15, color plate). Adult males average about 7 kg (around 16 lb), as opposed to a little over 5 kg (about 12 lb) for adult females (see Table 6-1). Howlers have an enlarged **hyoid bone** at the front of the neck, and this too is bigger in males. The inflated hyoid is used to amplify vocalizations (i.e., to make loud roars or howls) in conjunction with an air sac that extends from the vocal cords underneath the skin in the throat region. For this reason howlers, especially males, look jowly.

Mantled howlers are found in a variety of forests in southern Mexico, throughout most of Central America, and in parts of Colombia and Ecuador (Fig. 6-16). They are relatively slow-moving and slow-climbing quadrupeds that sometimes use their prehensile tails and bodies to help bridge gaps in the canopy so that younger individuals can cross. Howlers generally favor the higher parts of the canopy, and are more likely to simply drop to a lower level than to leap. They can also hang by their tails, and usually do so to reach out-of-the-way fruits and leaves. Although mantled howlers prefer to eat young leaves, like other howlers they have evolved the unusual ability to digest mature leaves (and immature fruit), which, although abundant, do not provide much energy. Enlarged salivary glands and hindguts aid in this process, as does an extremely slow rate of food passage through the gastrointestinal tract (described by Katharine Milton). The relatively low quality of their diet may account for why howlers expend less energy and rest more than the other atelines. Old World colobines also eat large quantities of leaves, but have different anatomical adaptations for doing so (see Chapter 7).

The social life of mantled howlers is complicated and fluid. They live in groups that average around 15 individuals, but range from 2 to 45. The ages and frequencies of each sex vary within these groups. Growing individuals of each sex are forced by their parent of the same sex to leave their natal troops. An expelled individual may spend some months or years alone (i.e., become solitary) before joining another troop. Some groups appear to have only one adult male (i.e., with testes descended at puberty). In other troops that are multimale, only the dominant male is usually reported as having access to receptive females. Male-male aggression may run high in these cases, which is not surprising for a polygynous, sexually dimorphic species. It should also be noted that, compared to some of the other neotropical primates, grooming and scent marking are of only minor importance to howler social life.

Figure 6-16 Although *Alouatta* is one of the most widespread tropical primates, mantled howlers are confined to the northern part of their range (i.e., from Mexico to Equador).

Kenneth Glander has spent more than 25 years studying mantled howlers at Hacienda La Pacifica in Guanacaste province, Costa Rica. He reports that, compared to the usual situation for primates that have dominance hierarchies, there is a curious reversal in linear dominance order. Instead of the oldest individual being the dominant, or alpha animal, the youngest adult holds this position while the oldest is the lowest-ranking animal. This holds for both males and females. However, as is often the case for other primates, all adult males appear to be dominant to all females and juveniles.

Glander also describes infanticide in *Alouatta palliata*, as has been reported for other howlers. This usually occurs when a solitary male attempts to join a new troop and confronts its **alpha male**. The two fight briefly, and if the new male is the winner he proceeds to kill all of the young infants in the troop. Once deprived of their nursing infants, mothers soon come into estrus and mate with the new alpha male. Although at first it may seem difficult to imagine the selective advantage of infanticide, such behavior is genetically advantageous for the invading male because it maximizes the number of his offspring. (This topic is explored in more detail in the next chapter's discussion about Hanuman langurs.) For their part, high-ranking females sometimes eliminate their offspring's competitors by kidnapping infants, which usually die or disappear in the process.

Although mantled howlers do not defend clear territories, they use loud vocalizations to maintain spacing between troops. Males produce the loudest roars, but females also join in. As with gibbons (see Chapter 10), loud vocalizations are most likely to occur at dawn and again in the afternoon. In addition to maintaining spacing, these calls provide information about the group's location and composition.

Unfortunately, despite the fact that it is at lower risk than some of the other howlers, habitat destruction and hunting by humans are taking their toll on this interesting species.

The Woolly Spider Monkey (*Brachyteles arachnoides*)

This chapter's final portrait is of the largest New World monkey, *Brachyteles arachnoides*, commonly known as the muriqui or woolly spider monkey. Muriquis have pretty golden or grayish-brown dense coats, bare black faces that may be splotched with pink, small canines, relatively long arms and legs, and prehensile tails (Fig. 6-17). Both males and females are large (see Table 6-1), with some adults reaching nearly 14 kg (about 30 lb). Interestingly, males have relatively huge testicles, perhaps the largest of any living primate, and a large bone in the penis. Female muriquis have elongated clitorises, as do female spider monkeys.

Confined to a few restricted, highly seasonal rain forests in southeastern Brazil (Fig. 6-18), the woolly spider monkey is the only species in its genus (i.e., is monotypic) and is one of the rarest and most endangered primates anywhere. Despite their large bodies, muriquis, like their close cousins the spider monkeys, are skilled, quick brachiators. As is frequently the case for bigger primates, the diet of muriquis

Figure 6-17 Muriquis or woolly spider monkeys (*Brachyteles arachnoides*) combine features of both woolly monkeys and spider monkeys.

Figure 6-18 Muriquis are found within very restricted habitats of highly seasonal forests in Brazil.

is heavily weighted toward leaves, although fruit and flowers are also eaten (and may even be preferred). Woolly spider monkeys resemble howler monkeys in having low metabolic rates and jaws and teeth that are specialized for processing hard-to-digest mature leaves. Unlike howlers, however, muriquis experience a relatively rapid passage of food through their digestive tracts. In fact, this species defecates about once an hour during the day and, curiously, its fecal matter gives off an odor that is described as smelling something like cinnamon.

Social organization in muriquis appears somewhat variable. Although they have been reported to live in cohesive multimale, multi-

REVIEW EXERCISES

1. Name the four genera of atelines and the five genera of primates in which adults have prehensile tails.
2. How do prehensile tails differ from nonprehensile tails, and what behaviors involve their use?
3. Which primates brachiate and what anatomical specializations facilitate this form of locomotion?
4. Which appears to be the most widely distributed of the platyrrhines?
5. How did the howler monkey get its name? Elaborate.
6. What is unusual about the dominance hierarchies of mantled howler monkeys?
7. What behaviors increase genetic fitness (i.e., relative number of surviving offspring) in male and female mantled howlers?
8. Which is the largest species of New World monkey?
9. Describe fission-fusion social systems, and name some species that have them.
10. How may large testicle size be related to the lack of male-male competition for estrous females in muriquis? What is the name for this phenomenon?

female troops that range from 8 to 45 individuals, in one group studied by Milton, individuals randomly came together and drifted apart, and small groups of two to four individuals were the norm. This loose fission-fusion social organization is similar to that of spider monkeys and chimpanzees. In muriquis, females migrate to other groups as subadults, so troops are focused around natal males, their mothers, immigrant females, and young individuals. According to Karen Strier of the University of Wisconsin, Madison, who has an ongoing study on muriquis at the Biological Station of Caratinga, migration is a stressful time for adolescent females, which leave only after being repeatedly threatened and chased by resident females.

Although woolly spider monkeys are large and polygamous, they do not appear to have dominance hierarchies, and male-male competition for mates is virtually lacking. When males detect (probably through olfaction) that a female is in estrus, they simply line up and peacefully take turns mating with her. Several workers have speculated that this affable mating behavior is related to a lack of sexual dimorphism in muriquis, but this is contradicted by the observation that, on average, adult males seem to be somewhat heavier than adult females (see Table 6-1).

Milton has a more interesting explanation. She hypothesizes that the large testicles of muriquis are associated with increased amounts of viable sperm. Milton suggests that males do not compete for estrous females because the competition to impregnate fertile females takes place between the sperm themselves, that is, *after*, rather than before, copulation! Thus, if Milton's hypothesis is correct, male-male competition is insignificant among muriquis because it is replaced by sperm competition.

Sadly, this fascinating species is highly endangered because of habitat destruction and illegal hunting for their flesh and hides. Only a few hundred woolly spider monkeys are thought to be alive today and it remains to be seen whether or not conservation efforts on their behalf will be successful.

SUMMARY

Rosenberger's evolutionary panels help structure this chapter's discussion of three of the four subfamilies of neotropical monkeys—the cebines, pitheciines, and atelines (callitrichines were the focus of Chapter 5). Rosenberger's classification (which differs somewhat from those of other workers) is particularly useful because it is based on shared locomotor and feeding adaptations that are key factors for interpreting the New World monkey pattern (see Chapter 5). This chapter begins by discussing the four genera of neotropical primates that primatologists have the most difficulty in classifying: *Saimiri* and *Cebus* are included in the same subfamily (cebines) because of suggestive molecular data, and because they resemble each other in their frugivorous-insectivorous diet, aspects of social organization, and certain features of the skull and teeth. *Cebus* is particularly interesting because of its tool-using skills (used in the wild to forage for food), which are viewed by some workers as convergent with those of chimpanzees (see Chapter 13).

The other two problematical genera, *Callicebus* (the titi monkey) and *Aotus* (owl monkeys), are placed not only in a different subfamily from the cebines (i.e., the pitheciines), but also in an entirely different family, the atelids. Since owl monkeys are the only nocturnal monkeys in the world, the possible causes of nocturnality are explored by comparing *Aotus* to the diurnal titi monkey. Three other genera in the pitheciine subfamily (*Pithecia*, sakis; *Cacajao*, uakaris; and *Chiropotes*, bearded sakis) share dental specializations and habitats that are linked to diets of extremely hard-to-open fruits and seeds. However, these monkeys remain somewhat mysterious because they live in areas that are frequently difficult for humans to access. Many workers regard the white uakari (*Cacajao calvus calvus*) as one of the most bizarre looking of all primates.

The ateline subfamily represents four large-bodied genera that span from slow-climbing leaf-eating howler monkeys (*Alouatta*) to the agile, brachiating, folivorous-frugivorous woolly (*Lagothrix*), woolly spider (*Brachyteles*), and spider (*Ateles*) monkeys. *Brachyteles arachnoides*, the largest of all the New World monkeys and the only species of woolly spider monkey, is extremely interesting because of its fission-fusion social organization, and its unusually peaceful, polygynous mating system that may be the result of sperm competition. Unfortunately, it is also one of the most highly endangered of the neotropical primates.

FURTHER READING

AYRES, J. M. (1990) Scarlet faces of the Amazon. *Natural History*, **99**(2):33–41.

BOINSKI, S. (1992) Monkeys with inflated sex appeal. *Natural History*, **101**(7):42–49.

NORCONK, M. A., ROSENBERGER, A. L. AND P. A. GARBER (1996) *Adaptive Radiations of Neotropical Primates*. New York: Plenum Press.

STRIER, K. B. (1993) Menu for a monkey. *Natural History*, **102**(3):34–43.

WRIGHT, P. C. (1994) Night watch on the Amazon. *Natural History*, **103**(5):44–51.

7

OLD WORLD MONKEYS: THE LEAF-EATING COLOBINES

Catarrhini
|
Cercopithecoidea
|
Cercopithecidae

Colobinae

Colobus (black-and-white colobus monkeys)
Nasalis (proboscis monkey)
Presbytis (leaf-monkey)
Procolobus (red and olive colobus monkeys)
Pygathrix (snub-nosed monkeys)
Semnopithecus (Hanuman langur)
Simias (pig-tailed monkey)
Trachypithecus (langurs)

INTRODUCTION

Although deeply rooted lineages of New World monkeys appear to have been conserved through time in the Americas (see Fig. 6-1), the fossil record indicates that primate evolution in the Old World was characterized by a series of replacements of successful species (or adaptive radiations). Ancestral Old World monkeys appeared in Africa by 20 mya, with some species reaching Europe and Asia by no later than 10 mya. Today, the Old World monkeys (for all practical purposes) are restricted to parts of Africa and Asia. As a consequence of their long separate evolution, cercopithecoids have evolved certain features that set them apart from New World monkeys. The Old World monkey pattern (discussed here) includes downward-turned noses, a dental formula of 2.1.2.3., nails that are flattened rather than clawlike, completely flexible thumbs in many species, and a variety of bodily decorations, natal coats, and sexual swellings. Catarrhine monkeys are generally (but not always) larger than platyrrhines. Although long tails developed in both New and Old World monkeys, Old World species lack the prehensile tails seen in some of the New World monkeys. However, stability in trees is facilitated by another anatomical feature that permits Old World monkeys (and gibbons) to sit on branches without falling off. These fibrous pads of fat, or **ischial callosities** (Fig. 7-1), are located in the rump region and are covered with toughened skin; that is, they reinforce the bottom of the pelvic bone that is used for sitting.

All cercopithecoids are diurnal. Various species live in the canopy where they have developed interesting strategies for avoiding aerial predators. In some cases, different species travel or feed together peacefully, although the costs (e.g., more mouths to feed) and benefits (e.g., safety in numbers) of such mixed-species associations are just beginning to be understood in Old World monkeys as well as other primates. Significantly, unlike New World monkeys, some cerco-

Figure 7-1 Pads of fat in the rump region, or ischial callosities, are found in Old World monkeys.

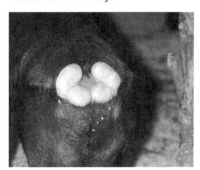

pithecoid species spend much of their time on the ground (i.e., are terrestrial) and have evolved adaptations that help compensate for the dangers of living away from the relative safety of trees. Because of this, and because Old World monkeys are genetically closer to humans than are the New World monkeys, terrestrial cercopithecoids are of great interest to researchers who ponder the origins of bipedalism in early hominids (see Chapter 14).

THE OLD WORLD MONKEY PATTERN

BONY FEATURES

1. Dental formula of 2.1.2.3.
2. Two crested (**bilophodont**) molars
3. Expanded ischial tuberosities on lower end of pelvis for attachment of callosities (see Fig. 7-1)

OTHER FEATURES

1. Nostrils that are closely spaced at their lower ends, downward-turned noses (catarrhine) (see Fig. 5-4)
2. All species diurnal
3. Some species terrestrial
4. Variety of sexual swellings and bodily decorations
5. Completely flexible thumb in some species
6. Derived frontal lobe sulcal pattern (see Neural Note 5)

Old World monkeys are divided into two subfamilies, the colobinae with eight genera, and the cercopithecinae with nine (Table 7-1). These subfamilies differ mainly in their digestive systems (Fig. 7-2). Millions of years ago the ancestors of cercopithecoids shifted their dietary preferences away from seasonally accessible, patchily distributed ripe fruits (eaten by the ancestors of apes) to more widely

TABLE 7-1 *Cercopithecoidea*	
Colobinae	**Cercopithecinae**
Colobus (black-and-white colobus monkeys)	*Alleñopithecus* (swamp monkey)
Nasalis (proboscis monkey)	*Cercocebus* (mangabeys)
Presbytis (leaf-monkeys)	*Cercopithecus* (guenons, vervets)
Procolobus (red and olive colobus monkeys)	*Erythrocebus* (patas monkey)
Pygathrix (snub-nosed monkeys)	*Macaca* (macaques)
Semnopithecus (Hanuman, grey langur)	*Mandrillus* (drill, mandrill)
Simias (pig-tailed monkey, simakobu)	*Miopithecus* (talapoin monkey)
Trachypithecus (langurs)	*Papio* (baboons)
	Theropithecus (gelada)

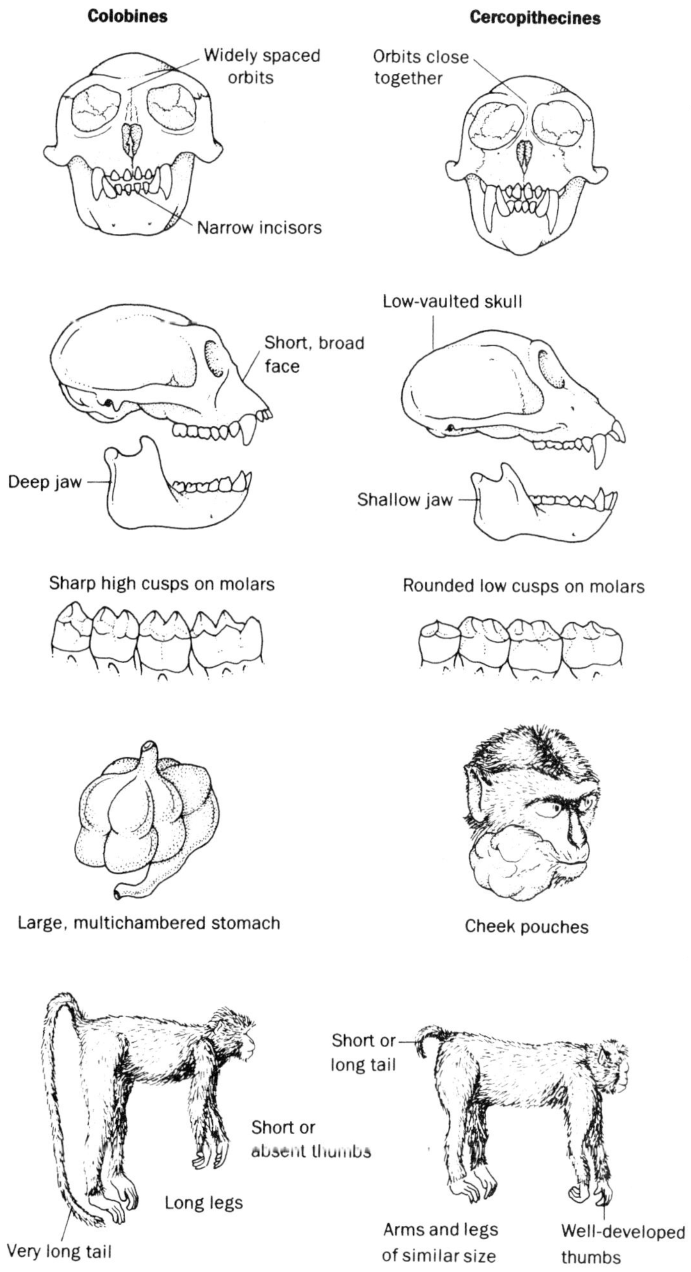

Colobines

Widely spaced orbits

Narrow incisors

Short, broad face

Deep jaw

Sharp high cusps on molars

Large, multichambered stomach

Very long tail

Long legs

Short or absent thumbs

Cercopithecines

Orbits close together

Low-vaulted skull

Shallow jaw

Rounded low cusps on molars

Cheek pouches

Short or long tail

Arms and legs of similar size

Well-developed thumbs

Figure 7-2 Anatomical features of colobines are on the left; those for cercopithecines are on the right.

available plant parts such as leaves, seeds, and unripe fruits. This pattern still holds today for the colobine subfamily, which has evolved a number of special features from sharp high cusps on the molars to adaptations of the digestive tract that are associated with consuming copious amounts of hard-to-digest leaves (see Fig. 7-2). For example, colobine salivary glands are relatively enlarged and produce ample saliva that helps to neutralize the effects of high levels of tannins in their diets. They also have sacculated or pouched stomachs and special microorganisms for digesting cellulose. Because leaves are not

Brain Evolution in Old World Monkeys

As we saw in Neural Note 6, sometimes specializations such as prehensile tails are reflected in cortical sulcal patterns. At other times, the functional significance of different patterns of convolutions is not as obvious, as is the case for the two subfamilies of Old World monkeys. Colobines usually have a small paroccipital sulcus (*par* in figure below) at the back of their brains. This feature is not visible in cercopithecine brains because it has been covered over by enlarged visual areas that have pressed forward from the back of the brain. The bottom parts of the frontal and temporal lobes of cercopithecines are also relatively expanded and therefore more likely to have extra sulci (e.g., the fronto-orbital sulcus and the occipitotemporal sulcus) than the comparable areas in colobines. These differences suggest that the two subfamilies of Old World monkeys might process visual stimuli differently, but this hypothesis is highly speculative and in need of testing. Meanwhile, comparisons with brains of prosimians suggest that these features are more derived or specialized in cercopithecines than in colobines.

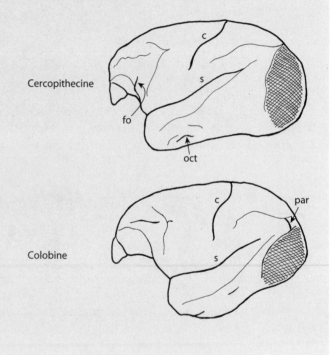

Left sides of brains of cercopithecine and colobine monkeys, with frontal lobes to the left of the figures and visual areas shaded (at the back ends) on the right. Abbreviations of sulci: **c**, central; **fo**, fronto-orbital; **oct**, occipitotemporal; **par**, paroccipital; **S**, Sylvian.

very nutritious, most colobines eat massive amounts in order to consume enough calories, which results in distended abdomens (and belching). According to David Chivers of the University of Cambridge, a few species also feed heavily on relatively nutritious, but potentially more toxic, seeds that have hard-to-digest coats. Because of their digestive systems, colobines are able to extract a good deal of water from leaves, although the frequently repeated suggestion that they can go for many weeks without drinking seems to be more myth than fact. In short, their highly specialized anatomies allow colobines to subsist on foods that would be unsuitable for most other primates.

The cercopithecine subfamily lacks the colobine specializations for digesting great quantities of leaves and seeds. Instead, cercopithecines have "herniated" cheek muscles within each side of the mouth that form little pouches which are handy for storing a wider variety of preferred foods. Generally speaking, of the two subfamilies the colobines are thought to have retained more ancestral (primitive) features, including some associated with the cerebral cortex (see Neural Note 7). The remainder of Chapter 7 describes colobines in more detail, with further discussion of the cercopithecines postponed until the next chapter.

THE COLOBINES

Approximately one-fourth of the more than 30 species of colobines live in Africa (colobus monkeys), while the rest dwell in parts of Asia (langurs and leaf-monkeys). Colobines take their name from the African species which have thumbs that are reduced or absent (*kolobus* means "mutilated" in Greek). Thumbs of Asian colobines, although somewhat larger, are still small compared to those of cercopithecines. Despite their large guts, colobines tend to be relatively slender in overall body build compared to cercopithecines (see Fig. 7-2). Most colobines inhabit tropical rain forests, and are predominantly arboreal and quadrupedal. Colobines are inclined to do a good bit of leaping, which is facilitated by a hind limb that is longer than the forelimb, long feet, and a long tail that acts as a balancing organ (see Fig. 7-2).

Colobines differ from cercopithecines, not only in how they digest food, but also in how they obtain it, according to John Oates of Hunter College (see In the Field) and Glyn Davies of University College London. Aided by well-developed thumbs, cercopithecines are able to pick up small morsels of ripe fruit or insects, often from the ground. Not so with colobines, which feed mostly aboveground by simply pulling leaves and unripe fruits from branches and stems. As discussed in Chapter 1, the folivorous diets of colobines are correlated with relatively large body sizes. Furthermore, because of the abundance of leaves and the fact that colobines have relatively small day ranges, the total weight of a given species per square kilometer, or **biomass**, is higher than that for neighboring cercopithecines. Needless to say, the continual browsing on leaves and other plant parts by colobines affects the growth of vegetation and is likely to have a significant selective impact on the evolution of plant reproductive strategies (e.g., timing of flowering, or level of toxins in seeds). Therefore, colobines are of special interest to evolutionary biologists interested in feeding ecology.

REVIEW EXERCISES

1. How do the overall patterns of primate evolution as represented in the fossil record differ in New and Old World monkeys?
2. Where in the world do Old World monkeys live?
3. What features comprise the Old World monkey pattern?
4. How do New and Old World monkeys differ?
5. Name the two subfamilies of the cercopithecoids. How many genera are in each?
6. What anatomical system best distinguishes colobines from cercopithecines? Enumerate the other features that distinguish the two subfamilies.
7. In what ways are the brains of colobines and cercopithecines known to differ?

Social organization is highly variable among colobines. Many species live in relatively small social groups of 10 to 15 individuals that contain several adult females and only one adult male. However, extremely large, possibly temporary, multimale groups have been observed for some populations of *Pygathrix*, *Colobus*, and the Hanuman langur (*Semnopithecus entellus*). Although the red colobus (*Procolobus badius*) is a notable exception, natal females usually form the stable nucleus of colobine social groups. At the other extreme, monogamous groups have been reported for the Mentawai leaf-monkey (*Presbytis potenziani*) and the pig-tailed monkey (*Simias concolor*). All-male groups also have been observed occasionally in a number of species such as the Hanuman langur, but solitary males are generally uncommon. Thomas Struhsaker of Duke University and Lysa Leland of the Kibale Forest Project nicely summarize the richness of colobine social life (1987:90): "The colobines provide us with a wide array of social systems, including one-male matrilineal groups, multi-male matrilineal and patrilineal groups, monogamous families, and enormous aggregations that may prove to be fusion-fission societies.... Breeding within these groups is usually promiscuous when the groups contain several males ... or polygynous in the one-male groups."

Colobine males that live in multimale groups are reported to have less obvious dominance hierarchies compared to multimale groups of cercopithecines, and aggression among colobine females is infrequent. Some workers attribute the general colobine affability to lack of competition for food owing to the abundance of more-or-less uniformly distributed leaves. However, it is not uncommon for neighboring groups of colobines to interact aggressively, and in some species (e.g., Hanuman langurs) females may play a role in these fights. As with other aspects of social life, the extent to which different colobine species are territorial is subject to variation, as is the size of their home ranges. Average day ranges, on the other hand, are consistently small in both African and Asian colobines, probably because groups need not travel far to find their next meals.

Female colobines reach sexual maturity at around 4 years of age; males, a little later. Sexual swellings related to estrus do not occur in female colobines to the extent that they do in cercopithecines. In particular, colobine females that live in one-male groups usually lack such swellings. Females solicit copulations with a variety of behaviors including head shaking, pursing the lips, staring, tongue smacking, pouting, or adopting a mating position (i.e., presenting). Depending on the species, mean gestation lengths vary from approximately 5 to 7 months, and birth peaks may or may not occur.

Infant killing by adult males in wild primates has been documented for one-male groups in several species of colobines, including Hanuman langurs (*S. entellus*), red colobus monkeys (*Procolobus badius*), and silvered langurs (*Trachypithecus cristatus*). As discussed in Chapter 2, this behavior is a male reproductive strategy (and a form of intrasexual selection) that is also found in a number of other primates and mammals. In colobines, infanticide emerges in situations where males typically have short tenures in female groups before being driven out by other males. As you will recall, this practice increases the likelihood that a male will reproduce because adult females that are deprived of their nursing infants come into estrus and mate with new resident males. Infanticide was first well documented for wild primates in Hanuman langurs (described later), and has since become an important focus of research for colobines and other primates.

Colobine mothers frequently allow other females to handle their young infants (allomothering or aunting), a practice that occurs in noticeably fewer cercopithecine species. Such baby-sitting not only provides time for the mother to feed independently, but also may provide valuable learning experience for future mothers. Furthermore, in the species where infanticide is practiced by invading males, it may be adaptive for infants to have bonds with (and presumably be defended by) more than one adult female. Whatever the evolutionary explanation for allomothering, colobine infants typically have coat colors that contrast markedly with those of adults, and this feature (like the cuteness of human babies) may help attract the attention of potential aunts. (Presumably this benefit outweighs the cost of attracting the attention of potentially infanticidal males.)

Many species of colobines are vulnerable to population reductions or even extinction for reasons that are sadly familiar by now: Humans continue to clear the forests that colobines inhabit, for commercial reasons related to harvesting timber, agriculture, and oil extraction. Colobines are also hunted for their meat, and some, such as the black-and-white colobines, for their beautiful coats. Colobines are generally less aggressive and wary than cercopithecines, which unfortunately adds to their vulnerability.

The African Colobines

The African colobines share a very reduced or missing thumb and a shortened ankle compared to Asian species. Despite the fact that they comprise only two of the eight colobine genera listed in Table 7-1 (*Colobus*, the black-and-white colobus; and *Procolobus*, red and olive colobus monkeys), these monkeys exhibit a surprising amount of ecological and social diversity. Consequently, the exact number of species in both the black-and-white and red groups is still a matter of debate. There is, however, only one species of olive colobus.

Colobines range across tropical Africa from the Atlantic Ocean coast to the Indian Ocean coasts of Kenya and Tanzania, in spite of the ongoing destruction of their wooded habitats by humans. In contrast to the Asian colobines, only one or two colobine species are usually found in any one African site, and cercopithecine species may not be as well represented in those locations as elsewhere. Some of these instances may be due to poor soils that yield difficult-to-digest plants with high levels of chemical defenses.

Within Africa, a number of anatomical and behavioral features distinguish *Colobus* from *Procolobus*, as enumerated by Oates: The black-and-white colobus has a three- rather than four-chambered

Colobus Monkeys, and a Few Hunters with Shotguns . . .

John Oates, a primatologist at Hunter College, is one of the world's leading experts on colobine monkeys. After decades of highly productive fieldwork, he now focuses much of his effort on primate conservation—and with good reason. In his own words:

JOHN OATES (LEFT) AND MICHAEL ABEDI-LARTEY OF THE GHANA WILDLIFE DEPARTMENT (RIGHT) ARE SHOWN HERE WITH A COLLEAGUE DURING A SURVEY OF ENDANGERED PRIMATES IN GHANA IN 1997. (PHOTO COURTESY OF JOHN OATES.)

I edged around the blind to get a better view of one of the small ponds which lay in the middle of a thickly vegetated, swampy hollow in Uganda's Kibale Forest. I was amazed to see a black-and-white colobus monkey (*Colobus guereza*) sitting in the center of the pond, up to its waist in water, scooping up water plants and eating them. It was April 1971, and some weeks previously I had built the blind out of elephant grass across one of the trails in my study area so that I could approach the pond without being immediately seen by any animals using it. Six months before this I had begun a study of the ecology and social behavior of Kibale's black-and-white colobus monkeys, and in the first weeks of my study I had noticed several colobus with gray or brown discoloration on their bushy tail tips, which are naturally white. I suspected that these large, tree-living colobus had been dragging their tails on the ground, and my Ugandan assistant told me that the monkeys were coming to the ground in the swamp in my study area. Since then I had managed to see the colobus going into the low undergrowth in this swamp and it sounded as if they were moving about on the ground and foraging near the pond, but I had not managed to see clearly what they were doing. Although I had suspected they were feeding, it was still a great surprise when I eventually saw them up to their waists in dirty pond water.

In the coming months I found that not only did my habituated group of 11 black-and-white colobus often go into the swamp to feed on aquatic plants, but also so did all the other black-and-white colobus groups living in this part of the forest, near the Kanyawara Forestry Station. On one occasion I saw six groups in the swamp at the same time, and one of these groups could only have got there by traveling half a mile from the center of its range through the territories of other groups. It seemed to me that for these monkeys to engage in such unusual and risky behavior, they must be obtaining something of considerable value. What could it be? To investigate this, I collected samples of the water plants they were eating, and on my return to London (where I was based for my Ph.D. studies) I had the plants analyzed. They proved to be rich in mineral nutrients (probably, these minerals were washed out of the soil in the high ground around the swamp and concentrated by evaporation in the ponds). Compared with the tree leaves that were the staple diet of the Kanyawara black-and-white colobus, the water plants were especially rich in sodium, and I hypothesized that this nutrient might be critically important to monkeys like these colobus, which ate no insects or other animals. Sodium is, of course, a vital element in the mammalian body, but it is often not a common element in plants (hence the value of salt to humans, especially in the days when it was not produced and distributed on an industrial scale).

This strange water-plant feeding by the black-and-white colobus was just one of several observations I made at Kibale that got me fascinated with the question of why primates select the foods they do, and the influence of this selectivity on other aspects of their lives. After carrying out several more studies of colobine food selection in Africa and Asia in the 1970s, (see figure below), my colleagues and I have gone on to investigate the community-wide implications of primate food preferences. We have found evidence that the total biomass (weight per unit area) of a range of forest primate communities in the Old World tropics is directly related to the quality of tree foliage as measured by the ratio of protein to fiber. But this finding only applies in the absence of human hunting.

Since I began my studies in Kibale (which is now a national park), the primates of tropical rain forests have become increasingly threatened with extinction, and in many places the greatest threat to the survival of primates is humans hunting them for their meat. This is a much more serious threat than logging alone, or other forms of forest disturbance. Although total removal of the tree canopy is obviously a devastating blow to forest primate populations, low-intensity logging can actually increase the quantities of available high-quality food. But logging roads often open the forest up to hunters and farmers, and a few hunters with shotguns are quite capable of driving populations of relatively large-bodied primates, like colobus monkeys, to extinction.

A FEMALE NILGIRI LANGUR EATS YOUNG LEAVES IN THE FORESTS OF SOUTH INDIA. THE REDDISH YOUNG LEAVES OF MANY FOREST TREES ARE TYPICALLY MORE DIGESTIBLE THAN MATURE FOLIAGE. (PHOTO COURTESY OF JOHN OATES.)

The challenge posed by the looming extinction of many tropical forest primates has increasingly influenced my fieldwork since the mid-1980s. I have put more and more effort into surveys of threatened populations and to the planning of protected areas. This work recently has taken me to Ghana, where one of my collaborators on field surveys has been Tom Struhsaker, who sponsored and advised me during my early fieldwork on the black-and-white colobus of Kibale. Our surveys in Ghana have led us to the conclusion that a local form of colobus monkey, Miss Waldron's red colobus, is probably extinct from human hunting. If we are correct, this is the first documented case of primate extinction in this century, and an omen of an extinction wave that will soon be upon us if effective protection programs for many threatened primates are not instituted in the very near future.

stomach and a large larynx with an associated air sac that is used to make loud resonant calls. In contrast, the red colobus has a more subtle, graded system of vocalizations. The male *Procolobus*, unlike male *Colobus*, has a sagittal crest on its skull, and separate rather than united ischial callosities. The female *Procolobus*, contrary to female *Colobus*, exhibits sexual swellings and rarely permits allomothering. The female red colobus transfers out of its social group more frequently than does the male, while the female olive colobus is the only monkey in the world that carries her infants by mouth.

Black-and-white colobus monkeys include five widely recognized species of the *Colobus* genus: *C. polykomos*, *C. vellerosus*, *C. satanas*, *C. angolensis*, and *C. guereza*. These monkeys have long black (*C. satanas*) or black-and-white coats that differ in their overall patterns. All species eat young leaves but differ in their ability to survive on mature leaves, seeds (highly favored by *C. satanas*), and even fruits when preferred food is scarce. The species are also distinguished by structural details of their teeth and skulls, with *C. guereza* being the most distinct. This species has also been the most widely studied of the black-and-white colobines.

THE GUEREZA (*COLOBUS GUEREZA*) With its stunning dark coat that is accented by dramatic white U-shaped mantles flowing along the sides (Fig. 7-3), the guereza (*C. guereza*, also known as the Abyssinian colobus) is truly beautiful—so much so, in fact, that the hides of this species were once intensely pursued by humans, which nearly led to its extinction. Guerezas have other, more subtle features. Compared to the other black-and-white colobines, *C. guereza* has relatively large canines in females, small incisors, long molars, and a wide nose. It is also the largest of the African colobines and is sexually dimorphic, with adult females averaging about 9 kg (around 19 lb), and males approximately 12 kg (26 lb) (Table 7-2).

Figure 7-3 Unfortunately guerezas (*Colobus guereza*) were once hunted for their beautiful black-and-white coats to such an extent that they almost became extinct.

TABLE 7-2 Statistics for Representative Old World Monkeys

	Colobus guereza (guereza)	Procolobus badius (red colobus)	Procolobus verus (olive colobus)	Nasalis larvatus (proboscis monkey)	Pygathrix bieti (Yunnan snub-nosed monkey)	Semno-pithecus entellus (Hanuman langur)	Simias concolor (pig-tailed monkey)
Adult mean body weight							
F	8.6 kg (18.8 lb)	8.2 kg (18.1 lb)	4.2 kg (9.3 lb)	10.0 kg (22.0 lb)	9.0 kg (19.8 lb)	11.2 kg (24.6 lb)	7.1 kg (15.7 lb)
M	11.8 kg (26.0 lb)	8.3 kg (18.3 lb)	4.7 kg (10.4 lb)	21.2 kg (46.7 lb)	13.0 kg (28.7 lb)	16.2 kg (35.8 lb)	8.8 kg (19.3 lb)
Adult brain weight (g)	82.0	77.0	57.8	97.0		135.2	59.0
Maximum life span (yr)	22.2			13.5		20	
Location (habitat)[a]	Af pf, sf, rif, wg	Af pf, sf, df, wg, gf	Af sw, rf, df	B m, rif, sw	C cf	As, I ts, pif	M ef, sef, sw
Locomotion[b]	qr, l	qr, l	qr, l	qr, c, l, sw	qr	qr	qr
Diet[c]	l, f, s, fl	l, f, s, fl	l, f, s, fl	l, f, s, fl, a	li, gr, l, f, b	l, f, fl, a, g, b	l, f
Seasonal breeders	0	2 peaks	Peak	0	+	+, peak	+
Gestation (days)		174	171	166	170	200	
Interbirth interval (mo)	25.2	25.5		12–24		15–26	
Social systems[d]	M-U, M-M	P-M, F-F	P-M?	M-U, F-F	?-M, F-F	M-U, M-M	Mo, M-U
Territorial	+	0		0		+, 0	
Ratio of adult females to males	1–4:1	2–6:1	1–2:1	2–4:1		2–20:1	1–5:1
Alloparenting	+	0	0	+		+	
Infanticide	+?	+	0	0		+	0
Group size	2–15	2–80	3–14	3–23	23–200	4–125	2–20
Density	10–315/km²	60–300/km²	7–78/km²	3–6/km²	Low	4–200/km²	7–220/km²
Home range	2–28 ha	9–114 ha	28 ha	130–900 ha	2500 ha	9–1300 ha	2.5–20.0 ha
Conservation status[e]	L	R-H	R	H	H	E	H

Note: Information in this table has been selected from numerous sources and should be viewed as tentative, i.e., subject to change in light of new information. Sources include but are not confined to Oates, Davies, and Delson, 1994 (body weights); Davies and Oates, 1994 (general statistics); Harvey et al., 1987 (most brain weights in Table 7-2 provided by D. Falk); Chivers, 1994 (diets); Newton and Dunbar, 1994 (social systems); Struhsaker and Leland, 1987; Burton, 1995; and Rowe, 1996 (conservation status and other statistics). Abbreviations: +, yes; 0, no. Blank entries indicate that no information is presently available. Dietary items are listed in order of decreasing importance to the species.

[a]Location (habitat): Af, Africa; As, Asia; B, Borneo; C, China; I, Indian subcontinent; M, Mentawai islands (Indonesian); cf, coniferous forest; df, dry forest; ef, evergreen forest; gf, gallery forest; m, mangroves; pf, primary forest; pif, pine forest; rf, rain forest; rif, riverine forest; sef, secondary forest; sf, seasonal forest; sw, swamp forest; ts, thorn scrub; wg, wooded grassland.

[b]Locomotion: c, climbing; l, leaping; qr, quadrupedalism; sw, swims.

[c]Diet: a, animal prey; b, bark; f, fruit; fl, flowers and buds; g, gums (sap); gr, grasses; l, leaves; li, lichen; s, seeds.

[d]Social systems: F-F, fission-fusion; M-U, matrilineal unimale; Mo, monogamous; M-M, matrilineal multimale, multifemale; P-M, patrilineal multimale, multifemale.

[e]Conservation status: E, endangered; H, highly endangered; L, lower risk; R, rare.

C. guereza occupies an unusually wide variety of forested habitats ranging from lowlands to mountains, and from the Nigeria-Cameroon border in the west to Ethiopia in the east (Fig. 7-4). Most field studies have been carried out in the eastern part of the range. Although the range of habitats and population densities vary greatly across the species, individual groups of *C. guereza* occupy very small home ranges at some sites, which is probably related in part to their diet. According to Oates, all species of black-and-white colobines prefer to eat young leaves or seeds, but will shift to harder-to-digest mature leaves when these are in short supply. *C. guereza* is better able to digest the high fiber content of mature leaves than are the other species of black-and-white colobines, however, and can even subsist on mature leaves. This ability, combined with a willingness to travel on the ground between food sources, may account for why *C. guereza* has achieved the greatest biomass recorded for any colobine in the highly seasonal gallery forest at Bole, Ethiopia. Oates concludes that their dietary adaptations constitute an evolved strategy which allows guerezas to subsist, on a seasonal basis, on low-quality diets.

The high-fiber, low-energy diet of guerezas is reflected in their day-to-day lives, as described by Oates for one group in the Kibale Forest. A typical day often begins with sunbathing, which is followed by a number of periods of movement and feeding interspersed with long periods of rest. Thus "food energy is subsidized by direct sun energy to maintain body temperature at a time when temperatures are lowest and the stomach is empty, and movement is minimized during long fermentation periods after [the] stomach fills" (Oates, 1994:100). (An association between a low-energy diet and a relaxed lifestyle also characterizes gorillas, as detailed in Chapter 12.) While relaxing, guerezas make friendly clucking sounds with the tongue and, curiously, engage in social belching. Despite their reputation for being

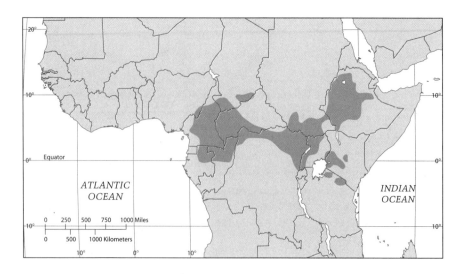

Figure 7-4 Guerezas are distributed from Nigeria to Ethiopia.

laid-back, however, guerezas are territorial and aggressively expel other groups from their core areas. Group spacing is further reinforced by low-pitched, widely spaced, resonating roars of adult males.

Although *C. guereza* varies greatly in its range of habitats and population densities, its social organization is remarkably stable. Groups with two to three adult females and only one adult male predominate, while multimale groups that are the norm for the other black-and-white colobines are less frequent and less stable in this species. One-male groups are matrilineal; that is, the males disperse to other groups when they mature, while females remain in their natal groups and reinforce their cohesiveness with much allogrooming. Interestingly, Oates speculates that the surprising lack of variability in guereza social organization may arise from an evolved genetic substrate in response to long-term food availability (or unavailability) and other environmental features. If so, "the black-and-white colobines provide some evidence in support of the idea that aspects of social organization can be among the least plastic elements of primate behaviour" (Oates, 1994:106).

Breeding is not strictly seasonal, although births may occur in peaks. Like most colobine females, *C. guereza* females lack sexual swellings. Around the age of 4, females solicit mating and, after an unknown period of gestation, infants are born with pink skin and white coats that darken to adult hues within a number of months. The striking coloration of infants appears to be extremely attractive to allomothers, which take and hold these infants for long periods of time.

Eagles have been observed to prey on *C. guereza*. Surprisingly, adult males are the most frequent targets, perhaps because they are sometimes solitary or because they tend to roar, which calls attention to themselves. As noted already, *C. guereza* was once intensely hunted for its beautiful coat. Logging and commercial agriculture continue to threaten this species, despite the fact that it is protected in a number of reserves.

THE RED COLOBUS MONKEY (*PROCOLOBUS BADIUS*) Like the black-and-white colobus monkeys, red colobus monkeys are extremely variable, occurring in up to 17 subspecies (some workers further divide *Procolobus badius* into several species). Their name is a bit of a misnomer because hair on the back, arms, crown, and tail may be any combination of black, brown, gray, or white. The hair on their lower limbs and stomach area is frequently light orange to red, however, which explains their name (Fig. 7-5, color plate). Compared to black-and-white colobus monkeys, red colobus monkeys are the same (in their basic digestive adaptations) and yet so different (behaviorally). The differences between these two groups (described below) may best be understood as socioecological adaptations related to different foraging strategies.

Red colobus monkeys are versatile in their habitat preferences, occupying moist rain forests, swamps, dry forests, and mountainous re-

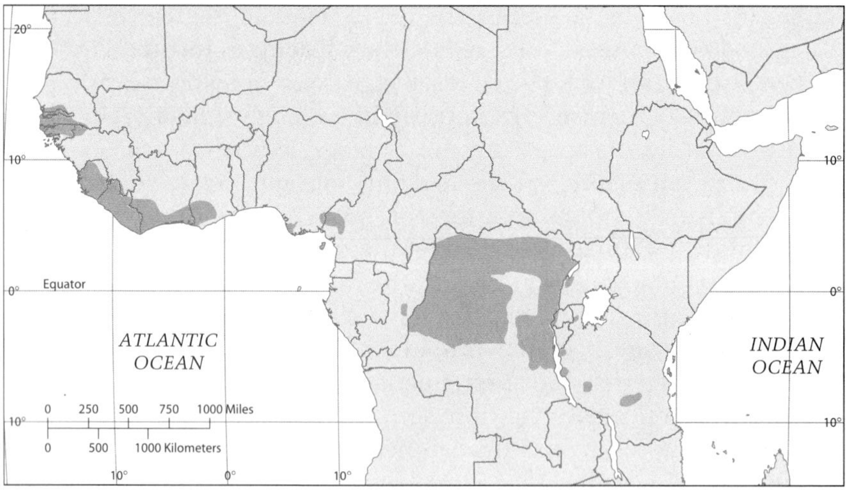

Figure 7-6 Red colobus monkeys occupy a variety of habitats in tropical West and Central Africa.

gions in parts of tropical West and Central Africa (Fig. 7-6). Whereas *C. guereza* seems content to consume relatively large amounts of low-quality mature leaves when preferred foods are in short supply, red colobus monkeys make a greater effort to find the high-quality foods they prefer. They frequently feed at various levels in the canopy in the largest available trees and, according to Oates, when studied by the same methods at the same site, red colobus monkeys spend more time feeding and less time resting than do black-and-white colobus monkeys. Consequently, their diet is more varied and contains a higher percentage of young leaves, flowers, and buds.

Because the trees in which *Procolobus badius* feeds are clumped in their distribution and ripen at different times, the patches that attract red colobus monkeys are widely distributed in both space and time. This pattern contributes to the relatively large, overlapping ranges of red colobus compared to neighboring black-and-white colobus monkeys (see Table 7-2). Red colobus monkeys also have generally larger social groups that contain numerous adult males, although single-male groups sometimes are found in marginal habitats. Also in contrast to other colobines, females are much more likely than males to transfer to other groups as they mature, leaving males to form the nucleus of their natal groups. For this reason, *Procolobus badius* groups are described as descended through the male line, or **patrilineal** (see Table 7-2).

Neighboring groups may be surprisingly tolerant of each other. At other times, aggressive episodes (that may or may not include females) occur and end when one dominant group supplants the other. Despite the fact that males cooperate to defend females or temporary feeding sites, however, *Procolobus badius* is not viewed as territorial because it does not defend fixed ranges (see Table 7-2). Lack of territoriality may be the result of the large (and therefore indefensible) ranges that are associated with red colobus diets. It is also interesting to observe how red colobus monkeys get along with each other within their social groups. They use a complex system of vocalizations in their social interactions, although males do not produce the roars that are typical of adult male *C. guereza*. Aunting behavior is rare compared to black-and-white colobus monkeys, which may be due to the fact that groups of adult females are not relatives because of the unusual patrilineal organization. The latter may also explain why adult females groom each other less, while adult males (that are relatives) groom each other more.

Another feature that distinguishes red from black-and-white colobus monkeys is that female *Procolobus badius* exhibit sexual swellings and produce quavering mating calls. Oates speculates that these unusual traits are related to the large multimale, multifemale groups that characterize red but not black-and-white colobines. Where there is more than one possible male to mate with, females may increase the fitness of their offspring by mating preferentially with the winners of aggressive interactions. Oates suggests that the sexual swellings and quavering "come hithers" of females are incentives that encourage male competition. Male-male competition and female choice therefore seem to be factors that determine mating patterns in red colobus monkeys, and it is possible that dominant males perform most of the copulations. As summarized by Oates (1994:118), "there may be a relationship between the seasonality of red colobus habitats, the seasonality of breeding, the size of female perineal swellings and the prevalence of copulation quaver calls in red colobus populations."

Red colobus monkeys are generally at greater risk of becoming extinct than are black-and-white colobines. One subspecies (*Procolobus badius temminckii*) is rare, while the others are endangered (see Table 7-2). Red colobus monkeys may remain still when frightened, which makes them easy targets for human hunters seeking food and pelts. Crowned hawk-eagles, chimpanzees (see Chapter 13), and (possibly) leopards are also threats, and humans continue to destroy some of the areas that red colobus monkeys inhabit for agricultural and commercial purposes.

THE OLIVE COLOBUS MONKEY (*PROCOLOBUS VERUS*) The olive colobus is composed of one species that is not further divided into subspecies; that is, it is **monotypic**. Although olive colobines have not been as thoroughly studied as some of the other colobines, investigations to date suggest that this species resembles red colobus monkeys in having a patrilineal social organization, female dispersion, female sexual swellings, lack of aunting behavior, and either multimale or unimale social groups (although group size is generally smaller in olive than in red colobus monkeys). This interesting monkey also has a number of unique features. For example, olive colobus monkeys take their name from their drab olive-brown coat. They are the smallest of all of the colobines, yet have the largest feet (Fig. 7-7).

Figure 7-7 *Procolobus verus* (olive colobus monkey) has been described somewhat poetically as a "thicket-haunter."

Figure 7-8   The olive colobus monkey is found only in the coastal forests of West Africa.

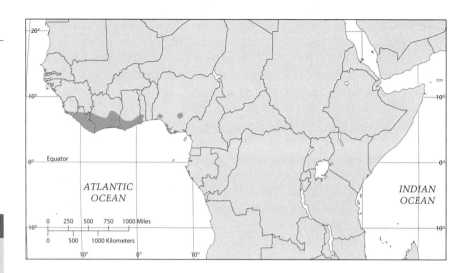

1. Where do colobines live?
2. Why are colobines of special interest to evolutionary biologists who are interested in feeding ecology?
3. What kinds of social groups do colobines live in?
4. Are sexual swellings common in estrous female colobines? Which usually initiates mating in colobines, males or females?
5. Describe the possible advantages of allomothering for colobines.
6. Name the two genera of African colobines. How do they differ from Asian colobines?
7. In what features is *Colobus guereza* unique among black-and-white colobines?
8. How does *Procolobus badius* (red colobus monkey) differ from *Colobus guereza*?
9. Of the unique behaviors discussed for olive colobus monkeys, which is the most unusual?

Procolobus verus is confined to coastal forests of West Africa (Fig. 7-8), where it moves and feeds in the lower part of the canopy and undergrowth, causing one early worker to describe it as a "thicket-haunter." Olive colobus monkeys do a good bit of leaping, and travel on the ground. Oates points out that the olive colobus has an unusual tendency to associate with other monkeys, particularly the Diana monkey (*Cercopithecus diana*). In one group, the association continued for over 5 years, although the interaction seemed to be one way, that is, with the olive colobus orienting toward, but being ignored by, the Diana monkeys. Oates speculates that *Procolobus verus* may associate with other forest monkeys as a defensive strategy, because its small body size combined with small group size puts it at greater risk from predators. Traveling time is relatively long for olive colobus monkeys, probably because they follow Diana monkeys.

Perhaps the most remarkable feature of olive colobus monkeys is the unique way in which mothers transport their newborns. When traveling from place to place, mothers carry their young in their mouths. During this procedure, the infant's body is tucked under the mother's neck and its tail is around her neck. No other anthropoid engages in this behavior.

Despite its cryptic coloration and habit of dwelling in thick vegetation, this small species is vulnerable to leopards, crowned hawk-eagles, chimpanzees, and humans. Sadly, human hunting and habitat destruction have caused the little olive colobus monkey to become quite rare.

The Asian Colobines

The Asian colobines differ from the two African genera in having longer thumbs and ankles (as noted earlier), but shorter faces. Six of the eight genera listed in Table 7-1 are Asian: *Nasalis*, the proboscis monkey; *Presbytis*, leaf-monkeys; *Pygathrix*, snub-nosed monkeys; *Semnopithecus*, Hanuman or grey langur; *Simias*, pig-tailed monkey; and *Trachypithecus*, langurs. Although this classification (by Oates and his colleagues) differs somewhat from those used in other textbooks, it is adopted here because it is based on the most authoritative and recent information.

The distribution of Asian colobines spans across Asia and Southeast Asia, with *Trachypithecus* (included in *Presbytis* in some classifications) being the most widely spread genus, and one that is sympatric with other genera in various regions. Some other genera are much more confined; for example, *Simias* is found only in the Mentawai islands. Each genus, of course, has its own distinctions: Monotypic *Nasalis* is known for its amazingly large and pendulous nose, while the seven species of *Presbytis* are distinguished by a number of cranial and dental features. *Pygathrix*, which contains five species (four of which are included in *Rhinopithecus* in some other classifications), has odd-looking flaps of skin at the top of the nostrils and greatly reduced or absent nasal bones. *Semnopithecus* (included in *Presbytis* by some) is another monotypic genus, and is known for a high degree of terrestriality as well as the occurrence of infanticide in many of its populations. The single species of *Simias* is the only colobine with an extremely short tail and the only Asian colobine in which females exhibit sexual swellings. This genus is sometimes monogamous. Finally, the nine species of *Trachypithecus* are distinguished by certain dental features and, as already noted, occupy a wide range of habitats. To illustrate the diversity of the Asian colobines, this chapter concludes with portraits of four species: *Nasalis larvatus*, *Pygathrix bieti*, *Semnopithecus entellus*, and *Simias concolor*.

THE PROBOSCIS MONKEY (*NASALIS LARVATUS*) The proboscis monkey (Fig. 7-9), named for its spectacular (some would say humongous) nose, is one of the most amazing-looking of all the primates. In males, the nose is enormous, protuberant, and fleshy and droops down over the mouth. It is less massive but still large in females, and somewhat upturned in infants. As if the size of the nose were not enough, it is accentuated even more by eyes that are narrowly spaced, and a bare pinkish face. No one is quite sure about how to explain the evolution of this dramatic anatomy, but it is possible that large noses in *Nasalis* are the result of sexual selection (see Chapter 2) by females for a trait which they perceive as particularly attractive in males. There is less mystery about the present-day function of such large noses—males use them to make loud aggressive honks that can be heard at some distance. Proboscis monkeys are large and highly sexually dimorphic, with males weighing about twice as much as females (i.e., averaging 21 versus 10 kg, or 47 versus 22 lb). Females have ischial callosities that are separated while those of males are fused, and body fur of both sexes is a blend of reds, browns, grays, and creams.

Figure 7-9 From this photograph, it is clear how the proboscis monkey (*Nasalis larvatus*) got its name.

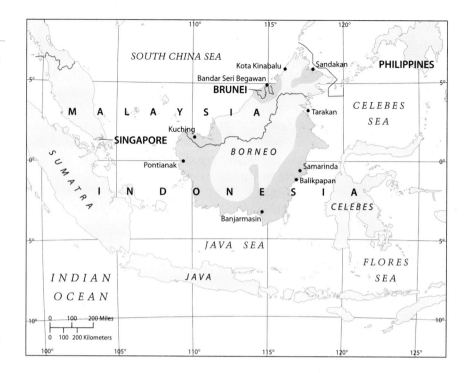

Figure 7-10 Proboscis monkeys are found only on the island of Borneo.

Nasalis lives only on the island of Borneo (Fig. 7-10), where it frequents both mangrove swamps and riverine forests. Although proboscis monkeys are basically arboreal quadrupeds, they have a remarkable repertoire of locomotor patterns that includes leaping with outstretched arms, hanging by the hands, and moving on the ground. They are also surprisingly good swimmers that can dive and swim under water. Despite its relatively large body size, *Nasalis* is reported to be a selective feeder that prefers young leaves, fruits, and seeds. The patchy distribution of these foods is probably related to the unusually large home ranges that have been reported for some groups.

The social organization of *Nasalis* is unique among colobines because it is a blend of two types, or a two-tiered system. At the most basic level, proboscis monkeys live in either all-male or unimale groups. The home ranges of these groups overlap extensively, and specific unimale groups frequently join together into larger bands (the second tier), for example, when they settle down next to a river for the night. Although some groups seem to avoid each other (possibly by using vocalizations), they are not territorial. The composition of one-male groups is somewhat flexible because males and, to a lesser extent, females may leave their natal groups. As Carey Yeager of Fordham University notes, fission-fusion between stable unimale groups is also found in hamadryas and gelada baboons (see Chapter 9). A proboscis male, like a gelada but unlike a hamadryas male, does not engage in herding the females in his group. Instead, according to Yeager (1991:84), "it calls softly and waits, or returns for female stragglers." Males seldom groom, but females groom each other and juveniles. Females may also compete with each other for the attention of the male.

Nasalis is typical of other colobines in its sexual behavior. Copulations are solicited by females through presentation of the hindquarters, head shaking, and pursed lips or pouting. As described by Yeager (1990:24), "males mount from the rear, grasping the female around the midsection with their hands and often clasping one of the female's rear legs with their foot. Repeated thrusting movements are made by the male, there is a brief pause, and the male dismounts." Same-sex mounts have been observed, and copulating couples may be harassed by infants or juveniles.

This interesting species is endangered because of deforestation associated with harvesting mangrove pulp, and because it continues to be hunted illegally for food and for sport. Its habit of sleeping in trees near rivers makes it especially vulnerable to hunters.

THE YUNNAN SNUB-NOSED MONKEY (*PYGATHRIX BIETI*) *Pygathrix bieti* is one of four recognized species of snub-nosed monkeys, none of which is very well known. However, the sparse information that is available for *Pygathrix bieti* is intriguing. These monkeys are extraordinary-looking (Fig. 7-11, color plate), with pale faces that appear as if they are made of porcelain. Other facial features include a rounded muzzle that projects in front of a tiny upturned nose. The latter is accentuated by two small flaps of skin that extend upward from the top of the nostrils. Rims of pale skin highlight the eyes, and a full mouth looks as if it has been painted with bright-pink lipstick. The back of the body and top of the head are jet black, which contrasts with the white to yellowish hair found elsewhere. Forward spiking bangs add to the dramatic look of this fairly large and sexually dimorphic species.

Living in China's Himalayas (Fig. 7-12), the Yunnan snub-nosed

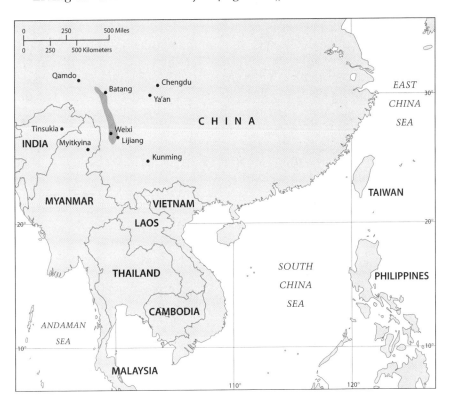

Figure 7-12 Yunnan snub-nosed monkeys live in the freezing cold Himalayas of China.

monkey is found at higher latitudes and experiences longer winters and colder temperatures than any other nonhuman primate in the world. This climate is extremely severe, with snow cover for half the year. However, *Pygathrix bieti* does not appear to migrate regularly to lower altitudes during the winter as *Pygathrix roxellana* does. Yunnan monkeys survive by eating lichens (mosses) that hang from trees year-round. This food is low in protein but high in carbohydrates, making it an ideal quick-energy food for the freezing Himalayan climes. *Pygathrix bieti* also eats grasses, leaves, fruit, and (in a pinch) bark.

A quick glance at Table 7-2 reveals that not much is known about the social life of Yunnan snub-nosed monkeys. Long-term research on this species only began in 1992, according to primatologist R. Craig Kirkpatrick who participated in the study. Group sizes vary tremendously, ranging from 23 to 200 individuals, and bands of *Pygathrix bieti* occupy unusually large home ranges (see Table 7-2). Kirkpatrick suggests that diet is the key to understanding this pattern because lichen is abundant enough to support large groups, but once an area has been denuded of this food, it will be at least two decades before it grows back. Unfortunately, not much else is known about the social organization or reproduction of this attractive but highly endangered species.

THE HANUMAN LANGUR (*SEMNOPITHECUS ENTELLUS*) Hanuman or grey langurs (*Presbytis entellus* in some classifications) are prominent in Indian mythology, and are called Hanuman after the Hindu monkey-god. Grey langurs are regarded as sacred in India, where they more or less have the run of towns and villages. Because numerous field studies have focused on *Semnopithecus entellus* since the 1930s, much more is known about this species than many others, such as the Yunnan snub-nosed monkey just discussed. There are approximately 15 subspecies of Hanuman langurs that vary in appearance and social organization (i.e., they are polytypic). As detailed below, grey langurs are best known for the light they shed on the evolution of infanticidal males.

Hanuman langur males range in size from a mean of 20 kg (44 lb) in the Himalayas to a much smaller 11 kg (24 lb) in Sri Lanka, and the species is highly sexually dimorphic (see Table 7-2). Coat color of adults varies in subspecies from brown to dark gray to buff (coats of young infants are blackish brown, however). The bare, dark face has a short muzzle that is dominated above by prominent brows and punky-looking bangs that project straight forward (Fig. 7-13). As is typical for colobines, the hands, feet, and tail are long. Grey langurs frequently appear to have potbellies.

The most western of the Asian colobines, *Semnopithecus* is widely spread across India, Nepal, Bangladesh, and Sri Lanka (Fig. 7-14), and its habitats range from desert, to dry jungle, to mountainous forests. They also live at altitudes that vary from sea level to 4000 meters. The climate is particularly harsh in the Himalayas, where grey langurs experience freezing temperatures and snow in the winter.

among primates, and is a by-product of more generally aggressive episodes. Hrdy and her colleagues (1994/95:152) counter that infanticide is widespread among primates and that "there is a range of male, female, and even infant behaviors that only begin to make sense when we assume that infanticide, like predation, is a recurring threat, even if the actual events are rarely witnessed." Some light is shed on this debate by Volker Sommer of University College London and his colleagues who monitored free-ranging Hanuman langurs at Jodhpur, India, for 13 years and were able to test predictions associated with Hrdy's hypothesis in a unimalc group in which all members are known individually. As predicted by the sexual selection hypothesis, they found that the loss of unweaned infants accelerated the resumption of menstruation and sexual receptivity, and reduced the spacing between subsequent births. The implication is that infanticidal males do indeed increase their fitness.

Sommer made other interesting observations about adult females. Estrous langurs signal their condition through their behavior, since they do not manifest sexual swellings. At Jodhpur, copulations were always solicited by females through head shuddering, presentation of the anogenital region, and lowering of the tail. In one-male groups, the probability of conception increased with the number of copulations, and females from Jodhpur apparently competed for the sperm of their leader by harassing other females as they copulated, and by mating during pregnancy (i.e., depleting sperm that would otherwise be available for others). Sommer's findings failed to support another suggestion, however, that postconception estrus is a female strategy for confusing paternity and thus deterring a new leader from killing her infant when it is born (i.e., since he might have fathered it).

Because it is widespread and protected for religious reasons in some areas, this species is not threatened with extinction, although like many other primates it is endangered. Some populations are under pressure from human encroachment, and golden jackals have been known to prey on immature grey langurs on the ground. Other predators include various cats and crocodiles.

THE PIG-TAILED MONKEY (*SIMIAS CONCOLOR*) The final portrait in this chapter is of a monotypic genus that is in need of further research. Nevertheless, *Simias concolor* (*Nasalis concolor* in some classifications) is described here because the little that is known about this species suggests that it is unusual. Pig-tailed langurs are so-named because, unlike other colobines, they have markedly short naked tails. Their fur is either creamy or dark gray or brown, and the bare faces are dark with upturned little noses (Fig. 7-16). *Simias* is proportioned like the Old World macaques, that is, with relatively short arms and legs that are approximately equal in length, and appears to be moderately sexually dimorphic (see Table 7-2).

Pig-tailed langurs live on the Mentawai islands off of the west coast of Sumatra, the largest of which is Siberut (Fig. 7-17). They inhabit evergreen, secondary, and swamp forests, and are known to de-

Figure 7-16 Not much is yet known about pig-tailed langurs (*Simias concolor*).

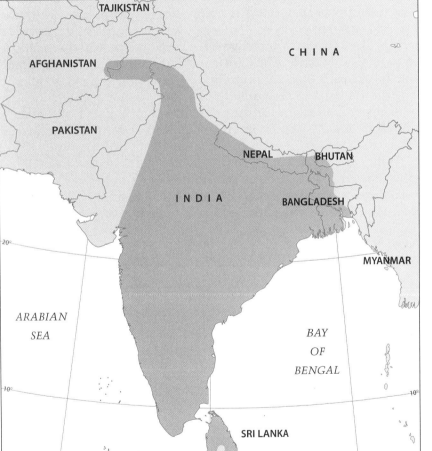

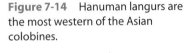

Figure 7-13 The Hanuman langur (*Semnopithecus entellus*) was the first Asian colobine to be studied.

Figure 7-14 Hanuman langurs are the most western of the Asian colobines.

This is probably why langurs in this region have bigger bodies (for heat conservation) and longer, thicker coats.

Diets of grey langurs are also quite varied. New leaves, fruits, and flowers are staples that may be supplemented with mature leaves, gum, soil, bark, and insects, depending on the season and site. *Semnopithecus* also raids human crops such as maize, wheat, and tubers, and can ingest toxic seeds and plants that are poisonous to humans. Depending on the distribution of foods, home ranges may be quite large. Although basically arboreal quadrupeds, grey langurs have been reported to spend a good deal of time traveling and feeding on the ground.

Social organization of Hanuman langurs is based on one-male groups, multimale groups, and all-male bands. The percentages of each type of group vary from site to site, however, as described by Paul Newton of Oxford University (Fig. 7-15). Heterosexual groups may include anywhere from 4 to 125 individuals, while all-male bands usually contain no more than 15 individuals. Dominant males lead their groups, and select sleeping sites and food trees. They also use vocalizations such as "whoops" to maintain spacing between groups. Although both sexes have been reported to migrate, males do so more than females. Thus, social organization in unimale and multimale groups is basically matrilineal. Allomothering is common in this species but L. S. Rajpurohit and S. M. Mohnot of the University of Jodhpur report that weaning of infants (which occurs by 13 months of age) is associated with hostility and indifference on the part of mothers.

Home ranges of the different types of groups may overlap. Intergroup encounters occur often and are frequently aggressive, involving vocalizations, chasing, other displays, and sometimes physical contact. As is well known, at some sites, adult males from all-male groups have been observed attacking one-male groups and killing infants (more often males) of adult females. In these cases, the group leader and adult females are unable to defend their offspring and the leader is usually driven out of the group altogether. Meanwhile, the invading males mate with the resident females until a new leader is established and drives them away. New leaders may have a short or long (e.g., 10 years) tenure before the cycle repeats itself. Such incidences of infanticide have been reported for a number of grey langur populations in South Asia, but by no means all of them. Interestingly, infanticide only occurs in populations that have a high percentage of one-male groups, whereas noninfanticidal populations are associated with a predominance of multimale groups (see Fig. 7-15).

Needless to say, the practice of infanticide in grey langurs (and some other primates as well) has generated a good deal of interest among primatologists. Some workers believe that this behavior is simply pathological and due to stress from encroachment by humans, habitat destruction, and overcrowded living conditions. However, infanticide has been observed in undisturbed areas, so this hypothesis seems unlikely. Another possibility (the sexual selection hypothesis)

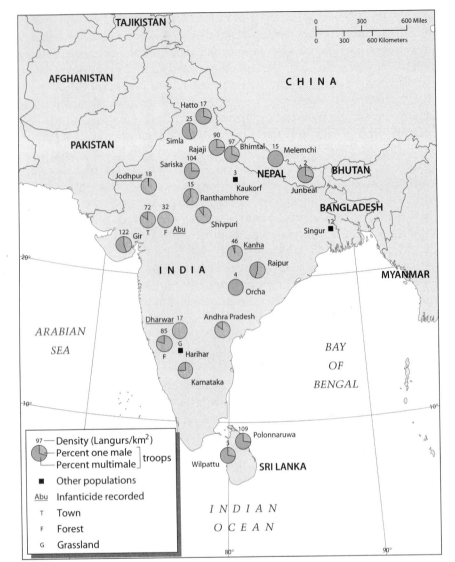

Figure 7-15 Hanuman langur organization varies from site to s shown in this map of South Asia

has been proposed by Sarah Blaffer Hrdy, who suggests that langur infanticide is a male reproductive strategy. According to Hrdy, invading males increase their chances of reproducing by killing nursing infants that they have not fathered, thereby causing their mothers to become sexually receptive (i.e., nursing inhibits ovulation, while its cessation stimulates it). Receptive females mate with and frequently conceive offspring of the new leaders. Thus, the invading male passes his genes into the future (increases his reproductive fitness) and, as an extra bonus, his infants do not have to compete directly with those of other fathers for resources.

The sexual selection hypothesis is consistent with socioecological theory (see Chapter 2). But is it true? Sussman, James Cheverud of Washington University, and Thad Bartlett of Dickinson College contend that it is not. Instead, they believe that infanticide occurs rarely

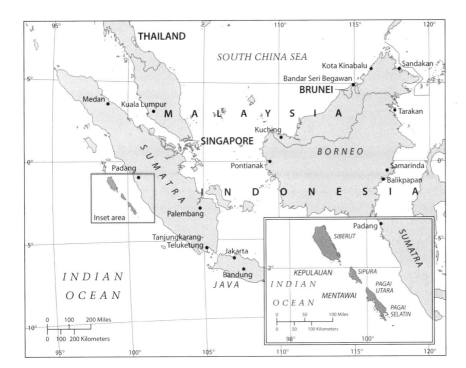

Figure 7-17 Pig-tailed langurs live on the Mentawai islands off of the west coast of Sumatra.

scend to the ground when fleeing danger. *Simias* is one of the few colobines that is known to live in small monogamous family units. In some populations, however, pig-tailed monkeys live in unimale groups that have up to five adult females. According to Richard Tenaza of the University of the Pacific, *Simias concolor* is one of the few colobine species (and the only Asian colobine) in which females have conspicuous sexual swellings. It is also the only colobine with sexual swellings that lives in unimale groups. As Tenaza notes, except for *Simias* and a few of the African monkeys that live in one-male groups, conspicuous swellings usually occur in primates that live in multimale groups. Tenaza hypothesizes that, when they do occur in unimale groups, conspicuous swellings might function in competition among females for attention of the male leader. He adds that swellings might also facilitate leaders' acceptance of females that attempt to transfer into their groups. These hypotheses remain to be tested, but are interesting because they focus on the females' fitness rather than emphasizing intragroup competition among males.

Tenaza has also studied intergroup loud calls produced by male pig-tailed monkeys that live in one-male groups. These vocalizations, which have been described as nasal barks by some, occur spontaneously throughout the day, but reach peaks in the early morning and late afternoon. Females rarely vocalize. Because loud calls provoke similar responses in other males, they may function to maintain spacing between groups, as hypothesized for various other primates who make such calls. Interestingly, males also emit loud calls in response to thunder, falling trees, or other startling sounds.

Simias concolor is extremely endangered. It is a timid primate that humans consider delicious to eat. This species has already been eliminated from some of the Mentawai islands as a result of logging and hunting.

SUMMARY

Old World monkeys lack the deeply rooted lineages of New World monkeys and, instead, emerged from a series of replacements of species through time. Today's cercopithecoids also differ from ceboids in having downturned noses, a 2.1.2.3. dental formula, ischial callosities, a variety of sexual swellings and bodily decorations, and terrestrial as well as arboreal species. There are two subfamilies of Old World monkeys that differ in their digestive systems. The colobines rely heavily on a folivorous diet and have sacculated stomachs and other features that aid in processing large amounts of hard-to-digest leaves. Cercopithecines, on the other hand, eat a more varied diet and have cheek pouches used for temporarily storing food.

Approximately 75% of the more than 30 colobine species live in Asia; the rest live in Africa. These monkeys are relatively large but slender, and have long hind limbs, feet, and tails. Their thumbs are reduced or even missing. Social organization is highly variable in colobines and includes unimale groups, multimale groups, all-male bands, large fission-fusion societies, and monogamous family units. Conspicuous sexual swellings occur less frequently in female colobines than in cercopithecines, but allomothering is more frequent.

As the portraits for three African (*Colobus guereza*, *Procolobus badius*, *Procolobus verus*) and four Asian (*Nasalis larvatus*, *Pygathrix bieti*, *Semnopithecus entellus*, *Simias concolor*) species suggest, diet is of key importance for understanding colobine social life. For example, the distribution and seasonal availability of preferred foods appear to influence group and range sizes, population densities, and other aspects of social organization. Because of their dietary adaptations and the impact they have on local vegetation, colobines are of special interest to evolutionary biologists interested in feeding ecology.

Colobines are also of keen interest to zoologists pondering sexual selection theory in general, and the evolution of infanticide in particular. Infanticide by adult males occurs in Hanuman langur populations that have a high percentage of unimale groups. Hrdy's sexual selection hypothesis suggests that infanticide in grey langurs is a male reproductive strategy. Predictions of this hypothesis have been tested in known free-ranging individuals and supported. Interestingly, females have been observed to compete for the attention of unimale group leaders in *Nasalis*, *Semnopithecus*, and *Simias* species. Hypotheses related to female competition are relatively new to primatology because they focus on females' rather than males' reproductive fitness.

FURTHER READING

HRDY, S. B., JANSON, C. AND C. VAN SCHAIK (1994/95) Infanticide: Let's not throw out the baby with the bath water. *Evolutionary Anthropology* **3**:151–154.

KIRKPATRICK, R. C. (1997) Search for the snub-nosed monkey. *Natural History* **106**(4):42–47.

OATES, J. F. AND A. G. DAVIES (1994) What are the colobines? In: A. G. Davies and J. F. Oates (eds) *Colobine Monkeys: Their Ecology, Behaviour and Evolution*. Cambridge: Cambridge University Press, pages 1–9.

SUSSMAN, R. W., CHEVERUD, J. M. AND T. Q. BARTLETT (1994/95) Infant killing as an evolutionary strategy: Reality or myth? *Evolutionary Anthropology* **3**:149–151.

8

OLD WORLD CHEEK-POUCHED MONKEYS: THE SHIFT TO TERRESTRIALITY

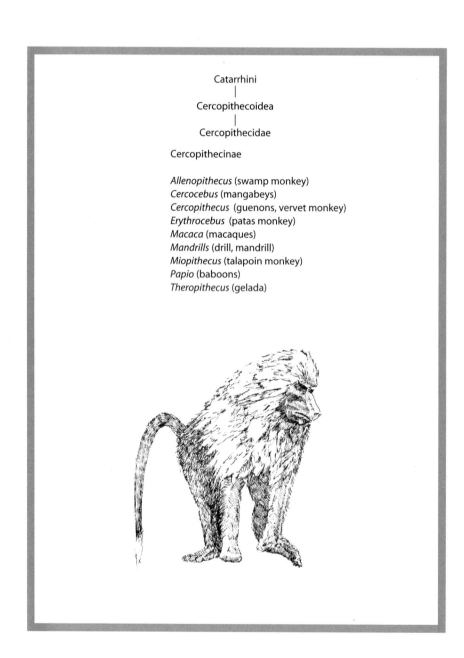

Catarrhini
|
Cercopithecoidea
|
Cercopithecidae

Cercopithecinae

Allenopithecus (swamp monkey)
Cercocebus (mangabeys)
Cercopithecus (guenons, vervet monkey)
Erythrocebus (patas monkey)
Macaca (macaques)
Mandrills (drill, mandrill)
Miopithecus (talapoin monkey)
Papio (baboons)
Theropithecus (gelada)

Figure 7-13 The Hanuman langur (*Semnopithecus entellus*) was the first Asian colobine to be studied.

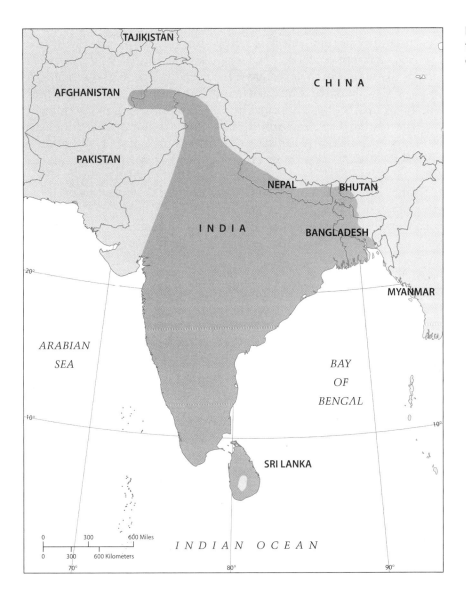

Figure 7-14 Hanuman langurs are the most western of the Asian colobines.

This is probably why langurs in this region have bigger bodies (for heat conservation) and longer, thicker coats.

Diets of grey langurs are also quite varied. New leaves, fruits, and flowers are staples that may be supplemented with mature leaves, gum, soil, bark, and insects, depending on the season and site. *Semnopithecus* also raids human crops such as maize, wheat, and tubers, and can ingest toxic seeds and plants that are poisonous to humans. Depending on the distribution of foods, home ranges may be quite large. Although basically arboreal quadrupeds, grey langurs have been reported to spend a good deal of time traveling and feeding on the ground.

Social organization of Hanuman langurs is based on one-male groups, multimale groups, and all-male bands. The percentages of each type of group vary from site to site, however, as described by Paul Newton of Oxford University (Fig. 7-15). Heterosexual groups may include anywhere from 4 to 125 individuals, while all-male bands usually contain no more than 15 individuals. Dominant males lead their groups, and select sleeping sites and food trees. They also use vocalizations such as "whoops" to maintain spacing between groups. Although both sexes have been reported to migrate, males do so more than females. Thus, social organization in unimale and multimale groups is basically matrilineal. Allomothering is common in this species but L. S. Rajpurohit and S. M. Mohnot of the University of Jodhpur report that weaning of infants (which occurs by 13 months of age) is associated with hostility and indifference on the part of mothers.

Home ranges of the different types of groups may overlap. Intergroup encounters occur often and are frequently aggressive, involving vocalizations, chasing, other displays, and sometimes physical contact. As is well known, at some sites, adult males from all-male groups have been observed attacking one-male groups and killing infants (more often males) of adult females. In these cases, the group leader and adult females are unable to defend their offspring and the leader is usually driven out of the group altogether. Meanwhile, the invading males mate with the resident females until a new leader is established and drives them away. New leaders may have a short or long (e.g., 10 years) tenure before the cycle repeats itself. Such incidences of infanticide have been reported for a number of grey langur populations in South Asia, but by no means all of them. Interestingly, infanticide only occurs in populations that have a high percentage of one-male groups, whereas noninfanticidal populations are associated with a predominance of multimale groups (see Fig. 7-15).

Needless to say, the practice of infanticide in grey langurs (and some other primates as well) has generated a good deal of interest among primatologists. Some workers believe that this behavior is simply pathological and due to stress from encroachment by humans, habitat destruction, and overcrowded living conditions. However, infanticide has been observed in undisturbed areas, so this hypothesis seems unlikely. Another possibility (the sexual selection hypothesis)

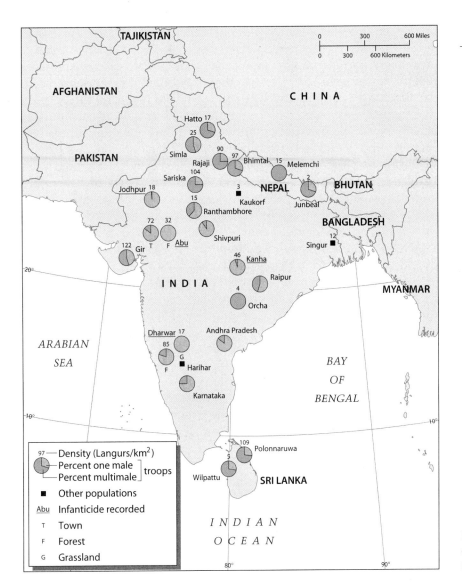

Figure 7-15 Hanuman langur social organization varies from site to site, as shown in this map of South Asia.

has been proposed by Sarah Blaffer Hrdy, who suggests that langur infanticide is a male reproductive strategy. According to Hrdy, invading males increase their chances of reproducing by killing nursing infants that they have not fathered, thereby causing their mothers to become sexually receptive (i.e., nursing inhibits ovulation, while its cessation stimulates it). Receptive females mate with and frequently conceive offspring of the new leaders. Thus, the invading male passes his genes into the future (increases his reproductive fitness) and, as an extra bonus, his infants do not have to compete directly with those of other fathers for resources.

The sexual selection hypothesis is consistent with socioecological theory (see Chapter 2). But is it true? Sussman, James Cheverud of Washington University, and Thad Bartlett of Dickinson College contend that it is not. Instead, they believe that infanticide occurs rarely among primates, and is a by-product of more generally aggressive episodes. Hrdy and her colleagues (1994/95:152) counter that infanticide is widespread among primates and that "there is a range of male, female, and even infant behaviors that only begin to make sense when we assume that infanticide, like predation, is a recurring threat, even if the actual events are rarely witnessed." Some light is shed on this debate by Volker Sommer of University College London and his colleagues who monitored free-ranging Hanuman langurs at Jodhpur, India, for 13 years and were able to test predictions associated with Hrdy's hypothesis in a unimale group in which all members are known individually. As predicted by the sexual selection hypothesis, they found that the loss of unweaned infants accelerated the resumption of menstruation and sexual receptivity, and reduced the spacing between subsequent births. The implication is that infanticidal males do indeed increase their fitness.

Sommer made other interesting observations about adult females. Estrous langurs signal their condition through their behavior, since they do not manifest sexual swellings. At Jodhpur, copulations were always solicited by females through head shuddering, presentation of the anogenital region, and lowering of the tail. In one-male groups, the probability of conception increased with the number of copulations, and females from Jodhpur apparently competed for the sperm of their leader by harassing other females as they copulated, and by mating during pregnancy (i.e., depleting sperm that would otherwise be available for others). Sommer's findings failed to support another suggestion, however, that postconception estrus is a female strategy for confusing paternity and thus deterring a new leader from killing her infant when it is born (i.e., since he might have fathered it).

Because it is widespread and protected for religious reasons in some areas, this species is not threatened with extinction, although like many other primates it is endangered. Some populations are under pressure from human encroachment, and golden jackals have been known to prey on immature grey langurs on the ground. Other predators include various cats and crocodiles.

THE PIG-TAILED MONKEY (SIMIAS CONCOLOR) The final portrait in this chapter is of a monotypic genus that is in need of further research. Nevertheless, *Simias concolor* (*Nasalis concolor* in some classifications) is described here because the little that is known about this species suggests that it is unusual. Pig-tailed langurs are so-named because, unlike other colobines, they have markedly short naked tails. Their fur is either creamy or dark gray or brown, and the bare faces are dark with upturned little noses (Fig. 7-16). *Simias* is proportioned like the Old World macaques, that is, with relatively short arms and legs that are approximately equal in length, and appears to be moderately sexually dimorphic (see Table 7-2).

Pig-tailed langurs live on the Mentawai islands off of the west coast of Sumatra, the largest of which is Siberut (Fig. 7-17). They inhabit evergreen, secondary, and swamp forests, and are known to de-

Figure 7-16 Not much is yet known about pig-tailed langurs (*Simias concolor*).

INTRODUCTION

The cheek-pouched monkeys are more numerous than colobines, with approximately 50 as compared to about 30 species. They include guenons, geladas, mandrills, drills, mangabeys, baboons, and macaques. Whereas the majority of colobines live in Asia, all but one of the nine genera of cercopithecines (see Table 7-1) live in Africa. However, that one genus, *Macaca*, has the widest geographical distribution of any nonhuman primate. For this reason, and because macaques and savanna baboons are closely related and have been especially well studied, the next chapter is devoted to them. In general, cheek-pouched monkeys are more diverse in their use of habitats and appearances than colobines, with various species sporting bright skin colors (some of them almost look tattooed), interesting hairdos, and a variety of cyclical sexual swellings in females.

As discussed in the previous chapter, Old World monkeys are divided into colobine and cercopithecine subfamilies on the basis of their dietary adaptations, and cercopithecines eat a much wider variety of foods than do the leaf-munching colobines. Cercopithecine dental features associated with a diet that contains varying amounts of fruit include relatively wide incisors and nonpointed cusps on the molars (see Box 1, Chapter 1). However, the most spectacular anatomical feature associated with the cercopithecine digestive tract is the cheek pouch (Fig. 8-1). When stuffed with food, these cavities form large, lumpy bulges that extend from the cheeks to the sides of the neck, giving the monkey an appearance of having a bad case of the mumps. The advantage of cheek pouches is that food may be stashed there for later consumption. To remove food, the monkey presses on a cheek with the back of the hand and then pushes the food out with its fists and chews. Another method is simply to use the fingers to scoop the chow from the mouth. In this manner, food can be collected in haste, and inspected and consumed at leisure a safe distance from other monkeys that might want it.

Colobines are basically forest monkeys, while cercopithecines may inhabit forests or savannas. The colonization of open grasslands by numerous cercopithecines (and a few colobines) is of supreme importance for understanding primate evolution because of its impact on the development of terrestriality. Compared to cercopithecines, colobines tend to have long, strong legs for leaping and long tails used for counterbalance during arboreal activities. Because of their relatively long hind limbs, when colobines are on the ground, their rumps tend to be the highest part of the body (see Fig. 7-2). Cercopithecines, on the other hand, have relatively longer arms (and shorter tails) and therefore tend to have their backs more horizontal when they are on the ground. As such, they reflect the beginning of an evolutionary trend that started with their ancestors and resulted in increasingly upright postures as new species adapted to more terrestrial lifestyles. The terrestrial great apes (i.e., chimpanzees and gorillas) carry the trend a step further because their shoulders are higher than their rumps when they are on the ground and they have completely lost the tail. The torsos of humans have undergone the most rotation, so much so that their bodies are completely vertical (i.e., they are **orthograde**). There is a certain irony in this trend toward verticality; having left the trees, terrestrial primates slowly but surely evolved

Figure 8-1 When their cheek pouches are stuffed with food, cercopithecines look like they have a bad case of the mumps.

Figure 8-2 Having left the trees, the terrestrial anthropoids that eventually gave rise to humans slowly evolved increasingly upright postures, as illustrated by this comparative sequence of living primates.

postures that returned increasingly larger portions of their bodies to greater heights (Fig. 8-2)! It should be noted, however, that whether they are arboreal or terrestrial, the Old World monkeys remain basically quadrupedal primates.

Hands also underwent important modifications associated with terrestriality. Arboreal primates use their hands to grasp the limbs and branches of trees, an activity that does not require skilled manipulation of individual digits. In fact, digits that might otherwise obstruct movement are greatly reduced in certain arboreal species that brachiate (the thumb of spider monkeys) or are strong graspers (the second digit of pottos). As noted in the Old World monkey pattern (see Chapter 7), cercopithecoids have evolved the ability to rotate their thumbs. (However, African colobus monkeys that have greatly reduced thumbs are an obvious exception.) Consequently, the pad of the thumb may be placed against that of one or more fingers; that is, the thumb is opposable. An **opposable thumb** (Fig. 8-3), which also is found in apes and humans, permits highly skilled manipulation of the fingers, as you may verify by self-examination. When you put a pen to hand or thread a needle, the tips of your thumb and next two fingers form a **precision grip** that allows fine movements. You would not be able to do this without an opposable thumb. The human's long, grasping thumb is also used in the less skillful, more forceful **power grip**, for example, when unscrewing the lid from a jar. As John Napier points out, despite the fact that Old World monkeys are capable of

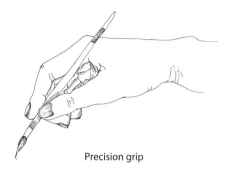

Precision grip

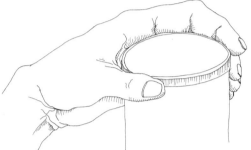

Figure 8-3 Because of their opposable thumbs, Old World monkeys, apes, and humans are capable of precision and power grips.

Power grip

Opposable thumb

both kinds of grips, some thumbs are more opposable than others. And the longest thumbs and best precision grips are found in terrestrial catarrhines including macaques, baboons, mandrills, geladas, and of course, humans, who have the longest thumbs of any primate.

The expansion of certain cercopithecine species to savanna grasslands and its associated terrestriality was complicated by increased pressure from new and dangerous predators. This is probably one reason why terrestriality is generally associated with larger body size and larger group size. For example, a number of species of baboons and macaques live in large multimale, multifemale social groups rather than the unimale units. However, the down side to living in groups with more than one adult male is that there is bound to be tension caused by competition for estrous females and other resources. As described in Chapter 9, such intragroup tension is controlled, and at times dissipated, by the interactions of individuals that recognize a particular pecking order, or dominance hierarchy. Social mechanisms have also evolved to protect the group, and these often involve singular or collective defense by males. This may be one reason why adult males have noticeably larger bodies and canines than adult females (i.e., terrestrial species usually exhibit a high degree of sexual dimorphism compared to arboreal species). A second (and some think more convincing) explanation for large male physiques is intrasexual competition for females (see Chapter 2). As we will see, social life in general is vastly complicated in terrestrial species, and the troop may be viewed as a kind of school in which individuals learn a complicated set of rules about social relationships.

GUENONS

Despite the fact that this chapter emphasizes the shift to terrestriality, not all cercopithecines are terrestrial or even semiterrestrial. Take the most common monkeys in Africa, the guenons, for example, in which most of the more than 20 species are wholly arboreal. Guenons are widely distributed across sub-Saharan Africa, and most belong to the genus *Cercopithecus*. The exceptions are monospecific *Allenopithecus* (Allen's swamp monkeys) and *Miopithecus* (talapoins), which are given generic distinction because their females are unusual among guenons in exhibiting cyclical sexual swellings, and *Erythrocebus* (patas monkeys), which have unique skeletal features associated with ground living. (It should be noted, however, that some workers include these monkeys in *Cercopithecus*, but identify each as a separate subgenus.) Of the three eccentric species, Allen's swamp monkey, which lives in the Congo basin, remains the most mysterious. Talapoins, the smallest of the Old World monkeys, are better understood. Interestingly, these little monkeys are unusual for forest guenons and similar to New World squirrel monkeys (with which they are frequently compared) because they live in large multimale groups. Patas monkeys have been thoroughly studied and are portrayed in the section on semiterrestrial guenons. Unfortunately, because most species are arboreal, guenons have not been as thoroughly studied as some monkeys living in savanna and woodland habitats that are more user-friendly for primatologists. Consequently, an authoritative volume about the evolutionary biology of guenons (edited by Annie Gautier-Hion of the University of Rennes in France and her colleagues) only became available in 1988, and information about the long-neglected arboreal species is just beginning to emerge.

Arboreal Guenons

Primatologists believe that the arboreal guenons are descended from monkeys that were once semiterrestrial, open-country animals, and that their proliferation today is the result of rapid evolution within forests during the last million years. They also think that new species of guenons may still be forming. Arboreal guenons are relatively small to medium-sized quadrupeds, with arms and legs of approximately equal length, and long tails. Different species are distinguished by a variety of coat colors, hairdos, facial hair, face patterns, and genital colors. As a result, some of these guenons are amazingly decorative (Fig. 8-4, color plate). In fact, a leading expert on this subject, Jonathan Kingdon of Oxford University in England, describes the faces of one

REVIEW EXERCISES

1. Where do most genera of cercopithecines live?
2. Which nonhuman primate has the widest geographical distribution and where does it live?
3. Which primates have cheek pouches and what are their function?
4. Compare the postures of colobines and cercopithecines when they are on the ground. How does this comparison reflect the beginning of the trend toward verticality in terrestrial primates?
5. Describe opposable thumbs. Who has them and how are they used?
6. Compare precision and power grips. Name an activity that requires each grip.
7. In what ways is ground living more complicated than arboreal life?
8. What is one probable function of the large body and canines in terrestrial male primates?

species (*Cercopithecus cephus*) as looking "like experiments in signal geometry." Colorful faces and hindquarters function to make guenons stand out in thick, dark arboreal habitats, whether they are coming or going. These distinctions also facilitate ready identification of a particular species in comparison with other guenons, mangabeys, and colobus monkeys with which guenons frequently associate. Kingdon also points out that these visual features may act as barriers to breeding between different groups of guenons and therefore could be prime movers of the reproductive isolation that is needed for speciation to occur.

All arboreal guenons are above-branch feeders that eat fruit as a main staple. However, when fruits dwindle seasonally, guenons eat other foods including insects. And as predicted (see Chapter 2), smaller guenons such as talapoin monkeys rely more heavily on insect nourishment than their larger cousins. Guenons are remarkably versatile and opportunistic. For example, little talapoins consume a fair amount of tubers, which have only recently been introduced into West Africa. Different species of arboreal guenons are often found associated. These mixed groups are thought to facilitate increasingly efficient foraging strategies that result in a better food supply for all parties concerned. Polyspecific groups also seem to be better at avoiding predators such as eagles and leopards.

Although the social systems of arboreal guenons are not as well studied as those of ground-living monkeys, it is generally acknowledged that forest guenons such as *Cercopithecus ascanius* and *Cercopithecus mitis* live in matrilineal unimale groups with one adult male that attempts to monopolize reproductive access to a group of adult females, a system known as **female defense polygyny**. According to Marina Cords of Columbia University (see In the Field), however, these groups are periodically disrupted by influxes of multiple males and mating by both sexes then becomes promiscuous. During these influxes, there is a good deal of aggression as males compete for estrous females. Nonetheless, active choice of mates has been observed in both sexes. Because there is no clear link between sexual receptivity and fertility in forest guenons, these promiscuous matings may or may not result in offspring. Thus, one benefit of promiscuous matings may be social; that is, they increase not only a female's odds of fertilization (a reproductive benefit), but also the likelihood that her future infant will be tolerated by numerous adult males (cases of infanticide have been reported for *Cercopithecus ascanius* and *Cercopithecus mitis*). Interestingly, during years when one-male groups remain relatively undisturbed, females sometimes mate with nonresidents. Thus the extent to which the resident male actually monopolizes reproduction is not yet known.

This is not to say that socioecological factors have no influence on reproduction in forest guenons. They do. Thomas Butynski of Makererre University in Uganda has shown that, although mating and birth seasons and peaks characterize all guenons, some populations that live in areas of high rainfall mate and give birth year-round. In fact, the lengths of mating and birth seasons are generally tied to the amount of annual rainfall and whether or not there are one or two wet seasons. The more wet seasons and greater amount of rainfall, the longer the mating and birth seasons. It is not surprising, then, that forest guenons living in the same area are highly synchronized in their mating and birth seasons. It seems that rainfall determines when food is most abundant, and this coincides roughly with birth and early lactation. In other words, Butynski has shown that rainfall affects food supply, which in turn is essential for determining the timing of reproductive events.

Semiterrestrial Guenons

Although most species of *Cercopithecus* are arboreal, a few are not. These include semiterrestrial species that live in forests such as *C. neglectus* (de Brazza's monkey), *C. hamlyni* (owl-faced monkey), and *C. lhoesti* (mountain monkey), as well as *C. aethiops* (vervet) which inhabits more open country. The semiterrestrial forest guenons are adverse to living in the mixed troops that characterize other forest guenons. According to Gautier-Hion, semiterrestrial guenons that avoid living in groups of mixed species prefer living in relatively small troops, tend to be highly sexually dimorphic, use their habitats intensively, and are likely to use silence and concealment to avoid predators. Curiously, except for *C. lhoesti*, there is a tendency among semiterrestrial species of *Cercopithecus* (as well as *Allenopithecus*) for adults of both sexes to mark objects in their environments with secretions from glands in the chest and the muzzle region. This is highly unusual behavior for Old World monkeys and probably serves some kind of olfactory signal function, as noted by Jean-Neil Loireau of Paimpont Biological Station in France and Gautier-Hion who also point out that such marking is totally absent in the arboreal guenons and may reflect a primitive retention in the semiterrestrial guenons that engage in it. Vervets (*Cercopithecus aethiops*) are the most terrestrial, widespread, and (some would say) successful of the guenons. Much is known about this interesting species, as revealed in the following portrait.

VERVETS (*CERCOPITHECUS AETHIOPS*) One might go so far as to say that vervets have been wildly successful. They are distributed over most of sub-Saharan Africa and may well be the most abundant of the African monkeys. Because of their wide presence, vervets have been studied in the field since the 1960s; Laurence and Linda Fedigan of the University of Alberta published an authoritative review that encompasses the ecology, demography, and social patterns of vervets. Vervets are also extremely flexible when it comes to diet and habitat, so much so that they seem to thrive even in marginal areas as their favored trees are removed by the timber industry. As one would expect in such a widely distributed and flexible species, vervets display a good deal of regional variation in their appearance. Consequently, some workers

Male Influxes in Blue Monkeys

Marina Cords has conducted field studies of wild guenons in a Kenyan rain forest since 1979. She is especially known for her surprising discoveries that challenge the basic assumption that social and breeding systems coincide in these monkeys. In her own words:

MARINA CORDS HAS SPENT MANY SEASONS IN THE FIELD STUDYING WILD GUENONS IN KENYA. (PHOTO COURTESY OF MARINA CORDS.)

When I began fieldwork on wild guenons in Kenya, my major previous experience studying primate behavior was 3 months with the rhesus macaques of Cayo Santiago. Even a beginner like me had little difficulty figuring out what made rhesus society tick. In a few months, I became quite good at predicting what was going to happen when two individuals or groups met. I therefore expected that it would be equally easy to "figure out" the societies of blue monkeys (*Cercopithecus mitis*), who are, after all, quite closely related to macaques.

But I was in for a lot of surprises. In the beginning it was hard to identify groups, let alone individuals. There were times when there seemed to be about 100 monkeys all around me: Individuals were widely spaced, not obviously moving in any particular direction, and there were no signs of hostilities between opposing factions. Earlier research had suggested that guenon groups were rather small and territorial, so this amoeboid sea of monkeys was puzzling. The key to figuring out what was going on was learning to recognize individuals: This took many months, but eventually I could reliably recognize individuals using natural markers like stiff fingers, tail thickness, and facial gestalt. Then it became clear that groups do have

stable female membership, and that there were imaginary lines in space that divided them when they happened to be in the same general area.

Because group members can be spaced over several hundred meters, an observer (and presumably a monkey) hardly ever sees the entire group at once. "Groupness" becomes most obvious when there are aggressive territorial encounters between groups: During these battles, which often occur near feeding trees, the group's females form a cohesive, threatening phalanx that takes on the opposing team. Males sit behind the front lines looking on, but are seldom directly involved. As the aggression dies down, all group members engage in what seems to be an intense display of togetherness, an orgy of concentrated grooming with rapid switching among partners, and often a focus on adult males. Perhaps 45 minutes later, grooming clusters disband, and the two groups drift apart and return to foraging.

But then you might notice a stranger in the group: You think you must be mistaken, as the previous fight has just shown how much antipathy there is between neighboring groups. But no—it really is a juvenile male from the neighboring group who's just hard to recognize out of context! It turns out that juvenile males don't adhere to the same pattern of group membership as females. When two groups are nearby, the young males often initiate long play sessions with each other off to one side. As the groups drift apart, a juvenile male from one group may remain with the neighbors for the rest of the afternoon. No one seems to mind this young intruder, who eventually rejoins his original group before nightfall.

Males moving in and out of groups whose identity is determined by mutually hostile females is a theme that carries on into adulthood. This theme jives with earlier descriptions of guenons as living in groups with only one male who would occasionally be ousted by another. Our observations confirmed that a typical blue monkey group includes just one adult male for most of the year, so there are many adult males who do not live together with females. However, we were surprised to find that these nonresident males form loose, flexible spatial associations, and occasionally even groom one another. Our long-term records allow us to determine that some male associates were peers in their natal groups: It seems very likely that relationships formed during youth influence social patterns in adulthood. Although these males do not live in cohesive, stable groups the same way that females do, it would be a mistake to call them "solitary."

For most of the year, nonresident males stay near the territorial boundaries of heterosexual groups, avoiding whichever group comes near by retreating into the neighboring territory. We were amazed, however, how dynamic male residence patterns can become during the breeding season, when nonresidents often visit female groups. In most years, these visits are rather inconspicuous and brief (though long enough for a female to mate with the visitor). In some years, however, several males arrive at once, so that a one-male group is rapidly transformed into a multimale group, and remains a multimale group for several months. The first time I witnessed such an influx, I wondered if it was just an aberrant transitional period in our study group. However, these breeding-season influxes of males occur regularly (every 4 to 5 group-years) at our site, and have been reported in other populations and other guenon species as well. We have come to realize that flexibility in male membership seems a characteristic of guenon society, not a phenomenon to be explained away as the result of anomalous events.

During a breeding-season influx, individual males come and go and there are varying numbers of males in the group on any one day. Some males are more regular and persistent visitors than others, and some lurk on the edges while others move boldly right to the center of the group. The male who was previously the sole resident usually tries to chase the intruders away, but in a thick forest, with limited visibility, it seems impossible for him to guard his widely dispersed females. His job is also made more difficult by the estrous females, who are attracted to the newcomers, approaching and following them, and soliciting them for mating with puckered lips. Some females even abandon their group for most of the day, to go off "on safari" with a newcomer. Many of the new males are in fact observed to mate with these eager females.

These observations challenge the oft-made assumption that social and breeding systems coincide. Clearly you don't need to be a resident male in order to mate, and intruders sometimes even mate more than residents. Residents even mate with the females from neighboring groups. However, to interpret these various patterns of male behavior from an evolutionary perspective, we need to know more than who mates: We need to know who is actually siring offspring, and how individual males move through the roles of resident and nonresident intruder over the course of a lifetime. We are trying to answer these questions now. I think they will keep us busy for quite some time to come.

regard *Cercopithecus aethiops* as a superspecies that contains a number of distinct species. For simplicity's sake, however, this text views vervets as one extremely varied or **polytypic** species. (Some workers now follow Collin Groves of the Australian National University by placing vervets in the genus *Chlorocebus* rather than *Cercopithecus*.) Again, because of regional variation, *Cercopithecus aethiops* has a variety of common names in different parts of Africa. Besides being called vervets, these monkeys may be referred to as green monkeys, grivets, or savanna monkeys.

Given that they are more terrestrial than most guenons, vervets are relatively small (averaging about 6 kg or around 13 lb) and not as sexually dimorphic in body size, contrary to the rule for most ground dwellers (Table 8-1). The canines, however, are large and sharp. Their bodies and faces are also not as dramatically decorated as those of the forest guenons. In general, vervets' coats range from gray to greenish gray to greenish gold, and they have black faces that are encircled with white fur. The abdominal skin is a faint eggshell blue in both sexes, while the scrotum of males is a much deeper blue. Males also have bright-red penises and their genitals are surrounded with white hair (Fig. 8-5, color plate). These colorful genitalia may be displayed to assert social dominance, a practice that primatologists (Fedigan and Fedigan, 1988:396) have dubbed as "flagging the red, white and blue." The Fedigans (1988:395) summarize *Cercopithecus aethiops* as "a rather elegant monkey with its delicate frame, rich fur, extravagant tail, and smooth, light-footed locomotor patterns both in the trees and on the ground."

Vervet habitats span from Senegal to Ethiopia, on down to the tip of South Africa (Fig. 8-6). Despite the fact that they prefer woodland areas near rivers, vervets have adapted to a wide range of African habitats including savanna woodland, the edges of rain forests, swamps, and (similar to rhesus monkeys in India, see Chapter 9) marginal areas that have been disturbed by humans. Vervets have also been transplanted to the Caribbean islands of St. Kitts, Nevis, and Barbados. In fact, they are so adaptable that, depending on the circumstances, vervets are shy and quiet in some areas, but boisterously raid crops and bother tourists in others. Although they are the most terrestrial species of *Cercopithecus*, vervets are equally at home on the ground and in trees and, for this reason, are officially regarded as both semi-terrestrial and semiarboreal. These little monkeys always sleep in trees, however. They use their long tails to brace themselves in trees, but hold them aloft at various angles while moving quadrupedally on the ground. Because vervets are able to subsist on a wide variety of foods, they are described as opportunistic omnivores. Foods are found both in the trees and on the ground, and may include fruits, seeds, flowers, leaves, grasses, roots, gums, bark, eggs, small birds, small mammals, lizards, insects, and handouts from tourists.

Unlike forest guenons that reside in matrilineal unimale groups, vervets live in multimale, multifemale troops that vary in size with

Figure 1-5   This is a preliminary reconstruction of *Eosimias*.

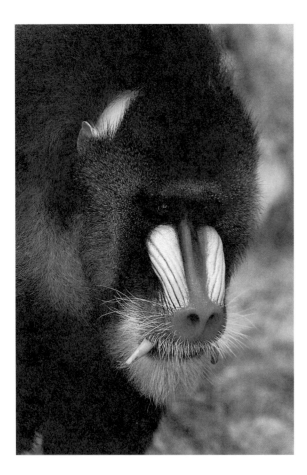

Figure 2-11   This male mandrill's brightly colored face could have arisen as a result of female choice.

Figure 3-14   Tarsiers are curious-looking little creatures.

Figure 5-11 The golden lion tamarin (*Leontopithecus rosalia*) is truly beautiful with its spectacular golden-reddish coat of hair.

Figure 5-13 Saddleback tamarins (*Saguinus fuscicollis*) belong to the hairy-faced group of *Saguinus*, and have three color zones along their backs. (Photo courtesy of Paul Garber.)

Figure 6-15 The mantled howler (*Alouatta palliata*) gets its name from the golden mantle of hairs on its back. (Photograph courtesy of Kenneth Glander.)

Figure 7-5 Red colobus monkeys (*Procolobus badius*) are named for the color of the hair on their lower limbs and stomachs.

Figure 7-11 The Yunnan snub-nosed monkey (*Pygathrix bieti*) looks like it's wearing lipstick.

Figure 8-4 Like this male Diana's monkey, arboreal guenons are characterized by a variety of interesting coat colors, hairdos, and facial hair.

Figure 8-5 Male vervets sometimes display their colorful genitals to assert dominance, known to primatologists as "flagging the red, white, and blue."

Figure 8-16 The top panel shows a fatted, group-associated male mandrill with fully developed secondary sexual characteristics; the male in the lower panel is a nonfatted nonsocial individual. (Photographs courtesy of Dr. Alan Dixson.)

Figure 11-7   Orangutans have feet that are built for
hanging on.

TABLE 8-1 Statistics for Representative Old World Monkeys (Cercopithecines)

	Cercopithecus aethiops (vervet)	Erythrocebus patas (patas monkey)	Papio hamadryas (hamadryas baboon)	Theropithecus gelada (gelada)	Mandrillus sphinx (mandrill)	Cercocebus torquatus atys (sooty mangabey)
Adult mean body weight						
F	5.6 kg (12.3 lb)	5.5 kg (12.1 lb)	12.0 kg (26.5 lb)	11.7 kg (25.8 lb)	11.5 kg (25.4 lb)	
						8.6 kg (19.0 lb)
M	7.0 kg (15.4 lb)	10.0 kg (22.0 lb)	21.3 kg (47.0 lb)	20.0 kg (44.1 lb)	26.9kg (59.3 lb)	
Adult brain weight (g)	60	107	143	132	159	
Maximum life span (yr)	31.0	21.6	35.6	19.2	46.3	18.0
Habitat[a]	sw, eh, mf, mh	aw, s	ad, sw	mg	pf, sef, gf, cf	pf, sef, ff, swf, m, gf
Locomotion[b]	qr, st, sa	qr, qfr	qr	qr	qr, c	qr
Diet[c]	f, s, fl, l, e, a	l, fl, f, s, a, i, g, gr	l, gr, s, r, t, a	gr, s, l, b, a	f, s, l, sh, p, i, a	f, s, a
Seasonal breeders	var	var	Peaks	Peaks	+	0
Mean gestation (days)	163	167	170	165	173	167
Interbirth interval (mo)	16	11.8	22	24	17	13
Social systems[d]	M-MM	M-U	F-F, P-MM, U	F-F, M-U	MM, U	MM
Territorial	var	0	0	0	0	
Ratio of adult females to males	1.5:1	1–19:1	1–9:1–3	1–10:1–3	5–10:1	
Alloparenting	+	+	+	+		+
Infanticide[e]	0	0	0	+		+
Group size	5–76	5–74	var	var	var	35
Density	14–154/km²	1.4/km²	1.9–3.0/km²	63–78/km²	1.5–7.5/km²	
Home range	13–178 ha	2340–8000 ha	2800 ha	344 ha	1000–5000 ha	Large
Conservation status[f]	L	L	R	T	H	H

Note: Information in this table has been selected from numerous sources and should be viewed as tentative, i.e., subject to change in light of new information. Sources include but are not confined to Fedigan and Fedigan, 1988 (*C. aethiops*); Harvey et al., 1987 (brain weights); Melnick and Pearl, 1987; Chism and Rowell, 1988 (*E. patas*); Stammbach, 1987; Rowe, 1996 (conservation status and other statistics). Abbreviations: +, yes; 0, no; var, variable. Blank entries indicate that no information is presently available. Dietary items are listed in approximate order of decreasing importance to the species.

[a]Habitat: ad, arid subdesert; aw, acacia woodland; cf, coniferous forest; eh, edge habitats near water (e.g., swamps); ff, floodplain forest; gf, gallery forest; m, mangroves; mf, montane forest; mg, montane grassland; mh, marginal areas disturbed by humans; pf, primary forest; s, savanna; sef, secondary forest; sw, savanna woodland; swf, swamp forest.

[b]Locomotion: c, climbing; qfr, quadrupedal fast running; qr, quadrupedalism; sa, semiarboreal; st, semiterrestrial.

[c]Diet: a, animal prey; b, bulbs; e, eggs; f, fruit; fl, flowers and buds; g, gums (sap); gr, grasses; i, insects; l, leaves; p, pith; r, roots; s, seeds; sh, shoots; t, tubers.

[d]Social system: F-F, fission-fusion; U, unimale group; M-U, matrilineal unimale; M-MM, matrilineal multimale, multifemale; MM, multimale, multifemale; P-MM, patrilineal multimale, multifemale.

[e]Infanticide: +, one or more reports.

[f]Conservation status: H, highly endangered; R, rare; L, lower risk; T, threatened.

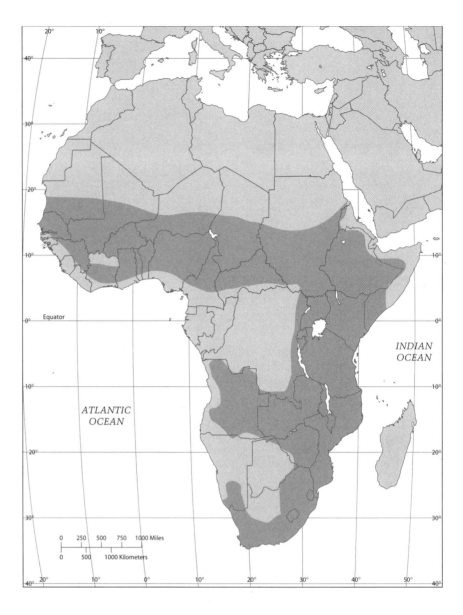

Figure 8-6 Vervet habitats are widely distributed in sub-Saharan Africa.

the availability of resources. Males tend to emigrate away from their natal troops as they mature, while females do not. As seen in previous chapters, this pattern, known as female philopatry, is associated with a social life that is centered around a stable group of core females. Both sexes have clear dominance hierarchies, but those of females are relatively stable because of their lifelong associations. Allomothering is widespread in this species, probably due to not only positive relationships among related females, but also the fact that mothers need help with relatively large, precocious infants that mature quickly. (Infanticidal males do not seem to be a problem in vervets, although females have been known to unite against males under other circumstances.) The extent to which vervets are territorial, again, varies from population to population. When it occurs, however, territoriality seems to be related to small, defensible ranges that have high population densities. Vervets are among those semiterrestrial guenons that engage in olfactory communication by rubbing secretions from glands in the chest on surfaces such as branches.

Because they are both semiterrestrial and semiarboreal, vervets have an interesting complex of communication systems. As noted, their habit of olfactory marking occurs in a number of other semiterrestrial guenons and may be a primitive retention that was lost in guenons that became more arboreal. Vervets also lack the variety of bright face and body decorations that characterize forest guenons. Such visual shouting might be more maladaptive (i.e., attracting predators) than helpful for monkeys that live in the open and, indeed, it is generally the rule that savanna primates are relatively dull compared to those that live in forests. The communication system that vervets are most famous for, however, is their repertoire of alarm calls and, as documented by Dorothy Cheney and Robert Seyfarth of the University of Pennsylvania, these too are suited for both arboreal and terrestrial conditions. Different alarm calls signify the presence of birds of prey, leopards, and snakes, and vervets react appropriately depending on the particular call. Upon hearing the alarm that signifies a raptor, troop members look up and take cover. The alarm for a leopard causes ground-dwelling vervets to run for the trees, while that for a snake results in vervets standing up and looking at the ground (and sometimes mobbing the snake). These findings are fascinating and show how complex nonhuman primate vocalization systems can be. They are also of interest to scientists studying the origins of language (see Chapter 14).

In their review, the Fedigans ask what it is that makes *Cercopithecus aethiops* such a successful primate. They suggest that a number of characteristics play a role, including the ability to be completely at home on the ground and in the trees, and a high degree of adaptability when it comes to foods and living in marginal habitats. According to the Fedigans, in vervets

> we have a primate that can live in almost any habitat which provides a few trees, water, and a minimal diet, and that can increase its population size rapidly when conditions are favorable. . . . So, while most *Cercopithecus* monkeys were evolving as splendid specialists of the African rainforest, *C. aethiops* seems to have specialized only in persistence under fluctuating conditions, and as a result, may prove to be one of the most successful "survivors" on the changing African continent. (Fedigan and Fedigan, 1988:411)

THE PATAS MONKEY (*ERYTHROCEBUS PATAS*) Although *C. aethiops* is the most terrestrial species of *Cercopithecus*, it is not the most terrestrial guenon. That honor falls to the closely related patas monkey (also known as the Hussar or military monkey), which has been thoroughly described by Janice Chism and Thelma Rowell of the University of California, Berkeley. According to these authors, patas monkeys have

REVIEW EXERCISES

1. Which monkeys are the most common in Africa? What is their genus? Are they mostly arboreal or terrestrial?
2. What two advantages do arboreal guenons appear to enjoy from living in groups of mixed species?
3. Describe female defense polygyny in arboreal guenons. Does it always work?
4. Among guenons, the durations of mating and birth seasons seem to rely on the amount of annual rainfall. What does rain have to do with reproductive events?
5. Which species of *Cercopithecus* are semiterrestrial? Is this the norm for the genus? Name some ways in which semiterrestrial guenons differ from arboreal guenons.
6. Which species of *Cercopithecus* is the most terrestrial?
7. In what ways are vervets extremely successful primates? To what do they owe their success?
8. Describe vervet communication systems and how they reflect the fact that these monkeys are both semiarboreal and semiterrestrial.

Figure 8-7 Because patas monkeys have white mustaches, they are sometimes called military monkeys.

retained a guenon lifestyle, but have moved away from the vervet habit of relying on large, tall trees in continuous woodlands. Instead, *Erythrocebus* has become a specialist in using patchy acacia woodlands and nearby open areas and, in so doing, has shifted even more toward terrestriality. In fact, patas monkeys may provide good comparative models for illuminating some of the adaptations that accompanied selection for terrestrial bipedalism in the earliest hominids (see Chapter 14).

Patas monkeys are medium-sized monkeys that are highly sexually dimorphic, with adult males averaging almost twice the weight of adult females (10 kg or 22 lb versus 5.5 kg or 12 lb, see Table 8-1). With long arms, legs, and tails, these slender quadrupeds have been likened to greyhound dogs. This comparison is not as fanciful as it may seem, because patas monkeys are the fastest runners of all the primates, capable of reaching speeds of up to 35 mph. Their coats are reddish brown with white to golden underparts, and the male's scrotum and penis are blue and red respectively, like those of the vervet. Adults of both sexes have white mustaches (hence the name military monkey), and black bands that stretch across the forehead to the ears (Fig. 8-7). Indeed, the latter resemble headbands, which is perhaps appropriate for the fastest running primate.

Erythrocebus is distributed in a broad band across the central part of Africa, from West Africa to Ethiopia (Fig. 8-8) and, as already noted, prefers to live at the edges of acacia woodlands, which are next to open areas and are not as densely covered with trees as the interior. Such a woodland habitat is somewhat surprising because this species was once thought to prefer living in more open grasslands. Patas monkeys have extremely large ranges (2340 to 8000 ha, see Table 8-1), however, and although they occasionally jump from tree to tree like vervets, they usually move from place to place on the ground. When moving quadrupedally, patas monkeys place weight on their fingers (which effectively lengthens the arms) rather than on the flats of their

Figure 8-8 Patas monkeys live in a broad band across the central part of Africa.

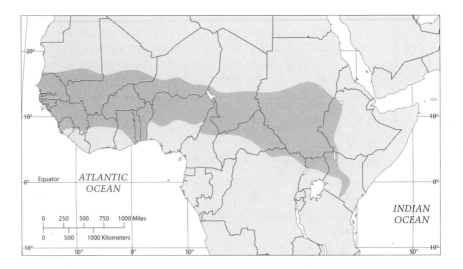

hands, an adaptation to terrestrialism that is not seen in vervets. As noted, these lanky animals are **cursorial**; that is, they are physically built for running. They also have an intriguing habit of standing bipedally, or propping themselves up like a tripod with both legs and the tail. As with other terrestrial monkeys, patas monkeys spend much of their day traveling on the ground in search of food, water, and resting sites. However, each monkey or mother with infant sleeps in a separate tree at night, which is unusual. This results in the troop being widely dispersed and, together with the fact that groups never sleep in the same trees on consecutive nights, may be part of the reason why home ranges are so large.

Another factor that may be responsible for large ranges is diet. Not surprisingly, patas monkeys rely heavily on acacia trees for food, eating many of their parts including new leaves, flowers, buds, seeds, gums, and insect galls. *Erythrocebus* also eats some herbs, fungi, invertebrates, and an occasional small vertebrate. Unlike baboons, grasses form a very limited part of the patas diet. On the other hand, Chism and Rowell report that insects are a primary source of food and that patas monkeys devote a good deal of time to finding and harvesting them. You will recall that high consumption of insects is usually associated with smaller-bodied primates. In fact, the extent of insect (and gum) ingestion in patas monkeys is unparalleled in other primates of their size and, as discussed in Chapter 1, such high-quality food would be difficult to obtain if home ranges were not relatively large.

Patas monkeys live in unimale social groups in which the resident male is somewhat more peripheral than is the case for many other primates that live in one-male groups. Males that do not reside in one-male groups may be solitary or part of temporary all-male groups. The core of patas society is therefore a group of related adult females and their offspring. In keeping with the peripheral nature of males, it is usually the adult females that determine the daily route, and females and juveniles are the chief participants when aggressive encounters occur with other groups of patas. Dominance relations exist among the adult females of a unimale group, but they do not seem to influence social interactions to a very large degree.

During mating season, a number of nonresident males may join a previously one-male group, resulting in intense competition between males for access to females. Multimale situations are especially likely to occur after takeover of the resident male's position, according to Hideyuki Ohsawa and colleagues of Kyoto University. When these situations occurred over a 3-year period at Kala Maloue National Park, Cameroon, resident males engaged in only 31% of the matings. Nevertheless, the resident males still sired more offspring than did sneakers, which Ohsawa suggests was probably due to additional matings on the part of the resident male diluting the sperm of the sneakers. Sneak matings resulting in births also occurred during one-male situations.

Curiously, patas monkeys give birth during the day, which is highly unusual for Old World monkeys. Furthermore, according to Chism (1986:49), patas monkeys have the fastest rates of sexual maturity of any Old World monkeys, which she views as "a response to a highly seasonal savannah environment in which there is a premium on ability to achieve nutritional, locomotor, and social self-sufficiency as quickly as possible." Thus, females in the wild reach sexual maturity at the remarkably young age of 2.5 years. As one would expect for primates that live in one-male groups, allomothering is an important part of group life. In fact, Naofumi Nakagawa of Kobe City College of Nursing in Japan reports that infants' distress calls are powerful stimuli that motivate females to locate and allomother even unrelated infants.

If there is one word that describes the patas personality, it is "cautious." Patas monkeys are extremely difficult to habituate (it takes months before primatologists can follow them), and with good reason: Their surroundings attract potential predators such as wild dogs, lions, leopards, cheetahs, jackals, hyenas, and birds of prey. According to Chism and Rowell, patas are vulnerable and clearly nervous if they stray too far from the safety of trees, which accounts for their strong preference for open acacia woodlands (Fig. 8-9). Even there, however, patas are vigilant in scanning their surroundings, doing so either from

Figure 8-9   Of the three woodland habitat types profiled by Chism and Rowell, patas monkeys have a strong preference for open acacia woodlands.

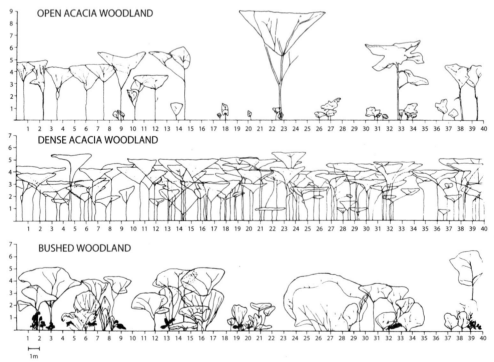

high perches or by standing up. Resident males play a special role in this regard, silently acting as lookouts at a distance from their groups and, if necessary, sounding an alarm and noisily diverting attention to themselves as the females and juveniles make their escape. Patas monkeys' extreme caution is nicely illustrated by Chism and Rowell's description of how they cross open areas:

> Patas usually crossed open grassland only after long periods of scanning it from the edge of the woods. They then crossed in a "leapfrog" fashion: one or two animals went ahead to a tree in the grassland, then the bulk of the group crossed to the tree and sat there, scanning again, while one or two animals moved ahead to another tree. This was repeated until the group again reached the margin of a wood. Often, one or two animals remained behind and scanned until the whole group had crossed, then raced across to catch up. The importance of the behavior of the lead animals in these crossings was underscored by attempted crossings in which an initial movement into open grassland was aborted when the leaders suddenly ran back to the wood. (Chism and Rowell, 1988:427–428)

OTHER PRIMATES THAT LIVE IN ONE-MALE GROUPS

Four closely related species of cercopithecines share the proclivity of *Erythrocebus* for living in one-male units. These species (loosely referred to as baboons by some) include hamadryas baboons (*Papio hamadryas*), geladas (*Theropithecus gelada*), mandrills (*Mandrillus sphinx*), and drills (*Mandrillus leucophaeus*). While the first two species have been fairly widely studied, much remains to be learned about the drills and mandrills. Nevertheless, Eduard Stammbach of the University of Zurich in Switzerland recently compared the four species, and it is clear from his analysis that the details of their social lives differ in important ways from those outlined for patas monkeys. In some sense, these four species have become even better adapted to the terrestrial way of life. Indeed, one of them (*Theropithecus*) is believed to be the most terrestrial of all of the nonhuman primates. These species also differ from *Erythrocebus* in that, unlike the latter, their unimale groups merge together and split apart on a regular basis that may be tied both to rather harsh environmental conditions (which vary with the species) and to patchy distributions of food and other resources. As discussed in the next chapter, in times of plenty, such multileveled or tiered societies may have given rise to the multimale, multifemale polygynous groups that characterize macaques and savanna baboons. Each of these species also has interesting and unique features, detailed below.

The Hamadryas Baboon (*Papio hamadryas*)

Hamadryas baboons, once considered sacred by the ancient Egyptians, are quadrupedal monkeys that resemble common baboons in having

Figure 8-10 Adult male hamadryas baboons have silvery whiskers and impressive mantles of fur around their shoulders.

rather long doglike muzzles. As befits terrestrial primates in which males attempt to monopolize a number of females (see Chapter 2), this species is highly sexually dimorphic (adult males average a little over 21 kg or about 47 lb, and adult females weigh about 12 kg or 26.5 lb, see Table 8-1). Faces tend to be pink, and coat colors range from dull shades of brown in females and juveniles to silvery gray in fully adult males. The latter also have silvery whiskers and impressive mantles of fur around their shoulders (Fig. 8-10), and are markedly longer in tooth than adult females.

Hamadryas baboons are also known as desert baboons because they are confined mainly to arid, rocky gorges of Ethiopia, Somalia, Saudi Arabia, and Yemen (Fig. 8-11) where they inhabit wooded areas or subdesert plains that are relatively treeless. This terrain is among the hottest and driest inhabited by any nonhuman primate. During the night, hamadryas baboons sleep in huge groups on the relatively few available cliffs and rock ledges. Each morning these large sleeping parties divide into smaller groups which go in search of food, shade, and scarce sources of water that are widely distributed. The long daily treks are broken up with rest periods, and conclude when the monkeys return to their sleeping sites at the end of the day. Various leaves, grasses, seeds, roots, and tubers are eaten and occasionally supplemented with insects or small mammals.

As already noted, the social organization of hamadryas baboons has multiple levels. A clan is composed of several one-male units, the males of which are likely to be genetically related. Members of clans regularly forage together. At a higher level, a number of one-male groups, clans, and single males form bands that also forage together and are remarkably stable over time. Finally, bands are likely to congregate into still larger troops (or herds) which contain up to 750

Figure 8-11 Hamadryas baboons are found in Ethiopia, Somalia, Saudi Arabia, and Yemen.

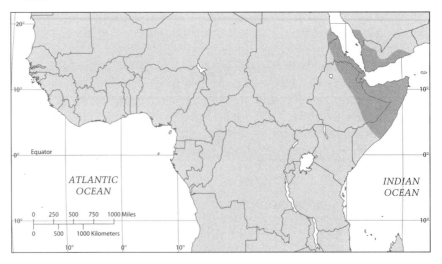

individuals that share sleeping sites. Significantly, unlike other cercopithecines, hamadryas males ultimately remain in their natal bands while females do most of the migrating.

Hamadryas unimale groups are more structured than those of many other primates when it comes to relationships between the sexes. As shown by research that was carried out during the 1960s by Hans Kummer of the University of Zurich, young males tend to baby-sit juveniles and, eventually, may kidnap a young female with which to start a new unimale group. As the male acquires more females in his group, he keeps them nearby with threats and bites to their necks and, at times, by actually herding them. This is probably a response to wariness about possible competition from other males. Interestingly, Kummer recently showed that this biting behavior has a genetic predisposition. Grooming the dominant male's silvery locks appears to be an irresistible urge for which females compete. One central female is likely to be higher ranking than the others, have a closer bond with the male, and may even help herd the other females. For their part, the more peripheral females seem to explore more and consequently locate new sources of food and potential predators.

According to H. Sigg of the University of Zurich and colleagues, male-female pair-bonds usually last several years, and exchanges of females within social units occur predominantly within bands. Females exhibit sexual swellings during estrus and are likely to bear their first young at about the age of 6. For their part, males acquire their first females between the ages of 8.5 and 11 years. As we have already seen for some other primates that live in unimale groups, group leaders have a difficult time monopolizing the mating activity of their females. D.G. Smith of the University of California, Davis, and colleagues note that brief and furtive matings occur between females and nonleader males within their groups or with male leaders of other unimale groups. In fact, recent genetic studies suggest that leaders of one-male groups may sire only about half of the offspring within their own groups.

As Stammbach points out, the social organization of hamadryas baboons must be viewed within an ecological context. Multilevel groups make it possible to tailor the size of the foraging unit to the size of available resources, so that smaller patches of food (e.g., one tree) may be exploited by smaller units (one unimale group), whereas more extended areas with more food can be used by larger clans or even bands. Sleeping cliffs are also likely to have been a selective factor promoting the multilevel social organization. In this particular species, it is a group of bonded males rather than unrelated females that make the decisions about how to divide up and where to travel during the day.

The Gelada (*Theropithecus gelada*)

Geladas are closely related to hamadryas baboons, which they resemble in a number of ways. For example, both species occupy arid

REVIEW EXERCISES

1. What factors may account for the large ranges of patas monkeys?
2. How is the diet of *Erythrocebus* unusual?
3. What animals are potential predators of patas monkeys, and what strategies have the latter developed to avoid them?
4. Describe the multilevel social organization of hamadryas baboons.
5. Which sex among *Papio hamadryas* migrates away at maturity and is this the norm for cercopithecines? What effect does this pattern have on bonding between individuals?
6. How do male hamadryas baboons control the females in their groups?

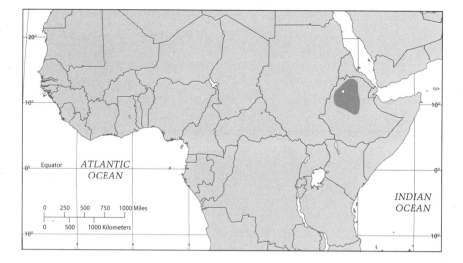

Figure 8-12 Geladas live in arid regions of Ethiopia.

Figure 8-13 Geladas have grooves along the sides of their rounded muzzles and little, turned-up noses.

regions of Ethiopia (Fig. 8-12), have multilevel societies based on unimale groups as the smallest groups, and form large sleeping parties at night in rocky areas or on cliffs. Both are relatively large bodied and highly sexually dimorphic (see Table 8-1), with males that have impressive shoulder mantles and large canines. However, there are also some intriguing differences between the two species. The mantles and whiskers of gelada males are brown rather than silvery, and both sexes have hollow cheeks, grooves along the sides of projecting rounded muzzles (although these are not as dramatic as those described for mandrills), and pert little turned-up noses (Fig. 8-13). Geladas are also said to have the most opposable thumbs of any Old World monkey.

Geladas are sometimes called "bleeding heart monkeys" because of the presence of naked, pink patches of skin on the chests of males (which may be triangular or heart shaped) and females (hourglass shaped). In females, the chest patch is surrounded by small bumps or vesicles that appear somewhat like a pearl necklace. The female's small pendulous nipples meet in the lower midline of her chest patch. During estrus, this patch becomes bright red and the surrounding vesicles swell. Interestingly, this display mimics a similar arrangement of swollen vesicles around the female's genitalia (Fig. 8-14). Kingdon suggests that chest patches of geladas convey important cues that indicate sexual receptivity of females or social submission by either sex (i.e., presenting the rear), signals that are normally communicated by the hindquarters of other monkeys. This redundant signaling system is necessitated by the fact that geladas spend inordinate amounts of time shuffling along on their bottoms (or ischial callosities) as they use their highly opposable thumbs to pluck the grass and small seeds that they love to eat. Geladas also send complex signals with their faces (see Neural Note 8).

Geladas are unique among primates in that grass comprises about 90% of their food. Such a **graminivorous** diet is surprising because

Figure 8-14 Female (top) and male (bottom) geladas have naked patches of pink skin on their chests. During estrus, the appearance of a female's genital region (middle) mimics that of her chest.

grass is extremely difficult to digest and geladas lack the sacculated stomachs and microorganisms for breaking down cellulose that are found in the leaf-eating colobines (see Chapter 7). By analyzing fecal samples from geladas and sympatric baboons that eat a much smaller proportion of grass (*Papio anubis*), Robin Dunbar and Utpaul Bose of University College London discovered that geladas break down their food into finer fragments than baboons do. Rather than spending more time feeding than baboons, however, geladas have relatively small in-

Of Lip-Flips and Endocasts of Fossil *Theropithecus*

In addition to their "bleeding heart" chests, geladas exhibit a fascinating combination of facial expressions that focus on both the eyes and the muzzle. When fearful, nervous, or intimidated, this species retracts its scalp, which reveals startling patches of white above the eyes. Simultaneously, the upper lip is turned inside out, and flipped up over the nose. This "lip-flip," which is unique to geladas, effectively exposes the bare gums and teeth (see figure). As one can imagine, the ability to voluntarily flip the lip in such a manner (i.e., without using fingers) is a muscular feat. In keeping with this, gelada endocasts have a unique sulcal pattern which suggests that the brain's motor representation for face is unusually enlarged in this species. Endocasts have also been studied from fossils of geladas that lived about 2 mya, but with bodies the size of gorillas. Amazingly, the endocasts of these giant monkeys reveal even larger face representations than those seen in endocasts of living *Theropithecus*. One can only wonder what it must have been like to encounter a several-hundred-pound gelada ferociously displaying a lip-flip and flashing its eyelids!

A

B

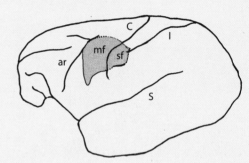

Macaque

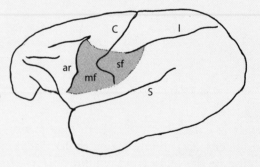

Fossil gelada
(note hook-shaped C
and enlarged mf and sf)

A. Rather than being an expression of aggression as a human observer might take it to be, the lip-flip of this gelada indicates that it is fearful or nervous. **B.** Left hemispheres of endocasts from fossil giant geladas show expansion in the areas that represent the face. **Ar**, arcuate sulcus; **S**, Sylvian; **C**, central; **I**, intraparietal; **mf**, motor face area; **sf**, sensory face area.

cisors and large molars that are adapted for grinding. Even so, geladas do not share the ability of some ungulates to digest dried grass and instead turn to roots and bulbs during the dry season.

Social organization is a little looser in geladas than it is for hamadryas baboons. For one thing, the one-male groups that make up larger bands may change over time, rather than remaining relatively constant. As is usually the case for cercopithecines, males rather than females migrate away from their natal bands as they mature. Contrary to hamadryas society, males do not herd (or bite the neck of) the females in their groups, and the relationships between females are strong and stable. The alpha female helps to maintain group cohesion among geladas, and the females rather than males are likely to initiate group movements during the day. The relationship between migration upon reaching sexual maturity and bonding patterns may be key to understanding the differences between these two multilevel societies. That is, the sex that remains in the natal band forms strong bonds. Thus, male-male bonding is strong in hamadryas baboons, whereas strong female-female bonding predominates among geladas.

The Mandrill (*Mandrillus sphinx*)

Not nearly as much is known about the drills and mandrills as for their cousins, the geladas and hamadryas baboons. This is no doubt because *Mandrillus* lives in extremely dense rain forests in West Africa (Fig. 8-15), that is, areas in which it is very difficult to conduct field research. Both species of *Mandrillus* are highly dimorphic and relatively large, with the mandrills being the biggest bodied (adult males average about 27 kg or 59 lb, and adult females average 11.5 kg or around 25 lb, see Table 8-1). While drills are rather dull looking, male mandrills are the most brightly colored monkeys in the world (see Figs. 2-11 and 8-16, color plates). In fact, they almost look like a

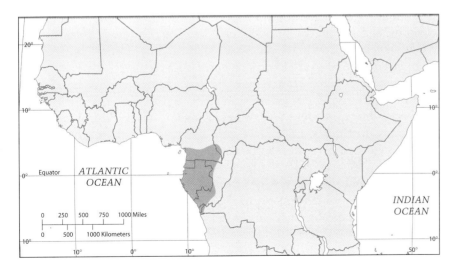

Figure 8-15   Mandrills live in rain forests in West Africa.

figment of someone's imagination, with a bright-red stripe of skin running down the middle of the face (from between the eyes to the nose), bordered on each side by a number of swollen vivid blue grooves. A bright yellowish to orange beard and whiskers complete the dramatic appearance, while females have less colorful beards. Further, just as the gelada's chest patch mimics the genital region, so too does the male mandrill's colorful face (see Fig. 8-16, color plate). While drill faces are boring by comparison, the male drill has genitals that are even brighter than the mandrill's. Females of both species have nasal ridges, but the color of their faces and perineal regions is generally drab. The tail is stumpy in both drills and mandrills, and it is thought that, like beacons in the night, the colorful rears of the males make it easier for others to follow their leads as groups proceed through the dense forests.

Mandrills feed throughout the day on fruit and tough-to-open seeds, which together comprise over 90% of their diet. They also eat leaves, pith, ants, termites, and other foods (see Table 8-1), according to Jiro Hoshino of Kyoto University. When food shortages occur during the dry season, mandrills form large hordes (up to more than 600 individuals) that travel through their extensive home ranges in search of fruit. As M. E. Rogers of the University of Edinburgh and her colleagues (1996:309) who studied mandrills in the Lopé Reserve in Gabon put it, "Our data suggest that the seasonal pattern of appearance of mandrill hordes in the forest edge at Lopé is a semi-migratory response to decreasing fruit availability in continuous forest, associated with maintenance of an eclectic diet involving extensive searching for scattered, high quality resources."

Although drills and mandrills are terrestrial, females and young feed in the trees. At night, groups sleep in trees, with dominant males on the lowermost branches. In this sense, *Mandrillus* resembles vervets, which are both semiterrestrial and semiarboreal. And like vervets, these monkeys have a complex system of communications that functions in both arboreal and terrestrial domains. The vivid coloration just described for the males is, as already noted, a form of visual shouting that is common in forest primates such as some of the guenons. Interestingly, *Mandrillus* is one of the few species of Old World monkeys in which males use chest glands to mark surfaces in their environments, as described earlier for semiterrestrial vervets. Other forms of self-advertisement include snarls, grunts, loud calls (including a rallying cry given by males), and yawning as a means of displaying large, impressive canines. Kingdon observes that such signals focus on the mouth and muzzle in drills and mandrills, but on eyes and colored eyelids in some of the other cercopithecines, such as baboons (see Chapter 9) or mangabeys.

When mandrills are not grouped in large semimigratory hordes, they live in one-male units or, in the case of males, may be peripheral or solitary. According to J. van Hooff of Utrecht University, Netherlands, an average one-male unit usually contains from 5 to 10 adult females and 10 juveniles. As discussed in Chapter 2, the degree of sexual dimorphism is largest in primates in which males defend a number of females. Dominant male mandrills vigorously defend their groups from other males and, in keeping with this, are up to three times the size of adult females. Mandrills are also extremely dimorphic in other features such as the coloration of the face and rump. In a fascinating study of semi-free-ranging mandrills conducted by E. J. Wickings and A. F. Dixson of the International Center of Medical Research in Franceville, Gabon, socially dominant adult males were found to have larger rumps with more vividly colored sex skins than were less dominant males; that is, they appeared "fatted" compared to peripheral or solitary males (see Fig. 8-16, color plate). The fatted males were also generally more massive than nonsocial males because of more fully developed nuchal crests, beards, and shoulder manes. Enhancement of these secondary sexual characteristics in social animals may be linked to size of the testes and plasma testosterone levels, since these were greater in social than in nonsocial males. Wickings and Dixson also report that dominance rank is related to sexual activity, since 70% of ejaculations observed during the 1991 mating season occurred in the highest-ranking group male. The authors conclude that social factors and reproductive development are interrelated in the male mandrill.

Because mandrills have been studied so little in the wild, much remains to be learned about their social organization. However, the fact that solitary males have been observed suggests to Stammbach that *Mandrillus* is more like *Theropithecus* than *Hamadryas* in its social organization. Hopefully, future studies will provide more information about the behavior and socioecology of this interesting species. Unfortunately, mandrills are endangered because of deforestation by humans. They are also hunted for their flesh in some parts of Africa.

A SPECIES THAT SPENDS TIME ON THE GROUND

The Sooty Mangabey (*Cercocebus torquatus* atys)

Cercocebus, like *Mandrillus*, is adapted to forests. Two of the four mangabey species spend a fair amount of time on the ground (*Cercocebus torquatus*, *C. galeritus*), while the other two are more strictly arboreal (*C. albigena*, *C. aterrimus*, both of which are placed in a separate genus, *Lophocebus*, by some workers). Mangabeys are large, slender monkeys with hollow cheeks, roomy cheek pouches, and long tails. Rather than having prominent muzzles like the drills and mandrills, these monkeys have decorative faces with startling white marks over the eyes (like poorly applied eye shadow) and flamboyant tufts of hair on the cheeks (Fig. 8-17). As with many forest monkeys, mangabeys produce a variety of loud calls that help to maintain spac-

Figure 8-17 The sooty mangabey, though rather drab looking compared to other mangabeys, has startling white marks over its eyes and tufts of hair on the cheeks.

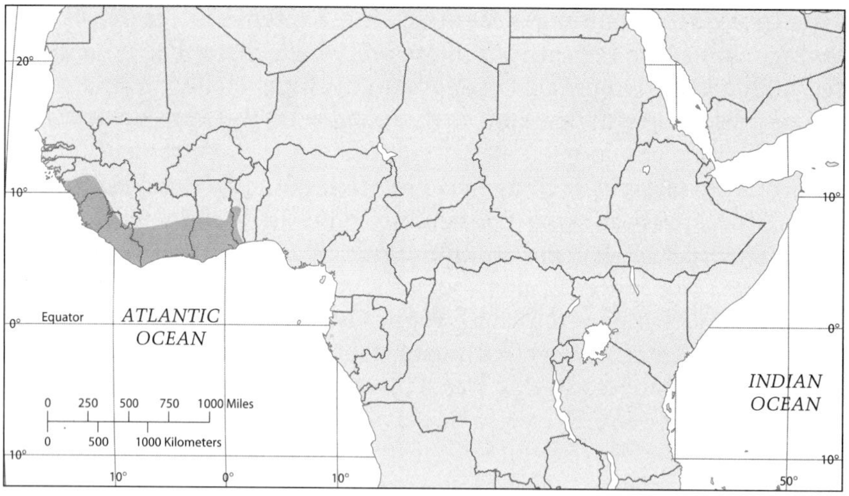

Figure 8-18 The sooty mangabey lives in a variety of forests in West Africa.

ing between groups. One of these, the "whoop-gobble," is given by mature males and can be heard over long distances. Most mangabeys live in multimale, multifemale groups, and are thought to be evolutionarily close to the subjects of the next chapter, the macaques and baboons.

Because of their uniformly brownish-gray coats, sooty mangabeys (*Cercocebus torquatus atys*) are drabber looking than many of the other mangabeys. The sooty mangabey lives in a variety of forests from Sierra Leone to Ghana in West Africa (Fig. 8-18), where it spends most of the day moving and feeding on the ground. Like mandrills, sooty mangabeys have strong jaws and teeth that are used to crack hard-to-open nuts and seeds. Although little is known about sooty mangabeys in the wild, captive studies carried out on descendants of a social group that was established in 1968 at Yerkes Regional Primate Research Center Field Station in Lawrenceville, Georgia, have provided some details about their behavior. However, there continue to be important gaps in our understanding of this primate (see Table 8-1).

An advantage of captive studies is that observational data, tissue samples (permitting genetic studies), and vital statistics may be collected from subjects more easily than can be done in the field. At Yerkes, over 80 sooty mangabeys are housed in a large outdoor compound with attached indoor quarters, and observational studies by Deborah Gust and Thomas Gordon of Emory University and others have been carried out from a tower that provides an unrestricted view of the compound. The results are fascinating. For example, unlike female macaques and baboons (see Chapter 9), social ranks of adult female sooty mangabeys (which live in multimale, multifemale groups) are not based on kin. This is true despite the fact that males and females rank just below their mothers until they are about 3 years old. After that, however, they usually rank over their mothers, and by the time they reach 5 to 6 years old, all males rank over all females. Gust speculates that such an individualistic dominance hierarchy might reflect ecological pressures that remain to be discovered for this species in the wild. (We will have more to say about this general theme in Chapter 9.)

Captive studies also permit more detailed studies of the reproductive lives of primates than do studies in the field. Based on observations of 671 mounts, Gust and Gordon describe the basic pattern of copulation in sooty mangabeys as consisting of

> a male mounting a female by grasping the female's ankles and supporting his upper body with his hands placed on the female's hips. After the male achieves intromission and begins thrusting, the female often turns her head, looking toward the male and occasionally will reach back with one hand to touch his leg. Subsequently the female vocalizes and darts away from the male, marking the end of the copulation. (Gust and Gordon, 1991:279)

At Yerkes, sooty mangabey males have turned out to be surprisingly precocious in their sexual development. According to Gust and Gordon, they exhibit erections within weeks after birth and are mounting females by as early as 9 months of age. Males as young as 4 years old have been observed to ejaculate. Although sexually immature males mount females almost three times as often as sexually mature males, the latter focus more on females that exhibit maximum sexual swellings (i.e., are ovulating). Despite their tolerance of copulations by immature males, dominant males attempt to socially inhibit (through aggression) copulations by other adult males. DNA-based paternity analyses for 78 sooty mangabeys were carried out by Gust and her colleagues. These studies revealed that, while higher-ranking males sire more offspring than do adult males of lower ranks, the same male is usually not the father of a given female's offspring from year to year. The paternity studies combined with behavioral observations also showed that adult males are not particularly attracted to their own offspring and vice versa. As can be seen, much has been learned about dominance rankings and reproductive behavior of sooty mangabeys in captivity. What is needed now are field studies that will elucidate their behaviors in the wild, as well as the socioecological contexts within which they occur.

SUMMARY

As discussed in this chapter, terrestriality occurs to a greater or lesser degree in a number of cercopithecine species. Ground living is associated not only with a number of anatomical adaptations, but also with a complex of communication systems based on olfaction, sexual swellings in females, other visual signals (coat colors, markings), and

REVIEW EXERCISES

1. In what ways are geladas similar to hamadryas baboons? How are they different?
2. Describe the gelada chest patch and discuss its likely function.
3. In what ways are gelada facial expressions unusual, and what part of the brain seems to be involved?
4. What appears to be the relationship between emigration patterns and intrasexual bonding among multilevel societies?
5. Which is the most colorful species of any primate? Describe and discuss the possible function of its appearance within an ecological context.
6. How does an adult male mandrill that is fatted differ from one that is not?
7. What have captive studies revealed about the development of social ranks in adult sooty mangabeys? About their reproductive lives?

vocalizations. The details of these systems vary, however, between semiterrestrial species that live in forested habitats and those that live in more open habitats. The approximately 50 species of cercopithecines include guenons, geladas, mandrills, drills, mangabeys, macaques, and baboons. These species are more numerous and varied in appearance than those of the colobines, and are distinguished from the latter by cheek pouches used to stash food. Unlike most colobines, a number of cercopithecines have moved away from forests to more open areas, and have experienced certain modifications in body proportions, posture, dexterity, and coloration that are related to ground living. Semiterrestrial monkeys also exhibit a greater degree of sexual dimorphism than arboreal species, partly in response to increased pressure from predators in more open areas. Not all cercopithecines have given up life in the trees, however. For example, the most common and widely spread monkeys in Africa, the guenons, are mostly arboreal. Colorful guenon faces and hindquarters stand out in the dark forests they inhabit, and may aid in species identification (by both primatologists and monkeys) as well as the prevention of breeding between different groups. Forest guenons tend to live in matrilineal one-male groups, with one adult male that attempts to monopolize access to a number of adult females. Only a few species of guenons are semiterrestrial and, of these, the vervets (*Cercopithecus aethiops*) are the most successful and the best known. Vervets are equally at home on the ground and in trees; live in multimale, multifemale troops that vary in size with the availability of resources; and are famous for their alarm calls. The most terrestrial of the guenons is the cautious patas monkey (*E. patas*), which inhabits the edges of acacia woodlands and appears to be continually on the lookout for various predators. Hamadryas baboons, geladas, drills, and mandrills all live in multileveled societies based on one-male units that merge and split apart on a regular basis. Captive studies of sooty mangabeys (*Cercocebus torquatus atys*) reveal interesting and complex behaviors related to dominance hierarchies and reproduction within their multimale, multifemale groups. A main theme of this chapter is that cercopithecine communication systems involving olfaction, visual signals, and vocalizations are often complex and tailored to particular arboreal or terrestrial habitats. We have seen hints that life on the ground may be associated with increased social complexity and, indeed, this will be even more apparent in Chapter 9.

FURTHER READING

CHENEY, D. L. AND R. M. SEYFARTH (1990) *How Monkeys See the World.* Chicago: University of Chicago Press.

GAUTIER-HION, A., BOURLIERE, F., GAUTIER, J.-P. AND J. KINGDON (1988) *A Primate Radiation: Evolutionary Biology of the African Guenons.* Cambridge: Cambridge University Press.

KUMMER, H. (1995) *In Quest of the Sacred Baboon: A Scientist's Journey*. Princeton: Princeton University Press.

ROWELL, T. E. (1984) Guenons, macaques and baboons. In: D. Macdonald (ed) *All the World's Animals: Primates*. New York: Torstar, pages 74–85.

9

MACAQUES AND SAVANNA BABOONS: SEXUAL POLITICS AND HUMAN EVOLUTION

Catarrhini
|
Cercopithecoidea
|
Cercopithecidae

Cercopithecinae

Allenopithecus (swamp monkey)
Cercocebus (mangabeys)
Cercopithecus (guenons, vervet monkey)
Erythrocebus (patas monkey)
Macaca (macaques)
Mandrillus (drill, mandrill)
Miopithecus (talapoin monkey)
Papio (baboons)
Theropithecus (gelada)

INTRODUCTION

Macaques (*Macaca*) and baboons (*Papio, Mandrillus,* and *Theropithecus*) are, respectively, Asian and African variants of closely related and highly successful species. These monkeys have 42 chromosomes, a number that was presumably inherited from a common ancestor that lived during the Miocene epoch. Both groups contain numerous species, including some that have made rather complete shifts to terrestrialism in ecological settings that resemble those in which early hominids are thought to have perfected bipedalism (see Chapter 14). In fact, both macaques and savanna baboons (*Papio*) have been used to speculate about the evolution of culture, complex social life, intelligence, and even the relationship between the sexes in higher primates including humans. Because macaques and savanna baboons continue to be of intense interest to primatologists and evolutionary biologists, this chapter is devoted to them. Even so, there is room to include only a small portion of what is known about these interesting cercopithecoids.

MACAQUES

Next to humans, *Macaca* is the most widely distributed primate on the planet. It is not surprising, then, that 19 species are currently recognized. These species are grouped into four groups (*fascicularis, sinica, silenus,* and *sylvanus*) based on anatomical, dental, and molecular data (Table 9-1). According to Dario Maestripieri of Emory University, cladistic analyses indicate that *M. sylvanus* represents the most ancestral lineage, while the *fascicularis* group is thought to have originated most recently. The two most widely studied species of macaque (rhesus macaques and Japanese macaques) happen to be from the *fascicularis* group. Consequently, some of their behaviors that traditionally have been attributed to all macaques (detailed below) are now known to differ from those of macaques in other groups.

TABLE 9-1 *Macaca Groups.*

Group	Species	Common Name
sylvanus	M. sylvanus	Barbary macaque
silenus	M. silenus	Lion-tailed macaque
	M. nemestrina	Pig-tailed macaque
	M. nigra	Sulawesi crested black macaque
	M. brunnescens	
	M. hecki	
	M. maura	Celebes moor macaque
	M. nigrescens	Sulawesi forest macaque
	M. ochreata	Booted macaque
	M. tonkeana	Tonkean macaque
sinica	M. sinica	Toque macaque
	M. arctoides	Stump-tailed macaque
	M. assamensis	Assamese macaque
	M. radiata	Bonnet macaque
	M. thibetana	Tibetan macaque
fascicularis	M. fascicularis	Crab-eating or long-tailed macaque
	M. cyclopis	Formosan rock macaque
	M. fuscata	Japanese macaque
	M. mulatta	Rhesus macaque

Figure 9-1 The rhesus monkey is the best known of the macaques.

Macaques are frequently described as largish, terrestrial monkeys that are generally dull gray or brown. Although the best known of the macaques, the rhesus monkey (Fig. 9-1), is indeed dull colored and relatively terrestrial (if not all that large), the above generalization ignores the marked variations that occur in the appearances of a number of the other macaques. For example, body weights vary a good deal, depending on the species. Rhesus monkeys and crab-eating macaques are among the lightest of species, while the Japanese macaque and pig-tailed macaque represent the heaviest (Table 9-2). Relative lengths of arms, legs, and tails also vary across macaques. In fact, tails range from very long (e.g., in the crab-eating macaque) to virtually nonexistent in two species (the Barbary macaque and the Sulawesi crested black macaque). The latter condition is very unusual for a monkey, the vast majority of which have tails. Indeed, lack of a tail is part of the ape pattern discussed in the next chapter, which is why the Barbary and Sulawesi crested black macaques have been incorrectly nicknamed apes at times. (These species are referred to as macaques in this text.) Interestingly, according to primatologist Joe Erwin of Bioqual Incorporated, the young of one group of crested black macaques resemble gorillas, an observation that may help perpetuate the ape misnomer.

Macaques have bare faces and exposed ears. Despite their reputation for dullness, some species such as the crested black macaque and the relatively arboreal lion-tailed macaque have hairdos that can only

	M. sylvanus (Barbary macaque)	**M. fuscata** (Japanese macaque)	**M. mulatta** (rhesus macaque)	**P. anubis** (olive baboon)	**P. ursinus** (chacma baboon)
Adult mean body weight					
F	10.6 kg (23.4 lb)	13.2 kg (29.1 lb)	7.7 kg (17.0 lb)	14.7 kg (32.4 lb)	16.8 kg (37.0 lb)
M	16.2 kg (35.7 lb)	14.5 kg (32.0 lb)	8.3 kg (18.3 lb)	28.4 kg (62.6 lb)	20.4 kg (45.0 lb)
Adult brain weight (g)	93	109	95	175	214
Maximum life span (yr)	22	33	29	45	45
Habitat[a]	NA, G: cef, of	J: f	As: sem, f, cef, of, w, sw	A: sem, ts, s, w, gf, rf	A: w, g, as, sem
Locomotion[b]	qr	qr	qr	qr	qr
Diet[c]	l, h, s, ac, ba, ce, m, b, a, i	f, s, ba, fu, a	f, s, l, g, bu, gr, r, ba, a	f, s, t, r, l, fl, a	f, s, l, fl, a
Seasonal breeders	+	+	+	0	0
Mean gestation (days)	165	173	164	180	187
Interbirth interval (mo)	16	14	18	26	21
Social systems[d]	M-MM	M-MM	M-MM	M-MM	M-MM, U
Youngest ascendancy	0	+	+		
Egalitarian or despotic	E	D	D		
Ratio of adult females to males	1–2:1	1–4:1	2–6:1	1–3:1	2–4:1
Alloparenting	+	+	+	+	+
Infanticide[e]	0	0	0	+	+
Group size	12–88	< 30–258	9–138	17–140	20–128
Density	1–70/km^2		7–120/km^2	4–37/km^2	2–4/km^2
Home range	127–901 ha	101–530 ha	0.01–2200 ha	75–4014 ha	210–3367 ha
Conservation status[f]	V	H	L	L	L

TABLE 9-2 Statistics for Representative Macaques and Savanna Baboons

Note: Information in this table has been selected from numerous sources and should be viewed as tentative, i.e., subject to change in light of new information. Sources include but are not confined to Menard and Vallet, 1996; Mehlman, 1989 (*M. sylvanus*); Harvey et al., 1987 (brain weights); Melnick and Pearl, 1987; Rowe, 1996 (conservation status and other statistics). Abbreviations: +, yes; 0, no; var, variable. Blank entries indicate that no information is presently available. Dietary items are listed in approximate order of decreasing importance to the species.

[a]Habitat: A, Africa; As, Asia; G. Gibraltar; J, Japan; NA, northern Africa; as, acacia scrub; cef, cedar forest; f, forest; g, grassland; gf, gallery forest; of, oak forest; rf, rain forest; s, savanna; sem, semidesert; sw, swamps; ts, thorn scrub; w, woodland.

[b]Locomotion: qr, quadrupedalism.

[c]Diet: a, animal prey; ac, acorns; b, bulbs; ba, bark; bu, buds; ce, cedar tree cones and needles; f, fruit; fl, flowers and buds; fu, fungi; g, gums (sap); gr, grasses; h, herbs; i, insects; l, leaves; m, mushrooms; r, roots; s, seeds; t, tubers.

[d]Social systems: M-MM, matrilineal multimale, multifemale; U, unimale group.

[e]Infanticide: +, one or more reports.

[f]Conservation status: H, highly endangered; L, lower risk; V, vulnerable.

be described as exciting (Fig. 9-2). It should also be noted that crested black macaques have facial features that are reminiscent of baboons, prompting some workers to view them as representing a transitional species along the macaque-baboon continuum. In terms of general ap-

Figure 9-2 Some macaques such as the lion-tailed macque and the Sulawesi crested black macaque have spectacular hairdos.

pearance, macaques have fully opposable thumbs, and the cheek pouches and ischial callosities that one expects in cercopithecines. This genus varies from moderately to highly sexually dimorphic in body size (see Table 9-2).

Although macaques were once native to Europe, this is no longer the case. (Indeed, you will recall that all Old World monkeys live in Africa and Asia, but not in Europe.) However, relics of earlier European macaques survive today as Barbary macaques (see Table 9-2) in parts of northern Africa and on the Rock of Gibraltar (a British fort at the entrance to the Mediterranean). The remaining macaques are Asian, ranging across Afghanistan, India, China, northern Japan, and Southeast Asia including its islands (Fig. 9-3). Some species, such as the rhesus monkey, have been successfully transplanted by humans to several Caribbean islands and to primate research centers in the United States. As expected in a widely distributed primate, macaques live in a variety of niches ranging from tropical rain forests to woodlands, swamps, mountains, grasslands, cliffs, and human-made structures such as temples and villages. Macaques are quadrupeds, and many species are equally at home on the ground and in the trees. A few species such as *M. mulatta* are highly terrestrial, however. As one would expect, diets are eclectic and vary with habitat. Depending on location and season, macaques may eat fruit, leaves, shoots, grass, tubers, herbs, seeds, berries, flowers, buds, gum, algae, insects, small vertebrates, cultivated crops, or (when times are hard) bark. Although a great variety of food may be consumed during the year, menus are more limited over shorter periods of time, and some workers regard *Macaca* as basically frugivorous.

The following description of social organization that theoretically typifies *Macaca* has emerged from numerous studies of rhesus and

Japanese macaques: Macaques live in multimale, multifemale troops that are centered around a stable core of females who remain in their natal groups for life. Group size, range size, and population density vary with habitat. For example, groups frequently average from 20 to 30 members, but can number in the hundreds. Males emigrate away from their birth troops, and may change troops several times during their lives. Females, frequently relatives, groom each other and have close social bonds, while relationships between males appear to be more aggressive. Males do less grooming than females. The sex ratio within troops is balanced in favor of females (see Table 9-1). Both sexes have strong dominance hierarchies, with mothers conferring their ranks on their daughters. Rhesus monkeys and Japanese macaques are usually characterized by "youngest ascendancy," in which the youngest daughter at least temporarily outranks her sisters and is second in rank to her mother (see In the Field). Adult males, on the other hand, are usually dominant to adult females, although there are exceptions. Troops travel and feed together during the day. When troops grow to be extremely large, they may permanently divide or fission (often along matrilines) into two groups.

There is a problem with the practice of generalizing this description of social organization to other macaques, however—much of it simply does not fit that of less well known species that are not in the *fascicularis* group. For example, female as well as male bonnet macaques and possibly stump-tailed macaques migrate from their natal groups. Furthermore, dominance hierarchies and kinship rela-

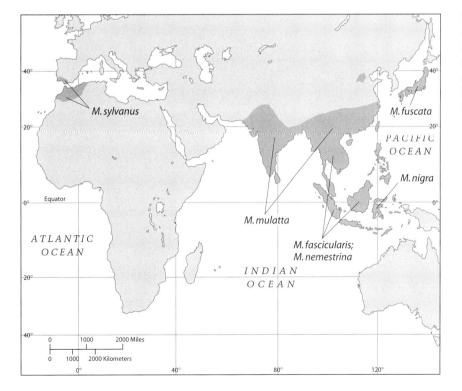

Figure 9-3 This map shows the distribution of 6 of the 19 species of living macaques. The only macaque that lives in Africa rather than Asia (*M. sylvanus*) is thought to be descended from the lineage that was ancestral to all living macaques.

The Missing Dyad

Donald Stone Sade of the North Country Institute for Natural Philosophy in Mexico, New York, helped establish the classic foundation for much of today's research on social interactions between nonhuman primates. When asked to recall what it was like to discover some of the basic dynamics of dominance hierarchies in monkeys, and to recount a particularly meaningful moment in the field, Sade realized that the satisfactions of fieldwork became more abstract as his experiences accumulated. In his own words:

THIS PHOTOGRAPH OF DONALD SADE WAS TAKEN IN 1970 AS HE WAS RETURNING FROM OBSERVING RHESUS MONKEYS ON THE ISLAND OF CAYO SANTIAGO. (PHOTO COURTESY OF D. SADE.)

Despite cabin fever, personality conflicts, administrative meddling, funding lapses, and academic politics, the work I began years ago on a colony of rhesus monkeys that lived on the small island of Cayo Santiago near Puerto Rico has provided many satisfactions. My first impressions were of a novel environment: feeling the wave of heat and humidity when first stepping off the plane, hearing the Coqui singing in the tropical downpour, seeing the flowering Flamboyan along the narrow road, learning the Andalusian dialect of Spanish spoken in the rural countryside, and even (perversely and in retrospect) recovering from new diseases ("la monga," dengue, malaria, bite of the back-fanged Dromicus snake, and lesions from the sap of the manchinella tree).

I initially focused my observations on a single small social group (later called group F). At first the monkeys all looked discouragingly alike, but after a certain time, as if through a transformation, I suddenly recognized individuals at a glance, and they and their personalities became regular visitors in my dreams. (In later years I had the pleasure of observing this same transformation in each of the many students that took part in the longitudinal study that grew out of my early work.) At first, the process of absorbing, systematically describing, and making sense of the monkeys' interactions required much effort. Within a few months of my arrival, however, the more obvious features of the communicative aspects of rhesus behavior were sufficiently recognizable so that patterns of social interaction could be systematically observed and recorded. This was especially true for the affiliative relations expressed by grooming and the agonistic relations expressed in dominance-subordination interactions (fights). Personal computers hadn't yet been invented so I naively tabulated my observations each day on matrix-ruled paper using colored pencils, a procedure I still advocate at the start of a study.

By the end of 1962 these matrices clearly showed the affiliative clustering of families ("genealogies") and the ordering of the families in a dominance hierarchy. Although the discovery (some would say imposition) of order satisfies the scientific aesthetic like nothing else, my records were still incomplete. In 1963 something called "academic requirements" (an impediment to fieldwork) left only a few months for observations. I was determined to find the dominance-subordination relations between every member of each age class to see if the rankings of the juveniles really corresponded precisely to those of their mothers. As my field time neared an end, most of the matrices had been completed, but I still lacked observations on one critical pair (or dyad) among the 2-year-old females because I had never seen them together. July 31, 1963, was my last day and I had to leave at noon. At about 11:30 I was walking down the hill toward the boat when I finally saw the two monkeys some distance apart. I had before successfully started fights among monkeys by presenting a small piece of platano and observing which monkey displayed threat and which displayed submission, irrespective of which got the treat. (I later came to abhor this technique because of its intrusiveness, but at the time it was effective.) At 11:48 I placed a small piece near one of the members of the missing dyad, and a series of fights broke out among a number of juveniles. Among them I observed three perfect agonistic interactions within the missing dyad, each in the direction predicted by their mothers' ranks. Because of this, I was able to complete the matrix at the last possible moment (the subordinate got the platano). It may be the remembered pleasure of this last-minute triumph that has kept my research focused on dominance hierarchies over many years.

The research up to that point was based on comparing a series of age classes (e.g., different groups of infants, juveniles, and adults) in order to model the general life history of rhesus monkeys. Although in subsequent years I was obligated to spend much of my time on academic duties, with the cooperation of a series of dedicated students my earlier studies were expanded. By 1972 we had collected individual life histories (i.e., data collected from individuals at different times as they grew up) and were able to document the very regular rise in rank of young females over their older sisters, to take their places just below their mothers in the adult hierarchy. These data represented the dynamics underlying the observed order. By 1976 other social groups had been included in the study, and we showed across the colony that family rank and reproductive success were correlated, thus placing dominance order within an evolutionary context.

We still lacked comparisons across different species of primates, however. That came years later when I had the opportunity to study emperor tamarins with Dr. Kerry Knox now of Rochester University. Despite the fact that social organization in emperor tamarins differs greatly from that of rhesus monkeys, we found that the important agonistic interactions were largely between siblings, suggesting a psychological commonality with the life histories of juvenile rhesus monkeys. Most recently I find myself speculating on the cognitive aspects of agonism, and the evolution of the ego—an abstraction far removed from the specifics of my earlier observations. I do wonder, though, if that last minute in 1963, when those two small females of the missing dyad squabbled over an insignificant piece of raw platano, was not pivotal in shaping the direction of my research.

THE 1990s FIND DONALD SADE IMMERSED IN RESEARCH ON NEW WORLD EMPEROR TAMARINS, ONE OF WHICH SITS ON HIS SHOULDER. (PHOTO COURTESY OF D. SADE.)

9 MACAQUES AND SAVANNA
BABOONS: SEXUAL POLITICS
AND HUMAN EVOLUTION

REVIEW EXERCISES

1. Next to *Homo*, which genus of primate is the most widely spread on the planet?
2. How many species of macaques are recognized?
3. Which of the four groups of macaques represents the most ancestral lineage? The most recent lineage?
4. Why are the "Barbary ape" and "Sulawesi black ape" misnamed? What part of their anatomy is responsible for their incorrect common names? What is their genus?
5. Where in the world are macaques found?
6. Describe the social organization that is theoretically typical for *Macaca*. What is the problem with this generalized view of macaques?
7. Describe youngest ascendancy. In which species does it routinely occur?
8. Name, compare, and contrast despotic and egalitarian species of macaques. What ecological factors may account for these different lifestyles?

tionships do not seem to be as strong in these two species or in Tonkean and Barbary macaques as they are in rhesus and Japanese macaques. Nor does youngest ascendancy routinely occur in bonnet or Barbary macaques, although, as expected, maternal rank is important for determining that of daughters. What accounts for this variety of social behaviors in macaques? Bernard Thierry of the University of Louis Pasteur in France offers an intriguing answer based on a model that views macaque social systems as "despotic" at one end of a continuum to "egalitarian" at the other. Thus, rhesus and Japanese macaques are despotic species characterized by extremely strong kinship relations; strong dominance hierarchies backed by intense, kin-supported aggression; a high degree of intolerance between unrelated individuals; and unidirectional aggression in which attacked animals flee rather than retaliate. Tonkean, stump-tailed, and, to some degree, Barbary macaques tend to be more egalitarian, on the other hand. According to Thierry, egalitarianism is characterized by weak differences in dominance ranks between individuals that are generally more tolerant of each other, and kinship relations that do not interfere with friendships between unrelated animals.

But why should some macaques be despotic and others egalitarian? Are there socioecological correlates that might help explain the adaptive significance of these systems? Although the jury is still out on this question, James Moore of the University of California, San Diego, offers interesting insights into this problem. Remembering that females go where there are resources and males go where there are females (see Chapter 2), Moore begins with female feeding strategies as an ecological starting point. Noting that kin selection may have been a primary force in the evolution of social behavior, he then asks why troop-living primates in which only males emigrate are more strongly matrilineal than species in which females also emigrate. He suggests that the former species tend to compete for patchy resources (e.g., fruit), and that such intergroup competition favors large, aggressive, kin-based troops. This fits well with the despotic rhesus and Japanese macaques. Apparently primate species in which adult females emigrate tend to be folivores which subsist on more evenly distributed resources that are not defended in a similar manner. Moore also notes that avoidance of predators and infanticidal males are likely to be factors that motivate female migration when it occurs.

In general, female macaques become sexually mature at between 3.5 and 6.5 years of age, whereas males are a little older when they mature, that is, between 4.5 to 7.0 years. Depending on the species, females exhibit more or less dramatic sexual swellings at estrus. Mating is promiscuous, although macaques form consort pairs, or **consortships**. In some species (e.g., *M. fascicularis*), dominant males copulate the most; in others (e.g., *M. fuscata*), they do not. Reports differ, however, on the extent to which a mother's rank correlates with her reproductive success. Single births tend to occur seasonally, and are probably governed by environmental factors (climate, rainfall, etc.) that are related to food supply. Male investment in offspring varies a good deal, reaching an extreme in *M. sylvanus* where male attention to nonrelated infants is surprisingly intense (see Box 9).

Animals such as dogs or raptorial birds may prey on macaques, but their worst enemies are humans who hunt them for food, fur, pets, supposed aphrodisiac properties of their bones, and export to scientific supply companies. Macaques continue to be used widely in biomedical research, for which they are sometimes bred. Because they raid crops, these monkeys are regarded as pests and killed in many areas. Macaques were once cherished as sacred in parts of Asia, but their religious significance is now in decline. This, combined with ongoing habitat destruction due to human settlements and commerce, has caused some species (e.g., the lion-tailed macaque) to become highly endangered. Nevertheless, as a genus, *Macaca* is highly successful, due at least in part to its high intelligence and complex social behavior (see Neural Note 9). These attributes have been especially well documented by Japanese primatologists who have studied *M. fuscata* since the early 1950s. We turn now to a portrait of this fascinating species.

Japanese Macaques (*Macaca fuscata*)

For half a century, monkeys have been revered and legally protected in Japan. Furthermore, researchers from Japan developed the now standard techniques of provisioning, habituating, and learning the individual identities of free-ranging primates (see Chapter 2). Consequently, a wealth of information is available about the lifestyle and inventiveness of Japanese macaques, some of which is truly amazing. Among the largest of the stump-tailed macaques, females and males average about 13 and 15 kg (29 and 32 lb), respectively (see Table 9-2). They have bare rumps and faces that tend to be red in adults, especially during mating season. These monkeys are especially attractive in winter when their fur, which varies from a grayish brown to almost white, is particularly thick and fluffy (Fig. 9-4).

Figure 9-4 Japanese macaques are among the largest of the macaques.

Box 9

An Odd Little Monkey . . .

The Barbary macaque (*Macaca sylvanus*) is an unusual monkey in several respects. Of the 19 species of macaques that are currently recognized, it is the only one that lives in Africa (i.e., in Morocco and Algeria, see Fig. 9-3) rather than Asia. This species also lives at the Rock of Gibraltar at the southern tip of Spain (see figure below), where it was introduced at least 260 years ago, probably by the British garrison, according to John Fa of the Jersey Wildlife Preservation Trust in the Channel Islands. Because legend has it that, as long as the monkeys remain at Gibraltar, so too will the British, the British Army provisioned and generally cared for the monkeys until 1992, when the task passed to the Gibraltar government. Today, as the only monkeys of any kind that live in Europe, Barbary macaques continue to be a huge tourist attraction.

Although *M. sylvanus* is generally regarded as the most primitive of the macaques, it lacks a tail, which is a derived feature that also appears in apes and humans. This may be one reason why the Barbary macaque was used as a substitute for human cadavers in Europe during the many centuries that it was illegal to use human bodies for medical studies. However, other than its lack of a tail, the plain hairdo and drab coloration of *M. sylvanus* seem unremarkable compared to some of the other macaques (see figure on right). Some of its behaviors, on the other hand, are quite unusual.

As is typical for their genus, Barbary macaques live in multimale, multifemale groups that are centered around matrilineal hierarchies (see Table 9-2). However, the pattern whereby females attain rank is surprisingly atypical. As noted earlier, the expected pattern for macaques is that younger sisters become

Barbary macaques live in northern Africa and on the Rock of Gibraltar, shown here, which makes them the only monkey that currently lives in Europe. (Photograph by John Guyer.)

Japanese macaques range from subtropical lowlands in the south to the mountains of Honshu Island in northern Japan (see Fig. 9-3). The macaques from Honshu Island experience extremely harsh winters and, for this reason, are known as snow monkeys. Other populations of *M. fuscata* are just as at home in human villages and temples as they are in forests, and are both terrestrial and arboreal. Japanese macaques are extremely adaptable, so much so that a large troop of 150 monkeys was successfully transplanted to a ranch in hot, dry southern Texas in the early 1970s, where they continue to flourish. The diets of *M. fuscata* are eclectic, as described for macaques in general. Snow monkeys, however, eat a good deal of bark during the winter when other foods are scarce.

Captive Japanese macaques are provisioned with food, which

This Barbary macaque lives on the Rock of Gibraltar. (Photograph by John Guyer.)

their natal groups when they are older than 5 years, which is unusually late for their genus. Compared to other species of macaques, a greater number of adult males reside in groups (reflected in relatively balanced sex ratios, see Table 9-2), and the relationships between them are surprisingly harmonious. One reason for this harmony seems to be that males frequently huddle together to pick up, examine, groom, pass around, and play with unweaned youngsters. This extent of male baby care is extraordinary for an Old World monkey. However, some of the males' behaviors toward infants, for example, the proclivity for picking up infants and presenting them to aggressors as appeasement gestures (or buffers), and for protecting infants from predators, are shared by other cercopithecoids. According to Maestripieri, such triadic interactions have been observed in baboons, geladas, and mangabeys, but appear to be limited to the *sylvanus* and *sinica* groups among macaques. He therefore suggests that intense male affiliation with infants and triadic interactions probably originated in a *sylvanus*-like common ancestor of *Macaca*, *Papio*, *Theropithecus*, and *Cercocebus*, and was eventually lost in the *fascicularis* group. In any event, as Fa (1984:93) says, "Baby care-taking is an important feature of troop life in any monkey. It has been elaborated in the Barbary macaque not only to promote survival of the infants by giving them constant attention but also to lower the level of tension between animals."

dominant to older sisters, but do not normally rise in rank above their mothers. In the case of *M. sylvanus*, however, younger sisters outrank neither their older sisters nor their mothers. D. A. Hill and N. Okayasu of Kyoto University suggest that, because they lack youngest ascendancy, Barbary macaques generally may be more egalitarian than some of the other macaques such as rhesus monkeys or Japanese macaques.

But what about the males? According to studies of wild groups in two different habitats in Algeria by N. Menard and D. Vallet of the University of Rennes in France, most male Barbary macaques emigrate from

sometimes results in huge troops that number in the hundreds. As is the case for other macaques, the social organization of *M. fuscata* centers around philopatric matrilines that are frequently descended from one female, and strong dominance hierarchies characterize both sexes. Interestingly, each newborn daughter becomes second in rank to her mother, which shifts the ranks of each of her sisters down a notch. Youngest ascendancy may be a temporary or variable situation, however, because other reports from the wild indicate that the eldest daughter attains the rank nearest her mother. Interestingly, Hill and Okayasu suggest that youngest ascendancy may be a by-product of the food that is provisioned to those groups in which it has been observed. Under these artificial feeding circumstances, they reason, mothers routinely support and protect their youngest offspring because they

Prime Mover Theories of Primate Brain Evolution

Since at least Darwin's time, scientists have attempted to identify one specific behavior that was targeted by natural selection and responsible for the dramatic increase in brain size that occurred in our genus, *Homo* (see Chapter 14). Proposed "prime mover" candidates have been numerous, and have included warfare, tool production, hunting, throwing, and language. Because they are known for being extremely intelligent and clever, macaques and baboons are among the primates used to speculate about the evolution of culture and social life in our earliest human relatives, and some of their behaviors have even been proposed as prime movers of brain evolution. As discussed in this chapter, the cognitive abilities of these Old World monkeys are partly due to their more-or-less complete shifts to terrestrialism with its correlated complex social organization and communication systems. The latter involve postural and facial expressions, and complicated vocalization systems with a variety of sounds including some that, rather than simply being emotional outbursts, seem to inform listeners about environmental features like the presence of predators. Richard Byrne and Andrew Whiten of the University of St. Andrews in Scotland have taken the interpretation of monkey (and ape) cleverness a step further. They note that higher primates are capable of deliberately misleading each other to their own gain. For exam-

ple, a baboon that is being aggressively chased might suddenly stop and adopt an alert posture that usually indicates presence of a predator, a deceptive act that effectively causes the chaser to stop and "look over there." Byrne and Whiten believe that the use of similar tactical deceptions to dupe others out of desirable resources such as food or mates are examples of "Machiavellian" intelligence in which seemingly altruistic acts are actually strategies that benefit the actors. They further suggest that Machiavellian intelligence may have been a prime mover that had a spiral (or runaway) effect of selecting for still more cleverness in the targets of actors. Other workers concur that social intelligence may have been an extremely important ingredient in higher primate brain evolution. For example, Robin Dunbar suggests that the enlarged human neocortex evolved primarily to keep track of multiple social relationships in increasingly large social groups. He notes that monkeys maintain group cohesion through social grooming and that this activity and brain size increase with group size. Human groups, however, are so large that there is not enough time for grooming to work as an effective bonding process. Interestingly, Dunbar proposes that language evolved to replace social grooming in hominids. In this sense, social intelligence and language may have together acted as the prime mover of hominid brain evolution (see Chapter 14).

are in constant danger from increased aggression that occurs in the vicinity of concentrated food supplies. Whether or not youngest ascendancy exists in a particular group, the kinship group of a troop's dominant female is dominant to the other matrilineages; that is, graded hierarchies exist between matrilines in general, as well as between individuals.

The reproductive life of Japanese macaques has been thoroughly studied by Linda Wolfe of the University of Florida who compared female sexual behavior in two troops that fissioned from one larger parent group. In 1972, one of the new troops was transplanted to a ranch in Texas and named Arashiyama West. The other troop, Arashiyama B, remained in Japan. Wolfe initiated her studies on the Texas troop in the early 1970s, and began studying the Japanese sister troop in 1977. Her findings, summarized here, are fascinating and important because of their comparative nature. Adult females do not have the dramatic sexual swellings that characterize estrous females of some species of macaques. Instead, their perinea and faces become reddened during the breeding season (so do those of males), which occurs during parts of autumn and winter. Consortships are common during the breeding season, and courtship involves simultaneous staring, flipping the eyebrows and ears, and smacking the lips. Either sex may approach the other, but females do not present their backsides in the typical primate manner. Rather, they sit on the ground and inch their way slowly toward the object of their desires. Copulating positions vary and include dorsal-ventral (i.e., back to front, with both feet of the mounted monkey clasped behind the partner's legs), ventral-ventral (face to face), sitting, and lie-on postures. Males of this species are series mounters; that is, the typical copulation averages about 13 minutes, during which the male mounts the female about 19 times before ejaculating. During the last mount, females often reach back and clutch their partners while turning to stare at their faces, a behavior that is believed to coincide with female orgasm.

While the above behaviors occurred in both Arashiyama B and Arashiyama West, the two troops differed in some telling respects. In Japan, estrous females mated not only with troop males but also with migrant males that appeared during the breeding season. This is not the case in Texas, however, because Arashiyama West was the only troop there. Compared to the Japanese troop, however, females in Texas engaged in much more female-female mounting (homosexual behavior occurs to some degree among all Japanese macaques) and more copulations with immature males. During the time of Wolfe's study, the sex ratio in Texas was approximately one mature male for every four or five adult females, instead of the more balanced ratio of one male for every two females found in Arashiyama B. Wolfe (1984:155–156) interpreted the significant increase in female homosexual behavior to the "quest for sexual novelty in the face of an altered sex ratio."

What Japanese macaques are best known for, however, is their inventiveness and proclivity for incorporating new discoveries into their social groups and passing them along to future generations. Curiously, a number of these "precultural" discoveries involved water, and in most cases the inventor was female. For example, some of the snow monkeys from Honshu Island are famous for their winter habit of keeping warm by soaking in natural hot springs (Fig. 9-5). The practice was introduced to the other macaques by a female who tried it after observing human bathers. Similarly, a female named Imo invented at

REVIEW EXERCISES

1. What enemies do macaques have?
2. Are any species of macaques highly endangered? Why or why not?
3. What species are Barbary macaques? In what ways do they differ socially from other macaques?
4. Describe triadic relationships that involve adult males and infants. Are these relationships more typical of macaques or baboons?
5. What is a "prime mover"? Name some candidates that have been proposed as possible prime movers of hominid brain evolution.
6. Define Machiavellian intelligence.

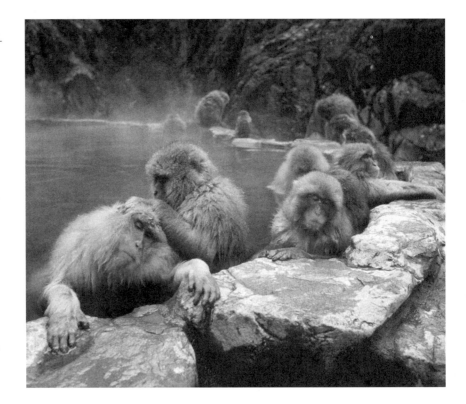

Figure 9-5 Snow monkeys may have learned the art of soaking in natural hot springs from observing humans.

Figure 9-6 A female named Imo was the first Japanese macaque observed washing sweet potatoes in the sea. As seen in this photograph, Imo's invention eventually spread to other monkeys.

least two new practices for preparing provisioned food on the southern island of Koshima. Primatologists there scattered sweet potatoes and wheat on the beaches for the macaques. At less than 2 years of age, Imo learned to wash the sand off of the sweet potatoes with water, and eventually started washing them in seawater, presumably because she liked the taste of the added salt (Fig. 9-6). Within 9 years, all except the youngest and oldest of the troop members had learned to wash and salt their potatoes, and this behavior was passed along to future generations. Imo later invented "placer-mining" washing of provisioned wheat by throwing handfuls of the sandy wheat into the sea, allowing the sand to sink, and scooping up the floating grains of wheat. This too was picked up by much of the rest of the troop (with a higher percentage of the 6- to 7-year-old monkeys learning the behavior than monkeys of any other age) and passed on to future generations. Led initially by another female, the macaques on Koshima have learned to wade, dive, and swim in the sea. Interestingly, new discoveries spread first among juveniles who teach them to their mothers, leaving the adult males as the last to acquire new habits.

Atsuo Tsumori from Japan and his colleagues studied problem-solving abilities of Japanese macaques experimentally by administering a sand-digging test to hundreds of monkeys in three troops. The test simply consisted of allowing a monkey to watch a human 6 feet away bury two or three peanuts in a hole in the sand and then smooth over the surface. The monkey was then given an opportunity to approach the location and dig out the peanuts. In this delayed-response experiment, the monkey needed to be motivated, remember the exact location, and keep the task in mind. Eighteen (44%) of the 41 monkeys in the Koshima troop that were tested were successful on the first trial, with the success rate rising with age until it peaked at 100% in the 6- to 7-year-olds and then dropping dramatically from ages 8 to 12 (Fig. 9-7). Monkeys older than 12 always failed on the first trial. What is particularly interesting is that females not only introduced new discoveries to the troop, but also outperformed males on the sand-digging test up to the 6- to 7-year age of peak performance. The theme of female inventiveness will be continued in Chapter 13's discussion about nut-cracking chimpanzees.

The above-described food-washing behaviors, first reported by Japanese primatologists in the 1950s and 1960s, were initially interpreted as examples of precultural or protocultural behavior because it was assumed that they were invented by one monkey, imitated by others, and passed along to future generations through imitative learning. This was startling at the time because cultural transmission of discovered knowledge was thought to be one of the characteristics that set humans apart from other animals. Although higher primates such as chimpanzees are now recognized as having cultural traditions (see Chapter 13), Elisabetta Visalberghi and her colleagues and Bennett Galef of McMaster University in Canada reject the idea that Japanese macaques do. Instead, they claim that dissemination of the techniques described above was too slow to be due to cultural trans-

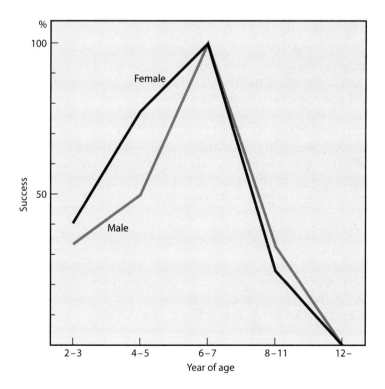

Figure 9-7 Female Japanese macaques outperformed males on a sand-digging test up to the 6- to 7-year age of peak performance.

REVIEW EXERCISES

1. What widely used techniques for studying primates were developed by researchers from Japan?
2. Describe the appearance of Japanese macaques. What is their species name?
3. Does youngest ascendancy always occur in *M. fuscata*? What ecological variable might it be related to?
4. How did female sexual behavior differ between Japanese macaques from Arashiyama B and Arashiyama West, according to Linda Wolfe's study? To what did she attribute these findings?
5. Describe some of the "precultural" discoveries made by Japanese macaques. In what ways are they cultural? How did they subsequently spread across age groups and gender?
6. Why do some workers reject the idea that *M. fuscata* has protocultural traditions?
7. Discuss recent evidence that supports earlier interpretations regarding protoculture in Japanese macaques.

mission; that is, rather than having learned the new behaviors by watching others, individual macaques probably discovered them independently.

Research on this fascinating species continues, however, and new reports reinforce and extend the earlier interpretations regarding protoculture. For example, Masayuki Nakamichi and colleagues from Osaku University report on the acquisition and transmission of new food-washing techniques and tool uses in the Katsuyama free-ranging group of Japanese macaques. Eleven adult females were observed carrying and washing grass roots. Most of the monkeys pulled the roots from the ground one at a time and set them aside to form a pile. Then the entire pile was carried to a river where each root was individually washed in water and scraped of dirt. Some of the females rubbed dirt off the roots by rolling them along flat rocks at the edge of the river. Interestingly, 6 of the 11 females belonged to the same matriline within the top-ranking Katsuyama kin group. This fact, together with a record of one infant watching her mother wash and eat roots, suggests that the behavior is not independently acquired by each animal. This is the first report of Japanese macaques washing a natural rather than provisioned food, and the authors believe that the economical piling of roots together reflects a relatively advanced cognitive ability. Perhaps they are right.

SAVANNA BABOONS

Four species of savanna baboons are recognized: the olive (*Papio anubis*), yellow (*P. cynocephalus*), chacma (*P. ursinus*), and guinea baboon (*P. papio*). All four species are relatively terrestrial and live in a variety of habitats that include savanna grasslands and woodlands, which, as already noted, cause them to be of special interest to paleoanthropologists interested in hominid evolution (see Chapter 14). Although they differ somewhat in size and color, the species resemble each other physically and behaviorally. Furthermore, their sub-Saharan distributions grade into one another, with guinea baboons living in the westernmost part of Central Africa, olive baboons occupying a band most of the way across the central part of Africa, yellow baboons living farther south, and chacma baboons inhabiting the southernmost part of the continent (Fig. 9-8). For this reason, some workers believe the four groups represent geographical variants, or subspecies, of one common species, *P. cynocephalus*.

Baboons are the largest of the monkeys. They have bare faces with long doglike muzzles that support large front teeth; indeed, *cynocephalus* means "dog head" in Latin. As expected for relatively terrestrial nonhuman primates, sexual dimorphism is marked in both body weight and canine size. In fact, adult males can be up to twice the size of adult females (Fig. 9-9). Hair may be somewhat long and, depending on the species, varies in color from reddish to olive to yellow to black.

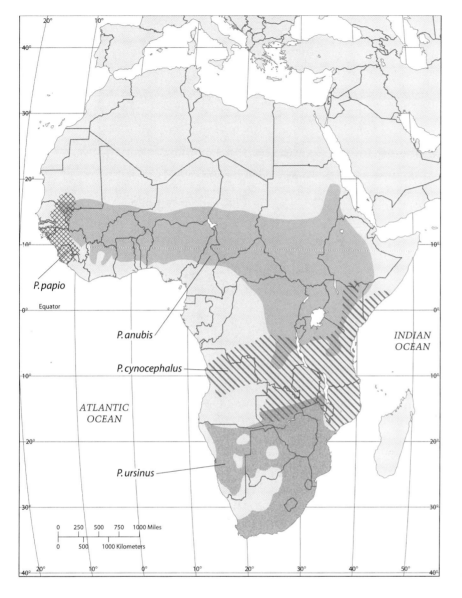

P. papio

P. anubis

P. cynocephalus

ATLANTIC
OCEAN

P. ursinus

Equator

INDIAN
OCEAN

0   250   500   750   1000 Miles

0   500   1000 Kilometers

Figure 9-8 The sub-Saharan distributions of the four species of savanna baboons grade into one another.

Baboons have moderately long tails and ischial callosities that are fused in males and separated in females.

Because they are big primates and food is widely dispersed in patches across their habitats, savanna baboons must range relatively far during the day to find enough to eat. Their diets are highly varied and may include grass, tubers, roots, bulbs, fruits, leaves, flowers, seeds, insects, gums, eggs, and even meat of small mammals that are sometimes hunted by males. Despite their omnivory, however, baboons are unusual because they can subsist on grasses and other foods found in open savannas. Don Melnick of Columbia University and Mary Pearl of the New York Zoological Society point out that, in ad-

Figure 9-9 It is not unusual for a male savanna baboon to be twice the weight of an adult female.

dition to being far-ranging, savanna baboons are characterized by consistently low population densities (see Table 9-2), and note that this combination of features is in keeping with the general trend for larger Old World monkeys. Baboons move across the savanna quadrupedally and, after a long day's trek, they bed down for the night in sleeping trees or rocky outcrops.

As already mentioned, the savanna is a dangerous place for primates. Humans find this out when they go on camera safaris in game reserves (e.g., in Kenya), from posted signs that strongly warn people to stay in their vehicles at all times. Potential predators include lions, hyenas, jackals, cheetahs, wild dogs, birds of prey, and leopards. Baboons are often found in the company of ungulates (hoofed mammals) such as gazelle, wildebeest, and zebra, and each responds to the startled reactions of the other upon encountering predators. The ungulates are thus able to take advantage of the baboons' keen eyesight, while the latter benefit from the enhanced sense of hearing and smell in their hoofed neighbors. This mutually beneficial or **symbiotic** relationship is not the only factor that contributes to the safety of savanna baboons, however. Their evolved, highly structured social life seems to be even more important.

The general outline of baboon social organization is like that described for their close relatives, the macaques. Savanna baboons live in multimale, multifemale groups that vary greatly in size from a handful to hundreds of animals, although troops of 40 to 50 monkeys seem to be the norm. (Some reports indicate that guinea baboons live in much larger groups that break down into smaller foraging units, however.) As with macaques, troop life is centered around a stable core of related females, and males emigrate away from their natal groups as they mature. Although the sex ratio is usually skewed in favor of females, males interact aggressively with each other, frequently competing for estrous females. Nevertheless, coalitions of individuals and friendships play an important role in baboon society. Despite their similarities, savanna baboons differ from macaques in some of the details of their reproductive lives. For example, all adult females exhibit dramatic swellings during estrus, and baboons tend to engage in more coordinated consortships than macaques. Baboons also lack the birth seasons that characterize macaques.

Savanna baboons have traditionally been one of the most widely studied of all primates. They are fascinating, not only because of their social interactions (which are reminiscent of humans'), but also because interpretation of their social behavior over the years has been subject to controversy and change in ways that parallel developments in human society. Consequently, the classic description of baboon social organization that originated in the 1960s is currently under revision. However, much of the early information about savanna baboons remains valid; details are provided in the next section.

The Classic Baboon Model

In the late 1950s and early 1960s, the late K. R. L. Hall and Irven DeVore of Harvard University and their colleagues carried out field research on chacma and olive baboons, respectively. Their findings continue to be extremely influential in primatology. Indeed, Hall and DeVore's observations have had such an impact that one may reasonably refer to their contributions as classic. This is not to say that every aspect of their research has gone unchallenged. As you will see, this is far from the case.

One of the main contributions of Hall and DeVore was to explore the nature of dominance hierarchies among male savanna baboons. At one level, males have a linear pecking order in which each individual has a specific status or rank. The most dominant of the baboons in a troop is the alpha male, which is identified by a number of criteria, including (1) he mates with (and theoretically fertilizes) most of the fully estrous females, (2) other monkeys present to him more than to others (as a submissive gesture), (3) he engages and prevails in more aggressive interactions than do other baboons, (4) he rushes to defend the troop when it is threatened by predators or strangers, and (5) mothers with infants stay close to him for protection. Hall and DeVore went on to delineate the functions of such dominance: The alpha male keeps peace by breaking up squabbles among other animals, and protecting mothers and infants. When faced with outside danger, the alpha male uses aggression to bring the troop together.

As fieldwork continued, primatologists studying rhesus monkeys as well as baboons began to focus primarily on the third criterion to assess dominance hierarchies; that is, they quantified and compared

Figure 9-10 Grooming and presenting are two friendly behaviors that savanna baboons frequently engage in.

Figure 9-11 Yawning may be either a threatening or a nervous behavior in savanna baboons.

the extent to which each male routinely prevailed over other individual males (see In the Field). In this way, primatologists could identify not only the alpha male for a given troop, but also which adults ranked second, third, and so on. Furthermore, it became clear that the emerging dominance patterns among savanna baboons could not be reconciled with the classic model in a strictly linear fashion. Instead, primatologists observed that individual friendships play an important role, so that two friends that constantly associate and support each other are often able to dominate another animal that, under other circumstances, is dominant to each. In other words, like human society, baboon troops have their cliques and politics. Because of this, it is extremely important for primatologists to observe the success of combinations of baboons against other individuals as an additional criterion when working out a troop's dominance hierarchy. Hall and DeVore found that troops frequently have a central hierarchy of males that act in concert and are thus able to control access to resources, and troop movements. The bottom line is that a male's success depends on his status as an individual (or linear rank) and his ability to enlist the support of other males in the central hierarchy.

The earlier fieldwork, summarized by Hall and DeVore, also provides details about how rank, fear, and friendship are expressed in savanna baboons. Grooming, presenting, and lip smacking are friendly behaviors (Fig. 9-10). So are greetings that may include embracing, nuzzling, and mouth-to-mouth contact. Threats, on the other hand, may be conveyed in a number of ways including staring, head jerking, raising the eyebrows, eyelid flashes, and flattening of the ears. Baboons may also threaten by shaking trees or pulling up rocks. Or they may grind their teeth, yawn (which displays the canines), or produce any number of shrieks, barks, roars, or screeches. Yawning may also be a sign of nervousness in baboons, as is jerky fidgeting (Fig. 9-11). Fright is often indicated with a fear grimace that involves retracting the lips (Fig. 9-12), or production of a yaking sound. Not surprisingly, fearful baboons are also known to run away. Clearly, although they lack the dramatic face paint and coat colors of their forest-dwelling relatives, savanna baboons have an extensive repertoire of visual and vocal signals with which to express themselves.

There is, however, a fly in the ointment of the classic baboon model, and it has to do with early misconceptions about the role of females in baboon society. Classically, the baboon troop was thought to be organized rather strictly around the dominance hierarchy of males, but not females. In fact, dominance in females was believed to be of fleeting concern because it seemed associated with being in estrus or having small infants, both of which are temporary states. To quote Hall and DeVore,

> it is possible that the female hierarchy is typically unstable, that it is individually based rather than partly organized around coalitions as in the males, and that it is expressed in more-or-less continuous minor bickering with very little real attacking and biting. (Hall and DeVore, 1965:67)

As you are aware, contrary to the classic model, the core of savanna baboon (and macaque) troops is now known to be a stable group of natal adult females, rather than the adult males who are immigrants. Just as society's expectations about the roles of women have changed since the 1960s, so too have primatologists' views about the social aspirations and roles of female primates.

Baboons' Lib

Since the classic view of baboon society was formulated in the 1960s, there has been a virtual revolution in the theoretical and methodological approaches to primate field studies. Consequently, primatology now offers a more balanced picture, not only of primate social life in general, but also of how social organizations have evolved in response to ecological constraints such as predator pressures, and availability of resources like food and mates (see Chapter 2). With respect to baboons and other cercopithecines, a particularly dramatic change has occurred in the interpretation of the role of females in troop social life.

Jane Lancaster of the University of New Mexico credits many of the recent advances in primatology to applications of sociobiological theory, population biology, and field ecology, as well as improved methods of collecting and statistically analyzing data. She finds that one generalization emerges from numerous studies of nonhuman primates (1984:3): "For all intents and purposes, male and female primates pursue different sex-specific strategies and tactics for successful reproduction through the course of the life-span." In contrast to earlier reports summarized previously, numerous field studies have now confirmed that females of many (but by no means all) species of Old World monkeys, including macaques and savanna baboons, have their own dominance hierarchies based on aggression, and that daughters are likely to inherit their ranks from their mothers.

Further, when resources are scarce, high-status females of some species begin reproducing earlier and have more surviving infants than do lower-ranking females (but see the portrait of *P. anubis* below). At least in these cases, females use their status to be successful mothers, while males use theirs for access to estrous females, that is, to become breeders. This seems to be particularly true of sexually dimorphic species and, as Lancaster notes, markedly different anatomies also fit with the above generalization about males' and females' different strategies for successful reproduction. That is, the males of dimorphic primates use their large bodies and canines for aggression, both to compete for females and for defense. Rather than growing big bodies, females on the other hand, use their calories to gestate and lactate (i.e., to grow and nurture offspring).

Another of the basic tenets of the classic model has also undergone revision. This involves the first criterion listed earlier for determining which monkey in a troop is the alpha male, that is, the assumption that estrous females initially mate promiscuously with any number of subordinate males, but end up mating with the most dominant male

Figure 9-12 This savanna baboon is baring its teeth in a fear grimace.

when they are at the peak of estrus (which coincides with ovulation) and are therefore most likely to conceive. In this manner, it was traditionally thought that alpha males contributed a relatively large proportion of their genes to future generations. As noted earlier, however, it is now known that the extent to which rank correlates with reproductive fitness (for either sex) varies between different groups and sometimes appears tied to ecological factors such as availability of food. It may also be tied to social variables. For example, Shirley Strum of the University of California, San Diego, and Barbara Smuts of the University of Michigan have shown that some estrous baboons prefer to mate with special male friends rather than high-ranking individuals. Consequently, primatologists are now generally more aware of female choice when it comes to matters of sex. This is not to say that females are exclusive in their preferences. In fact, rather than being choosy about mates, estrous females in multimale, multifemale groups usually seek numerous sexual partners. Interestingly, Sarah Blaffer Hrdy suggests that such behavior may add to the reproductive fitness of females by motivating males to be nurturing and protective of subsequently born offspring, which they might have fathered. That is, the more promiscuous the female, the greater the number of potentially investing males.

As this chapter shows, the core of the baboon or macaque troop is a stable matriline of females that has a dominance hierarchy separate from that of the immigrant males. As we have seen, the sexual dimorphism that characterizes these largely terrestrial species as well as their orchestrated social lives are both adaptive for surviving, and even thriving under, the rigors of savanna life. But the liberated image of female baboons and macaques goes further than simply thinking in terms of reproductive fitness. Older females are respected in their troops by all members, not just those in their matrilines. In times of controversy, it is they who lead the others, as beautifully illustrated by Jeanne Altmann's observation of yellow baboons in Amboseli National Park in Kenya:

> The elderly baboon, known to us as Handle, stared intently toward the distant grove of trees. The five dozen members of her group sat tensely in clusters nearby. Indecision was in the air. . . . A short while before, as they were heading southward toward a grove of trees in which to spend the night, the group had spotted a leopard—the baboons' major predator. The sun was dropping rapidly now, and the short equatorial dusk would give the baboons little time to make a decision: stick to the original plan or strike out for one of the other groves—all farther away—scattered across their East African savanna home. If they didn't act fast, darkness would overtake them far from the safety of tall trees. . . . After ten long minutes, in a single, smooth motion, Handle stood, her infant still clutching her sides but now riding under her, and began to move decisively westward toward a grove still hidden in the dusty, dry-season haze. . . . Handle paused as she looked back; throughout the group, baboons responded with soft grunts and moved to follow her. The rippling motion rapidly grew into a wave, and soon, all the baboons followed, silent except for the protests of tired youngsters for whom first the tension and now the sudden rapid pace were too much. (Altmann, 1992:48).

The Olive Baboon (*Papio anubis*)

Of the four species of savanna baboons, olive baboons are the most widely distributed (see Fig. 9-8) and have been relatively well studied since the classic years of baboon watching. The purpose of this brief portrait is to highlight some of the more interesting details about the politics and possible adaptive significance of interbaboon relationships. As noted earlier, the formation of friendships is important for olive baboons. This holds not only for male-female pairs, but also for relationships between males. For example, among olive baboons that live on the Eburru Cliffs in Kenya, coalitions of older, lower-ranking males are able to take estrous females away from younger, higher-ranking males. According to Smuts and John Watanabe of Dartmouth College, older males are able to do this because, unlike younger males, they have formed stable, cooperative relationships among themselves.

Male olive baboons are usually tense and competitive with each other, and fights occur almost daily. The exception to this is when they are exchanging friendly greetings that occur in neutral contexts. These greetings, which are unique to males and also occur in other savanna baboons, entail a fascinating sequence of behaviors. As described by Smuts and Watanabe, a greeting begins when one male uses a rapid, swinging gait to approach another male that is usually (but not always) less dominant. The approacher often smacks his lips, narrows the eyes, and has the ears back—all friendly gestures. If the second animal accepts the approach, two or more behaviors occur, including posterior presenting, grasping the posterior, mounting, and fondling of the penis or scrotum (Smuts and Watanabe term this last behavior "diddling"). For example, baboon A might present to, and be mounted by, baboon B. Occasionally these behaviors are mutual, for example, when each male simultaneously touches the other's scrotum. Older males carry out greetings more easily than younger males (Fig. 9-13), which have difficulty completing greetings because of unresolved tensions in their relationships. Typically, an entire greeting is extremely ritualized and takes only a few seconds, after which one or both males move rapidly away using the same swinging gait that began the greeting.

Figure 9-13 These two male olive baboons are greeting each other in a ritualized manner. (Photo courtesy of Fred Bercovitch.)

Smuts and Watanabe hypothesize that males use greetings to de-fine and negotiate aspects of their relationships, including coopera-tion. They (1990:169) note that, with respect to genital touching, "lacking articulate speech, and unable to swear oaths, perhaps male baboons make a gestural equivalent by literally placing their future reproductive success in the trust of another male." The authors also draw provocative parallels with humans when they (1990:169–170) observe, "In *Genesis* (xxiv 9), Abraham's servant swears an oath to him while placing his hand under his master's 'loins'. Furthermore, the *Oxford English Dictionary* suggests that the words 'testify,' 'testi-mony,' and 'testicle' may all share a common root, 'testis,' which originally meant 'witness.' In other words, the formal parallels here suggest an evolutionary continuity. . . ."

But what about friendships between adult male and female *P. anu-bis* baboons? Not surprisingly, the particular males that sexually swollen females selectively groom and present to are the ones with which they are most likely to consort. Fred Bercovitch of the Caribbean Primate Research Center has observed that special friend-ships between adult males and females frequently develop after the fe-male has conceived. Baboon infant mortality is very high during the first 2 years of life. Most infants have at least one adult male friend, and Bercovitch thinks that females may develop friendships with males that mated with them as a means of fostering future male care of infants. Whether or not this male care hypothesis eventually proves to be correct, it is fascinating to ponder Smuts's description of the be-ginning of a friendship between two baboons:

> Alex and Thalia sat about fifteen feet apart on the sleeping cliffs. It was like watching two novices in a singles bar. Alex stared at Thalia until she turned and almost caught him looking at her. He glanced away immedi-ately, and then she stared at him until his head began to turn toward her. She suddenly became engrossed in grooming her toes. But as soon as Alex looked away, her gaze returned to him. They went on like this for more than fifteen minutes, always with split-second timing. Finally, Alex man-aged to catch Thalia looking at him. He made the friendly eyes-narrowed, ears-back face and smacked his lips together rhythmically. Thalia froze, and for a second she looked into his eyes. Alex approached, and Thalia, still nervous, groomed him. Soon she calmed down, and I found them to-gether on the cliffs the next morning. . . . Six years later, when I returned to Eburru Cliffs, they were still friends. (Smuts, 1987:43)

Having male friends is not the only way that female olive baboons increase their reproductive success, as shown by decades of research by Jane Goodall and her colleagues at the Gombe Stream Research Centre in Tanzania. High-ranking females have easier access to food and other scarce resources, and suffer less stress because they are not as harassed by their companions as the low-ranking females are. Analyses of data on rank, menstrual cycles, births, and deaths by Craig Packer of the University of Minnesota and his colleagues reveal that, as expected, high-ranking females at Gombe have shorter inter-birth intervals and higher infant survival rates than do low-ranking mothers. This finding raises a theoretical problem, however. Since rank is scored largely by success in aggressive interactions between pairs of individuals, if high-ranking females are characterized by higher reproductive fitness compared to lower-ranking females, what is to prevent olive baboons from becoming a hyperaggressive species? Packer's team provides a surprising answer to this dilemma: A number of high-ranking females appear to have more miscarriages and reduced fertility than their lower-ranking cohorts. Packer hypothesizes that the higher-ranking females may have higher amounts of male sex hormones that interfere with conception and pregnancy compared to other females. If so, this provides a balance against the potential for runaway selection resulting in hyperaggressive females.

SUMMARY

Two related groups of large-bodied, sexually dimorphic, and relatively terrestrial Old World monkeys are the macaques (*Macaca*) and savanna baboons (*Papio*). There are 19 recognized species of macaques, which are the most widespread of the nonhuman primates. They live in a variety of habitats ranging from tropical islands to snowy mountains, located in parts of northern Africa and across Asia. On the other hand, savanna baboons, of which there are four species, live in sub-Saharan Africa. They are the largest of all the monkeys and are of particular interest to paleoanthropologists because of their adaptations to harsh savanna habitats that characterize those in which our earliest human ancestors are thought to have perfected bipedalism. The high degree of sexual dimorphism and terrestriality is probably one of the ingredients that accounts for similarities in macaque and savanna baboon social organizations. Both groups live in multimale, multifemale troops that are centered around a stable core of females who remain in their natal groups for life. Dominance hierarchies are found in macaque and baboon males and females, and under some circumstances, rank may be correlated with reproductive fitness within each sex. The classic model of baboon social organization has been updated in light of recent findings regarding the roles of female primates.

FURTHER READING

ALTMANN, J. (1992) Leading ladies. *Natural History* **101**(2):48–49.

FA, J. E. (1984) Baby care in Barbary macaques. In: D. Macdonald (ed) *All the World's Animals: Primates*. New York: Torstar, pages 92–93.

LANCASTER, J. B. (1984) Introduction. In: M. F. Small (ed) *Female Primates: Studies by Women Primatologists*. New York: Alan R. Liss, pages 1–10.

NAKAMICHI, M., KATO, E., KOJIMA, Y. AND N. ITOIGAWA (1998) Carrying and washing of grass roots by free-ranging Japanese macaques at Katsuyama. *Folia Primatologica* **69**:35–40.

SMUTS, B. (1987) What are friends for? *Natural History* **96**(2):36–44.

REVIEW EXERCISES

1. How many species of savanna baboons are recognized in this text? In what country and type of habitats do they live?
2. Why are savanna baboons of particular interest to paleoanthropologists?
3. What do savanna baboon and macaque social organizations have in common? How do their reproductive lives differ?
4. How were female baboons viewed in the classic model of baboon society? In what ways have beliefs about the role of female monkeys in troop social life changed since the 1960s?
5. Compare the benefits that male and female terrestrial Old World monkeys may derive from high status.
6. How do different body sizes of sexually dimorphic monkeys contribute to reproductive fitness in each sex?
7. Why might female savanna baboons form special friendships with certain males?
8. Describe the ritualized greetings of male olive baboons. What might their function be?

10 GIBBONS:
THE LESSER APES

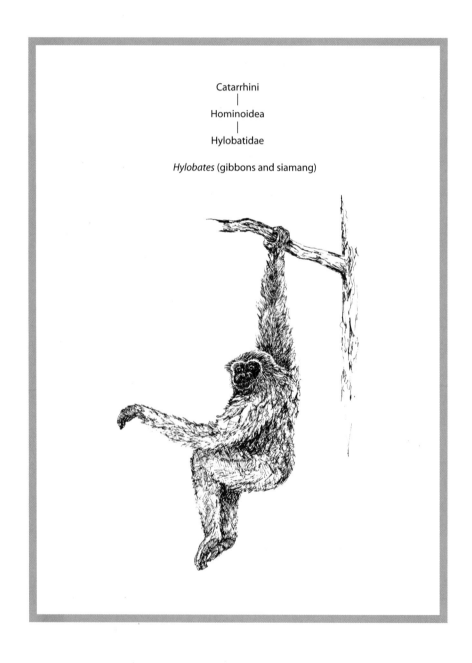

Catarrhini
|
Hominoidea
|
Hylobatidae

Hylobates (gibbons and siamang)

INTRODUCTION

We turn now from monkeys to apes, which belong to the superfamily Hominoidea. You will recall that hominoids are divided into three families (see Fig. i-6), the hylobatids (gibbons and siamangs, known as the lesser apes), pongids (the three genera of great apes), and hominids (humans and their bipedal relatives). Altogether, there are four genera of living apes, great and small: *Hylobates* (gibbons and siamangs) and *Pongo* (orangutans) from Asia, and *Pan* (chimpanzees and bonobos) and *Gorilla* (gorillas) from Africa. Thus, there are few genera (and species) of living apes compared to the numbers that flourished during the Miocene epoch, and the three great apes are best regarded as precious and endangered relics that, today, survive precariously in relatively tiny isolates. Apes are very different from monkeys, as is clear from the ape pattern outlined below. Nevertheless, it is not uncommon to see footage of chimpanzees or gorillas during television news programs accompanied by the announcement "stay tuned for a monkey story"—a distressing situation for those who take their primates seriously.

THE APE PATTERN

BONY FEATURES

1. Y-5 cusp pattern on lower molars (see Fig. 1-11)
2. Wide chest (thorax)
3. Flexible shoulder, elbow, and wrist joints
4. Long slender hands
5. Short back with no tail
6. Long legs, and very long forelimbs and arms
7. Robust big toe

8. Mainly frugivorous or vegetarian
9. Brachiators
10. Able to rotate forearms 180°
11. Biologically closer to humans than are monkeys

At first glance, it may seem difficult to interpret the ape pattern from a functional point of view. However, most of the traits that characterize apes are related to how they move in arboreal habitats. Rather than scampering along branches the way quadrupedal monkeys do, apes tend to hang by their arms and move by swinging arm over arm underneath branches. This form of locomotion, known as brachiation, also occurs in some of the New World monkeys (see Chapter 6). The relatively small gibbons and siamangs are strictly arboreal and therefore practice brachiation more extensively than the largely terrestrial African great apes. Although highly arboreal, orangutans (see Chapter 11) tend to use both hands and feet to clamber cautiously through the trees. Despite their different locomotor preferences, however, the three great apes share the gibbon's anatomy and, to a lesser degree, proclivity for brachiation (i.e., they are structural brachiators).

To brachiate, a primate must be all arms and shoulders and, indeed, this is reflected in most of the traits listed for the ape pattern. For example, a wide chest places the arms out at the sides of the body, which facilitates suspension from branches. Flexible shoulders, elbows, and wrists help swing the body into brachiation, and long hands are good for hanging onto branches. Longer arms permit bigger arcs and therefore more distance to be covered per swing, which is energetically efficient. Swinging a long body is not efficient, however. Instead, brachiators tend to tuck the lower limbs up as they swing along, which reduces their drag. This streamlining effect is enhanced by a relatively shortened trunk compared to monkeys, which have longer more flexible backbones. This is not to imply that apes never use their feet in the trees. They do, and a large big toe is useful for grasping branches, especially since apes do not have a tail, unlike New World monkeys (e.g., spider monkeys) that make spectacular use of their prehensile tails when brachiating (see Chapter 6).

Besides getting from point A to point B, a brachiating anatomy permits apes to suspend themselves by one arm, around which they are able to pivot their bodies. This is handy for reaching fruits clumped on branches that are too small to support an ape's weight. Indeed, fruit may be the key to the Y-5 lower molar pattern (see Fig. 1-11) that characterizes apes (and hominids). As discussed in Chapter 1, frugivores require large, relatively flat molar surfaces upon which to pulp fruit. Indeed, this is just what the Y-5 ape molars have, compared to the steeper bilophodont molars of monkeys that eat a higher proportion of leaves (see Box 1).

Apes are of particular interest to primatologists because they are

---

REVIEW EXERCISES

1. How many genera of apes are there? What are their names and where do they live?

2. Describe the anatomical features that are associated with the ape pattern.

3. What functions are associated with the ape pattern?

4. What is the Y-5 pattern? What dietary preference may it reflect?

5. In what order did the four genera of apes probably appear during their evolution? What evidence supports this branching pattern?

6. Which ape is the closest cousin of humans?

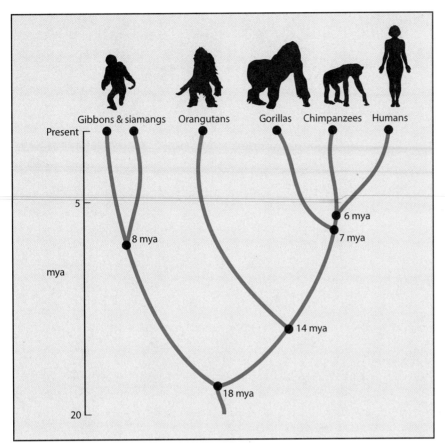

Figure 10-1 Comparative molecular studies suggest that the direct ancestors of living hominoids radiated within the last 20 million years and that ancestral gibbons were the first to branch off. The relative times of divergences are based on data from Goodman et al. (1998).

Figure 10-2 The bony features of this skeleton from a male siamang are typical of hylobatids.

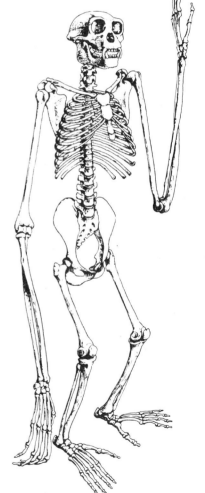

more closely related to humans than are the other nonhuman primates. However, not all apes are equally related to *Homo sapiens*. Comparative molecular studies suggest that gibbons branched off first from the common stock that eventually gave rise to the great apes and hominids. This makes the lesser apes our most distant hominoid cousins. The next to separate were the other Asian apes, the orangutans, followed later by gorillas in Africa (Fig. 10-1). Finally, somewhere around 5 to 6 mya, the remaining stock split into two lines that gave rise to chimpanzees and hominids. The remainder of this chapter is devoted to gibbons. The next three chapters cover the great apes in the order of their evolutionary appearances (or degree of relatedness to hominids), that is, orangutans, gorillas, and chimpanzees.

THE HYLOBATID PATTERN

BONY FEATURES (FIG. 10-2)

1. Small, gracile skeleton
2. Extremely long forelimbs and thumbs
3. Saber-like canines in both sexes
4. Lack of sexual dimorphism in body size

TABLE 10-1 Statistics for Representative Hylobatids

	H. syndactylus (siamang)	H. lar (white-handed gibbon)	H. concolor (crested gibbon)	H. hoolock (white-browed gibbon)
Adult mean F	10.6 kg (23.4 lb)	5.6 kg (12.3 lb)	5.8 kg (12.8 lb)	6.1 kg (13.4 lb)
body weight M	13.5 kg (29.8 lb)	6.3 kg (13.9 lb)	5.6 kg (12.3 lb)	6.9 kg (15.2 lb)
F + M			4.5–9.0 kg (9.9–19.9 lb)	
Adult brain weight (g)	122	108	132	109
Maximum lifespan (yr)	35	44	46	42
Habitat[a]	SA: pf, sf	SA: pf, sf, df, rf	SA: trf, mf	SA: pf
Locomotion[b]	br	br, bp	br	br, l, bp
Diet[c]	l, f, fl, a	f, l, i, fl, st, sh, bu	bu, sh, f, l, fl	f, l, bu, fl, a
Seasonal breeders	peak?	0		+
Extra-pair copulations?	+	+		
Mean gestation (days)	214	205	206	
Interbirth interval (yr)	> 3 (var)	2.5		3.0
Social systems[d]	M	M	M	M
Intensive male parenting	+	0	+?	
Infanticide[e]	0	+ (zoo)		
Group size	2–6	2–6		2–6
Groups/km²	1–2	1–2		2
Home range	15–48 ha	12–58 ha	40–500 ha	18–400 ha
Conservation status[f]	H	H	H	H

Note: Information in this table has been selected from numerous sources and should be viewed as tentative, i.e., subject to change in light of new information. Sources include but are not confined to: Harvey et al., 1987 (brain weights); Palombit, 1995, 1996 (*H. syndactylus, H. lar*), Leighton, 1987 (groups/km²), Reichard and Sommer, 1997; Bleisch and Chen, 1991 (some home ranges); Rowe, 1996 (conservation status and other statistics). Abbreviations: +, yes; 0, no; var, variable. Blank entries indicate that no information is presently available. Dietary items are listed in approximate order of decreasing importance to the species.

[a]Habitat: SA, Southeast Asia; df, deciduous forest; mf, montane forest; pf, primary forest; rf, rain forest; sf, secondary forest; trf, tropical forest.

[b]Locomotion: bp, bipedal walking or hopping; br, brachiation; l, leaping.

[c]Diet: a, animal prey; bu, buds; f, fruit; fl, flowers; i, insects; l, leaves; sh, shoots; st, stems.

[d]Social systems: M, monogamous.

[e]Infanticide: +, one or more reports.

[f]Conservation status: H, highly endangered.

OTHER FEATURES

5. Strictly arboreal
6. Good precision grips
7. Ischial callosities
8. Lack of sexual swellings in females of most species
9. Vocal duetting
10. Monogamous

In addition to the features that gibbons and siamangs share with other apes, they have other characteristics that set them apart, as just outlined in the hylobatid pattern. Some of these traits can be seen in the skeleton of the largest as well as one of the most distinctive of the hylobatids, the siamang (see Fig. 10-2). Although the siamang is big compared to gibbons, its adult weight averages only about 11 to 14 kg (roughly 23 to 30 lb, Table 10-1), which is within the monkey range. Perhaps because of their small size, hylobatids spend the vast majority of their time in trees and engage in much more brachiation than do the other apes. Indeed, they are the most graceful and skilled brachiators among all of the primates. As discussed below, features 1, 2, 5, 6, and 7 of the hylobatid pattern reflect adaptations that are useful for brachiation and life in the trees. Features 3, 4, 8, 9, and 10, on the other hand, form a package that is related to an unusual and fascinating social system in both gibbons and siamangs.

Hylobatids are, in a word, beautiful. They have large eyes and bare black faces shaped something like a three-leaf clover (Fig. 10-3), and their coats are thick and luxurious. Their colors range across the nine species recognized in this text (see Table 10-2), from blonde to black to silver, and may change with age. Although gibbons are not particularly sexually dimorphic in body and canine size, adult males and females may be two different colors, or **sexually dichromatic** (Fig. 10-4). Markings on the face and head also vary with species. For example, the delicate face is completely framed with a band of white hair in *Hylobates lar* (Fig. 10-5), but is only highlighted above by thick white eyebrow flashes in *Hylobates hoolock*. Other species exhibit contrasting color in the hair on their cheeks, temples, or tops of their heads, while the entire coat of *Hylobates moloch* is a striking silvery gray. On the other hand, siamangs of both sexes (*Hylobates syndactylus*, described later) are totally black. The lesser apes really are monkey sized. Most adult gibbons weigh between 5 and 7 kg, although as already noted, siamangs are heavier and may weigh up to 15 kg (see Table 10-1).

Figure 10-3 Gibbons are beautiful little apes, as shown in this photograph of a *Hylobates lar* mother and infant.

Figure 10-4 Crested gibbons (*Hylobates concolor*) are sexually dichromatic and sometimes appear to have pointed heads because of their crests.

TABLE 10-2	*Hylobatids*	
Genus	**Species**	**Common Name**
Hylobates	*agilis*	Agile gibbon
	concolor	Crested gibbon
	hoolock	White-browed gibbon
	klossii	Kloss gibbon, beeloh
	lar	White-handed, or common, gibbon
	moloch	Silvery gibbon
	muelleri	Gray gibbon
	pileatus	Capped gibbon
	syndactylus	Siamang

Figure 10-5   The face of the white-handed gibbon (*Hylobates lar*) is framed with a band of white hair.

Figure 10-6   Except for the siamang, which overlaps with *Hylobates lar* and *H. agilis*, gibbon species live in different parts of Southeast Asia. (**1**) Siamang. (**2**) Concolor gibbon (**a**) black-cheeked and (**b**) white-cheeked phases. (**3**) Hoolock gibbon. (**4**) Kloss gibbon. (**5**) Pileated gibbon. (**6**) Müller's gibbon. (**7**) Moloch or Silvery gibbon. (**8**) Agile gibbon (sexes similar in one population): (**a**), (**b**) forms in Malay Peninsula and southern Sumatra; and (**c**) southwest Borneo. (**9**) Lar gibbon (sexes similar in same population): Thailand, dark phase (**a**) and (**b**) light phase: (**c**) south of Malay Peninsula: (**d**) northern Sumatra.

In terms of numbers, hylobatids are highly successful. The nine species of *Hylobates* are distributed widely across Southeast Asia, including Sumatra, Borneo, and Java (Fig. 10-6). Although the diploid number of chromosomes varies between 38 and 52 depending on the species, different species have been known to interbreed in captivity. In nature, however, there is little geographical overlap between species, except for the siamang, which lives in the same region or is **sympatric** with the smaller-bodied *H. lar* and *H. agilis*. (In some areas, gibbons are also sympatric with orangutans and macaques.) The hylobatids are restricted primarily to evergreen forests, and they prefer to spend time in the higher parts of the canopy. Because they live in small families that are territorial (see below), gibbons are not as densely distributed as many primates. According to Donna Leighton of Earthwatch, however, home range sizes and population densities may vary markedly within species, perhaps in response to greater or lesser availability of trees at different sites. Leighton also notes that

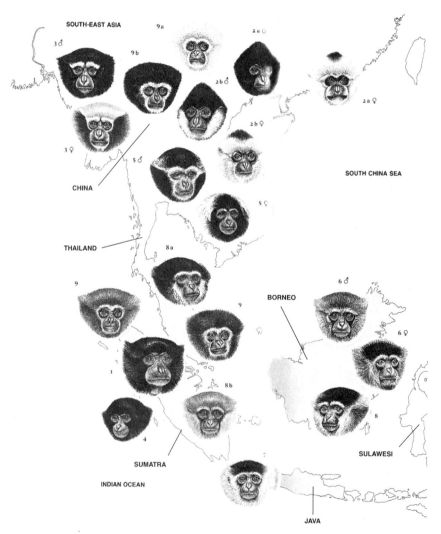

Figure 10-7 This montage shows the arm-over-arm swinging motion of one gibbon as it brachiates.

the more folivorous diet of siamangs permits smaller territories than would otherwise be expected for primates of their size.

Gibbons and siamangs are a joy to watch as they gracefully brachiate arm over arm through the trees or along ropes, as illustrated in Figure 10-7. At other times, they leap between branches with arms outstretched like a ballerina (Fig. 10-8), or walk bipedally along branches like a tightrope walker (Fig. 10-9). In fact, *Hylobates* means "tree walker," and hylobatids are regarded by some as **preadapted** for bipedalism because their anatomy is adapted for some of the movements it entails. This raises the possibility that protohominids with some of the same anatomical features as living hylobatids (or bonobos, see Chapter 13) became modified for terrestrial bipedalism as they shifted toward ground living during hominid evolution. Besides their agility in moving from place to place, hylobatids are the apes that engage in the most suspension, often hanging by one arm to feed (Fig. 10-10). Such hang-feeding permits gibbons to reach fruit at the ends of small branches, known as terminal-branch feeding. Thus the brachiating anatomy of hylobatids facilitates both fluid and efficient

Figure 10-8 Gibbons sometimes leap through the trees.

Figure 10-9 Hylobatids sometimes walk bipedally along branches.

Figure 10-10 Hylobatids often feed while suspended by one arm.

Figure 10-11 Hylobatids frequently sleep sitting vertically in trees with their heads down on their knees.

locomotion between scattered food sources, and hang-feeding at those sources.

The postural behavior of hylobatids is interesting in another regard, as discussed by Russell Tuttle of the University of Chicago. The lesser apes have specialized pads of hairless, cornified skin, or ischial callosities, that are attached to the bony pelvis. These hardened areas cushion the buttocks and function as sitting pads, as is also the case for Old World monkeys. At night, hylobatids curl up on branches to sleep or, more often, they sleep in trees sitting vertically with legs folded, and heads down on their knees (Fig. 10-11). As noted by Tuttle, hylobatids are the only apes that do not build nests in which to sleep. Instead, they sleep comfortably sitting on their ischial callosities.

Hylobatids are basically frugivorous and spend up to a third of their waking hours feeding. Most species prefer ripe, juicy fruits such as grapes, mangoes, plums, and figs, and these are most often picked using a precision grip. A smaller but still considerable portion of leaves is also consumed, and insects and eggs are sometimes eaten. According to Leighton, siamangs eat a greater proportion of leaves than gibbons do, and consume more of the difficult-to-digest mature leaves. Because they are less choosy about their diets, siamangs need not travel as far afield to find enough food and, as already noted, they frequently have smaller territories than gibbons.

GIBBON SOCIAL LIFE

Gibbons have not been studied as much in the wild as the other apes. Because of this, and because the statistics that are available are similar across species (see Table 10-1), the best known species, *H. lar*, frequently has been accepted as a model for all gibbons. The classic portrait of hylobatids goes like this: They are one of the few primates, and the only ape, that is monogamous. The typical social group consists of a bonded adult male and female and up to four of their offspring, which are spaced over 2 years apart in age (Fig. 10-12). Males and females are often said to be codominant. A gibbon family occupies and defends a territory, both with loud morning calls that reinforce the spacing between different groups and, to a lesser degree, physically. Relationships between neighboring groups are hostile. Mated pairs tend to stay together for decades, during which time they produce up to 10 offspring. When one of these offspring approaches sexual maturity at around 8 to 10 years of age, it leaves its family to seek a mate that has also left its natal group. Parents sometimes encourage the departure of mature offspring by interacting aggressively with them, that is, by booting them out of the territory (if not the nest), and it is not unusual for the parent of the same sex to do the booting.

This traditional view of hylobatid behavior is currently undergoing revision, however, thanks to new research by a number of primatolo-

Figure 10-12 The typical gibbon group consists of a mated pair and their young offspring.

gists. Although much of the traditional view still holds, gibbon social life is neither as inflexible nor as uniform across species as once thought. Take monogamy, for example (see Box 10). Some years ago, John Mitani of the University of Michigan studied the responses of mated gray gibbons (*H. muelleri*) to recordings of songs of solitary females and males. When songs of female strangers were played back from the centers of the mated females' ranges, they reacted aggressively by beginning duets (see Duetting) while leading their groups toward the sites of the played-back songs. Males, on the other hand, led silent approaches toward the sites of the played-back songs from strange males. Mitani concluded that males are forced into accepting monogamy by the range defense of adult females, and that monogamy is maintained by intrasexual aggression that is supported by the opposite sex. The fact that monogamy requires such vigilant regulation suggests that it might not be as inflexibly ingrained as traditionally believed; in other words, if hylobatids weren't guarded by their mates, they would mate with outsiders.

Indeed, this recently was shown to be the case for siamangs and at least one species of gibbon. In 1994, Ryne Palombit of the University of California, Davis, reported that a female siamang mated with three males from a neighboring group. The following year, Ulrich Reichard of George August University in Göttingen, Germany, published the first report of similar **extra-pair copulations** (EPCs) in white-handed gibbons (*H. lar*), which occurred between one adult paired female and two adult paired males from neighboring groups. The EPCs accounted for 10% of all copulations reported for the three groups, and Reichard noted that one could only speculate about which copulations resulted in fertilization. Infanticide has been reported for *H. lar* in a zoo (but not the wild), and Reichard speculates that EPCs may be part of a strategy to forestall infanticide and to breed with a partner of superior quality to the current mate. According to Reichard (1995:108), "although, gibbons can still be characterized as monogamous, the observations in this study suggest a more flexible reproductive strategy for gibbons than expected from earlier descriptions of an obligate

REVIEW EXERCISES

1. What features set gibbons and siamangs apart from the other apes?
2. Why are hylobatids called lesser apes?
3. Describe the appearance of hylobatids.
4. Where in the world do gibbons and siamangs live?
5. What does the name *Hylobates* mean?
6. In what ways are hylobatids apparently preadapted for bipedalism?
7. What activity do gibbons frequently engage in when they hang by one arm?
8. How and where do gibbons sleep?
9. What anatomical feature is associated with the fact that, unlike the great apes, gibbons do not build nests in which to sleep?
10. What do gibbons eat? How does this compare to the diet of siamangs?

Box 10

Monogamous Primates—Why Be That Way?

Although in our society monogamy is the ideal norm, this is not the case for most other mammals including primates. In fact, monogamy such as that seen in gibbons and siamangs is rare in other anthropoids and prosimians. Among the latter, this mating pattern is found in only the indri (a lemur) and the spectral tarsier (see Chapter 3). Although, in the New World, cooperative polyandry is common among callitrichids (see Chapter 5), monogamy shows up only in the pitheciines (titi monkeys, owl monkeys, and some of the sakis; see Chapter 6) and possibly Goeldi's monkey. Old World monkeys are even less monogamous, with this mating pattern appearing regularly in only the Mentawai leaf-monkey (an Asian colobine, see Chapter 7). Monogamy sometimes appears as a variant in a few other primates, but the species noted above are the only ones that have been reported to regularly engage in this mating system—albeit with some extra-pair copulations in white-handed gibbons, siamangs, and possibly other species. Since it is so rare, one wonders why *any* primate practices it.

To begin with, the above-mentioned species all share certain features: They are arboreal and lack sexual dimorphism in body size, but may be dichromatic. Monogamous primates occupy and defend relatively small territories that contain dispersed patches of food. Duetting is used to defend the territory as well as to reinforce the pair-bond (see Neural Note 10). Females of most species lack dramatic sexual swellings during estrus (they don't need to advertise for a mate), and males are usually good and sometimes excellent parents (there are no extra adult females around to provide allomothering). Finally, social life may be somewhat subdued within family units, per-

haps as a natural result of youngsters growing up with few if any same-aged playmates.

But why monogamy? Put simply, this system allows its practitioners to use their energies to defend a relatively small territory (dietary requirements permitting) and therefore corner the market on its resources. Part of those resources are the sexually mature adults themselves, and the mate guarding associated with monogamy minimizes infidelity of mates (hardly the norm in polygynous primates). According to Palombit, relatively long-lasting pair-bonds in some white-handed gibbons appear to facilitate shortened inter-birth intervals and thus increase reproductive fitness.

Body size is usually restricted in highly arboreal primates, but sexual dimorphism (big male bodies) is not needed among monogamous primates because combat is usually done vocally. In fact, duetting is all important for maintaining monogamy because it is acoustically designed for long-range propagation; that is, one of its functions is to serve as a "Keep Out" sign. As Elliott Haimoff of the University of California, Los Angeles (1986:58), notes, "There have been no observations of any stable monogamous and territorial primate species in which duetting does not occur, or any polygynous primate species (either territorial or non-territorial) in which duetting does occur." (The owl monkey, *Aotus*, may be an exception to this rule [see Chapter 6]. It should also be noted that the spectral tarsier [see Chapter 3], which has a social system that is monogamous rather than polygynous 80% of the time, duets.) The bottom line is that monogamous primates may be somewhat lonely, but they don't have to worry as much as some other primates about paternity, daycare, or where to find their next meal.

monogamy, where sexual intercourse occurs exclusively between pair-mates."

Daily Activities

An average day in the life of a gibbon is relatively restrained, compared to most primates. Beginning at dawn, gibbons are active only for about 8 to 10 hours, at which time they enter the upper canopy of trees located in the central parts of their territories, where they may sit for hours before falling to sleep. They spend most of their active hours traveling and foraging for food in the main canopy, as well as a good deal of time resting. The typical family travels about a mile a day, but group members (except for dependent young) are likely to forage independently and to sleep in different trees (siamang pairs, on the other hand, usually sleep in the same tree). Leighton remarks that, given all their leisure time, it is surprising how little gibbons interact with one another. There are few vocal or visual signals between individuals, and although grooming and play are important, they occur less frequently than is the case for most other anthropoids. Part of the reason for their low level of social activity may be that, unlike many primates, gibbons do not grow up with playmates of a similar age (see In the Field, however).

DUETTING

Probably the most exciting time of day for gibbons is the morning when adults (primarily) engage in loud, ritualized song bouts with their neighbors. Hylobatid songs are said to be among the most beautiful of any land mammal and are species-specific. In most species, males sing long solos just before dawn and females sing in duet with their mates between dawn and mid-morning. Females rarely sing solo. According to Leighton,

> duets begin spontaneously or in response to calls by neighboring groups. They last about 15 minutes on average and include several "great calls" by the female, produced as she moves quickly about the crown of a tall tree. Great calls are introduced by and interspersed with exchanges of notes between the male and female. Different species show varying degrees of male participation in duets. . . . Other family members may also contribute vocally and behaviorally. Infants often squeal during a mother's great call, and juveniles may join duets with imperfect, sex-appropriate sounds. As they pass through adolescence, male offspring seem to be increasingly inhibited from singing with their parents. While subadult females usually give great calls simultaneously with their mothers, subadult males (except for siamangs) rarely sing in the presence of the adult male. (Leighton, 1987:140)

Haimoff has studied gibbon vocalizations in detail, with interesting results. He notes that gibbons produce calls that are short responses to specific external stimuli (such as alarm calls upon seeing predators), and songs that are more spontaneous (i.e., not driven by ex-

REVIEW EXERCISES

1. Describe the typical gibbon social group.
2. How do gibbons defend their territories? Do they frequently come to blows?
3. How common is monogamy among primates? Name the primates that are regularly monogamous.
4. Do monogamous hylobatids ever engage in extra-pair copulations (EPCs)? What might be the function of EPCs?
5. What package of physical traits do monogamous primates share, and how can they be explained?
6. What ecological variables are associated with monogamy?
7. What are the wider social ramifications of a monogamous lifestyle for primates?
8. What advantages are associated with monogamy? Why do you think it is so rare?

Close Encounters of the Friendly Kind

Thad Bartlett of Dickinson College is a primatologist who is helping to revise the traditional view of gibbons. For his dissertation research, Bartlett studied white-handed gibbons in Thailand, which resulted in a number of surprising and important revelations. In his own words:

WHILE IN THAILAND, THAD BARTLETT DISCOVERED A NUMBER OF SURPRISING THINGS ABOUT WHITE-HANDED GIBBONS. (PHOTO COURTESY OF THAD BARTLETT.)

Often, even the best preparation in the library does little to prepare you for the reality of field research. For many years, social interactions between gibbon groups were thought to be almost exclusively hostile or aggressive. All species of gibbon are territorial; that is, they defend their range against gibbons from other groups. Before going into the field, I had read a great deal about territorial interactions among gibbons. These encounters were said to include either vigorous chasing between males or "vocal battles," where opposing males hang below tree branches within view of one another and sing loud complex calls. As a result, I was completely unprepared to see friendly interactions between neighboring gibbon social groups. But this is exactly what I did see shortly after I began the research for my dissertation on the behavioral ecology of the white-handed gibbon in Khao Yai National Park, Thailand.

One morning a few months after I had arrived in Thailand, I was following a gibbon group previously labeled "group A." I knew group A was composed of five animals in all: two adults, two juveniles, and a newborn infant. I had been observing the group as they fed in a large fruiting tree. After feeding for 15 to 20 minutes on fruit, the group began to travel away from the tree down a very steep ridge. Rather than tumble down the slope after them, I opted for a less direct, but also less harrowing, route to the base of the ridge. In doing so, I lost contact with the animals for several minutes.

When I rejoined the group, their number had grown. White-handed gibbons are asexually dichromatic. This means that individuals have one of two coat colors, tan or black, irrespective of their sex. In other words, sometimes females are black, and sometimes they are tan, and the same is true for males. Moreover, the proportion of these two color phases within a gibbon social group helps to keep track of which particular group you are following. Some groups contain all tan animals, some all black, and others have a mixture. I knew that the group I had been following had three tan individuals (including an adult female and her infant), and two black ones. But once I had reestablished contact with the gibbons, I counted at least three black gibbons (one more than expected). I was stumped. Though I had been apart from the group for several minutes, it seemed unlikely that I could have lost them completely. Still, my first response was to conclude that this must be a different gibbon group altogether. So, I watched quietly in hopes of not disturbing them. If this wasn't an habituated group—one used to human observers—it would flee as soon as any of the animals saw me.

From my hiding place I could see four juvenile gibbons—two black and two tan—playing together low in the trees about 50 feet away from me. Still more gibbons watched from above. The juveniles chased one another, swinging and leaping from tree to tree in the forest understory. As they swung from one tiny trunk to another, their feet would almost touch the ground. Again, I was shocked. To see these renowned brachiators so close to the ground surprised me. What's more, most articles I had read indicated that gibbons rarely engaged in social behavior of any kind. Yet, here these juveniles were engaged in a long bout of playful chasing. A haze of confusion enveloped me.

Then slowly, as I began to decode what I was seeing, my confusion turned to disbelief. It is not easy to make visual contact with all the members of a primate group when they are dispersed in the trees overhead; this is true even of small groups. But by sinking to my knees, craning my neck, and crawling under branches, I was finally able to confirm that at least some of the gibbons I was watching were indeed from group A. The most important clue was the tan adult female with her tan infant clinging to her belly. Near her was a black adult male, her mate. These individuals looked down from above, as the four juveniles played below. The presence of group A accounted for two of the juvenile gibbons—one tan and one black. But there were still two juveniles to account for. Where did they come from? To which group did they belong? And if there were two social groups present, why weren't they fighting?

I checked my trail map. I knew from the information that I had gleaned from other researchers at the park that my location was near the border of two gibbon groups. Because gibbons defend their home ranges, the borders between groups remain relatively stable over time. With this information, I finally concluded that the other animals present must belong to group C. But it wasn't until I saw the second adult female that I was sure. Like group A, group C had five animals—three black and two tan. The female of group C, however, was black, as was her infant. Because gibbons live in monogamous social groups, you would never see two females with infants in a single group. So by identifying these two females, I had determined not only that two different gibbon groups were present, but also that these two territorial rivals were interacting peacefully—despite the fact that I had seen the males of these two groups chase each other on a number of occasions in the past.

Subsequent observations would reveal that these friendly interactions between neighbors were not limited to juveniles alone. In fact, I once watched an adult male groom a juvenile from a neighboring group for nearly 20 minutes. We now know that amicable interactions are a regular part of the social lives of gibbons. While the majority of group encounters are hostile, friendly encounters do occur regularly among neighbors. This significant discovery means we will have to reconsider the meaning behind territoriality in gibbons. Yet, it is not surprising that such encounters were missed in the past. Hostile encounters between gibbons groups tend to be easy to detect. Male gibbons sing loud calls and shake branches during territorial displays. Often I was able to find gibbons precisely because they were engaged in a territorial dispute. Friendly encounters, on the other hand, are eerily quiet. More importantly, only animals that are very comfortable with human observers are relaxed enough to exhibit these behaviors. Many of the animals in these two groups have spent much of their lives being followed by inquisitive primatologists plodding along below them. These gibbons accept humans as a normal part of their environment.

Much of what we have learned about gibbons, and primates in general, has emerged gradually over time. The social lives of primates, probably more than any other group of mammals, are complex and highly variable. It is certain that many currently held views and assumptions will have to be modified in the future. To be sure, there are many questions still to be answered.

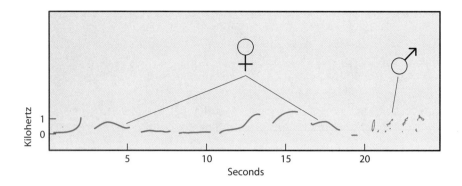

Figure 10-13 This sonogram shows the great call of *Hylobates lar*, which is begun by the female and joined by the male toward the end.

ternal stimuli such as predators). Songs are also pure in tone, and highly musical. In duets, both partners sing in an organized and mutually coordinated manner. Although the details of songs differ between the nine species of hylobatids, the duets that eight of the nine species (all but the Kloss gibbon) sing share certain general features. Each begins with an introductory sequence that Haimoff interprets as a warming-up phase. This is followed by alternations of great calls by the female with interludes in which both partners sing as described by Leighton (Fig. 10-13). The interludes, according to Haimoff, may allow the couple to organize and synchronize their duet to produce the great-call sequences that transmit information to other groups about the status (e.g., pair-bonded) and vigor of the pair.

Indeed, there are a number of reasons why Haimoff believes that the primary function of gibbon duets may be for intergroup communication. For one, duets are designed acoustically to be heard at a distance. For example, their notes are loud and pure in tone, which causes them to carry farther. In addition, duets are sung in the morning when atmospheric turbulence is minimal and temperature gradients are favorable for long-range propagation of songs through the forest.

Despite extensive research on gibbon duetting by numerous researchers, its exact function remains mysterious, as summarized by Warren Brockelman of Mahidol University in Thailand. One interpretation, noted already, is that gibbon duets convey information to other groups about characteristics of the singers (age, sex, health, pair-bond status) and their occupancy of a specific territory. In this sense, duetting may act as a spacing mechanism, as a warning to "keep out." On the other hand, duetting is repetitive to the point of appearing almost ritualistic and some workers have questioned whether such tiring repetitions might not be overkill (i.e., taking an excessive amount of time and energy) if their function is merely to convey information to outsiders. This second school of thought views duetting as a mechanism for reinforcing the pair-bond (see Neural Note 10). There is no reason to favor either the intergroup or intragroup explanation. Both could easily be true.

TERRITORIALITY AND SEX Gibbons are territorial, some would even say fiercely so, and there is no doubt that calling and singing play an important role in this regard. Calls may be given from the center of a

Singing in the Brain

Warren Brockelman (1984:286) notes that "there is a growing realization that most loud calling serves, in an ingenious way, intra-group as well as intergroup functions" and that "duets may well help to maintain the pair bond." However, Brockelman continues, "it is not clear how exactly duets would do this, or what kind of evidence would support the idea." In an effort to answer these questions, one can explore the relationship between singing and emotion in the human brain, and investigate whether there is any specific evidence related to how these activities are processed by the brains of hylobatids.

Neural Note 4 explained that speech and right-handedness are dominated by the same (left) side of the brain in most people. Certain aspects of both emotion and music, on the other hand, are strongly lateralized to the right hemisphere (which therefore probably provides the main neurological underpinnings for our species's predilection for emotion-laden love songs). For example, a song's melody usually is processed mostly by the right hemisphere. The right hemisphere is also normally responsible for tone of voice, which adds emotional nuance to the words produced by the left hemisphere. Indeed, one way to treat brain-damaged people who are no longer able to speak is to encourage them to sing words that otherwise cannot be uttered (i.e., singing flips the neurological processing of word production from the left to the undamaged right hemisphere).

Is there any evidence that hylobatid brains are lateralized for handedness or emotional vocalizations, similar to the brains of humans? Donna Stafford and her colleagues at the University of Memphis have addressed this question, with intriguing results. Not surprisingly, hylobatids are far from being right-handed in the human sense. However, adult female hylobatids have a propensity to pick up food with their right hands that increases with age, unlike males who show no particular pattern. This is interesting because human females tend to be more strongly right-handed than human males. Stafford also notes that three of nine hylobatids in her sample prefer to lead with their right limbs when they begin to brachiate, while the rest have no preference. Furthermore, the leading limb tends to shift to the other side when vocalizations accompany the initiation of brachiation, which Stafford and her colleagues (1990:407) attribute to "the arousal effects of species typical vocalizations." Thus, it appears that singing in hylobatids may be emotionally charged and controlled more by the hemisphere that is opposite to the hemisphere that normally dominates limb preference in at least some individuals, as is the case for the vast majority of humans. Perhaps the ducts of gibbons and those of humans use similar neurological pathways that communicate with evolutionarily old parts of the brain, that is, those which process the emotions that (among other things) are important for establishing pair-bonds.

territory, or its boundaries. As noted earlier, playback experiments indicate that the adult of the same sex usually responds aggressively to the song of an invading stranger, which suggests that vocalizations are used to defend mates as well as space and resources. Upon encountering each other, contesting males engage in conflicting vocalizations (hoos) and brachiating displays, and chase each other back and forth across boundaries. In this way, they dissipate tension and are often able to avoid direct physical contact (but see below). According to Leighton, the rates of intergroup encounters are highly variable across species, varying from once every 2 days to once a month. Actual fighting is relatively rare, but it does happen, as indicated by a limited number of direct reports, as well as suggestive observations of facial scars, wounds, and broken teeth—especially among males. Palombit, for example, reported a serious injury inflicted by one adult male white-handed gibbon on another that disappeared within 24 days and presumably died. Most encounters, however, involve only duetting. The extent to which females engage in aggressive territorial disputes is not yet clear. Nevertheless, they are generally regarded as less likely to be involved in actual fights but, again, it does happen (particularly in the case of widowed or recently paired females). Despite their high degree of territoriality, gibbons do not appear to patrol the borders of their territories the way that chimpanzees do (see Chapter 13).

Females first menstruate at about the age of 8 years, after which they cycle for only a few months every couple of years. Sexual swellings are usually lacking during their infrequent periods of estrus, although Palombit notes that very slight swelling and color changes have been reported for *H. lar*, *H. hoolock*, and possibly *H. syndactylus*. The traditional view that subadult gibbons that leave or are expelled from their natal groups form pair-bonds, establish territories, and start nuclear families with other expelled subadults needs revising in light of recent work by Palombit challenging the nuclear family model. Instead of finding other lone subadults, it now appears that dispersing white-handed gibbons usually replace adults in neighboring groups, thus creating nonnuclear families:

> In one case, forcible displacement of the resident male resulted in a group which included a young juvenile presumably fathered by the previous male, two younger juveniles (probably brothers) from the new male's original group, and (later) offspring of the new pair. Social relations within this heterogeneous group remained harmonious: the adults groomed all the young and play occurred between all preadult members. In only two out of a total of seven cases of dispersal seen did two subadults pair and disperse into new territorial space. (Brockelman et al., 1998:329)

However, the first observations of the formation of a new *H. lar* group happened to involve two subadults that had left their natal groups (which may explain the origin of the accepted nuclear family model), as delightfully summarized by Tuttle:

> Initially they noticed a male gibbon with a male siamang at fig laden trees. They often sang together in the morning and travelled together during part of the day. . . . One day, 25 minutes after they had called, a female gibbon appeared. The male gibbon attacked and briefly grappled with her and then followed her away from the siamang. The next morning the new pair sang together, including female climactic territorial great-calls. . . . They fed and travelled together for 3 hours. The honeymoon was abruptly busted up by a group of 3 gibbons that viciously attacked them and left with the female (MacKinnon and MacKinnon, 1977).

Both subjects were found alone the next day. The male called. But the fickle female began to travel with the lone siamang. She even sang with him. Three days after their separation, the 2 gibbons were reunited and resumed territorial morning song bouts. During their next battle with the homewreckers, the new pair held their own. The female was an active combatant, which is unusual among lar gibbons. The new group became firmly established in the area and produced an infant 32 months after their pairing. (Tuttle, 1986:245–246)

Either sex may initiate mating. The few matings that have been observed in wild hylobatids have been either from the back or face to face (the latter while suspended). Ellefson notes that both adults appear irritable when the female is in estrus, and that gibbons make whining squeals at the end of copulations that probably coincide with ejaculation. There is no apparent birth season for most species, and the occurrence of birth peaks is variable. Gestation length is approximately 7 months, and interbirth intervals are more variable than previously believed. For example, Palombit showed that intervals for *H. lar* and *H. syndactylus* exceed the 2 to 3 years commonly attributed to these two species. In some species, males are reported to spend more time playing with and grooming their offspring than do females. In keeping with this, females have sometimes been described as the more socially aloof parent.

A Further Word about Siamangs

We have already discussed a number of ways in which siamangs (*H. syndactylus*) are distinguished from gibbons. In fact, they are so different that some primatologists formerly placed siamangs in a separate genus, *Symphalangus*. The names *syndactylus* and *symphalangus* refer to joined digits, which is appropriate since the second and third toes of siamangs are frequently webbed together. Some workers have speculated that siamangs might be closer to the great apes than to gibbons because of their larger, sturdier bodies. However, as David Chivers notes, siamangs have specialized guts that reflect dietary preferences (e.g., a high reliance on young vines, figs, and mature leaves), which probably accounts for their relatively large bodies, smaller territories, shorter day ranges, and closer spatial relations between family members. Thus, most workers are comfortable placing siamangs in the genus *Hylobates* and viewing their unique features as reflections of their distinct ecological niche.

Siamangs of both sexes are striking-looking because of their uniformly dark, shaggy hair and bare, pinkish throat sacs that become inflated and function as resonating chambers during singing (Fig. 10-14). Some of the gibbon species also have throat sacs, but these are usually less impressive than those of the siamangs. Like all hylobatids, siamangs are monogamous. As noted earlier however, unlike other monogamous primates that lack sexual swellings during estrus (including most, but not all, female gibbons), female siamangs exhibit

Figure 10-14 The siamang uses its throat sac to vocalize.

slight swelling and change in coloration of the genitals when they ovulate.

Siamangs also differ socially from gibbons. Although they are territorial, siamangs have fewer interactions with neighboring groups. They also call less frequently, even within family units. Part of the reason for this may be that siamang families stay in fairly close visual contact during the day, whereas gibbon groups separate to forage for food. Siamangs seem generally more sociable than gibbons; for example, Palombit has shown that grooming is more important and more reciprocal between the sexes in siamangs than in white-handed gibbons where females groom their mates less than they are groomed by them. Subadult siamangs are better tolerated, and relationships with other species (including humans) appear friendlier. They also engage in a bit of role reversal. The female leads the group as it ranges, and subadult females take more initiative in the formation of new groups than is the case for gibbons. Adult male siamangs, on the other hand, carry infants after they have become weaned from the mother at about 1 year of age, until they become independent. Males also play with and groom their offspring more frequently than do mothers, and they engage in more peripheralizing of subadults when it is time for them to leave and form their own family units. Such intensive male parenting is highly unusual among primates. As Palombit (1996:322) summarizes it, "greater intra-group feeding competition in the gibbon and substantial paternal care in the siamang may account for the evolution of more reciprocal and stronger pair bonds in the latter."

HYLOBATID CONSERVATION

Despite their wide distribution, success in terms of numbers, and the fact that they are not especially vulnerable to natural predators, siamangs and most species of gibbons are endangered by ongoing deforestation, in some cases highly so. According to Brockelman and Chivers (1984:3), "the future of wild gibbon populations will depend entirely on the fate of the tropical forests of South-east Asia, on which gibbons rely totally for food, shelter and movement." They propose two strategies to secure the long-term survival of hylobatids. The first is to promote the gibbon's image along with forest conservation. The second tactic is to directly manage the forests beyond those located in national parks and sanctuaries, and to establish breeding research centers and programs for releasing captive animals back into natural habitats. According to Chivers, because of their resemblance to humans, gibbons are highly regarded by forest peoples, and even revered by some as representing good spirits. Sadly, without immediate and successful conservation efforts, these good spirits might cease to be.

SUMMARY

One lesser (*Hylobates*) and three great ape genera (*Pongo*, *Gorilla*, and *Pan*) share a distinctive pattern of features that reflect a brachiating anatomy and a high component of fruits, vegetables, or both in the diet. Apes differ greatly in body size, however, and in their degree of genetic similarity with humans. Molecular data are consistent with an early evolutionary appearance of the direct ancestor of gibbons and siamangs, followed by ancestors for orangutans and then gorillas, and finally (about 5–6 mya) a split between the ancestors of chimpanzees and hominids. Gibbons are much smaller than the great apes, spend virtually all of their time in trees, and engage in much more brachiation. Thus, many of the hylobatid traits reflect adaptations that are useful for brachiation and arboreal life in general (e.g., ischial callosities for roosting in trees). Other features are related to their monogamous lifestyle and territoriality.

The nine species of hylobatids are widely distributed across Southeast Asia, with little geographical overlap between them (except for the larger, more folivorous siamang, which overlaps with two species of gibbons). Coat colors and markings on the face and head vary between species, and they often differ between the sexes within species. All hylobatids are gorgeous brachiators that engage in much one-armed suspensory feeding, which allows them to pick fruits located at the tips of branches. They are the only ape (and one of the few primates) that is monogamous. A gibbon or siamang family consisting of two adults and up to four differently aged offspring lives in and defends a relatively small territory that contains sufficient, dispersed resources. When it occurs regularly in primates such as *Hylobates*, monogamy is associated with a suite of features including arboreality, lack of sexual dimorphism, territoriality, duetting, good parenting on the part of fathers, and generally subdued social interactions between family members. Together, this configuration allows a family to exclusively occupy and harvest the resources in a small territory, while strengthening the pair-bond (by duetting) and decreasing the likelihood of extra-pair copulations. It remains to be seen whether or not gibbons and siamangs will ultimately survive in light of the ongoing deforestation of their habitats.

Further Reading

BROCKELMAN, W. Y., REICHARD, U., TRESUVON, U. AND J. J. RAEMAEKERS (1998) Dispersal, pair formation and social structure in gibbons (*Hylobates lar*). *Behavioral Ecology and Sociobiology* **42**:329–339.

PALOMBIT, R. A. (1996) Pair bonds in monogamous apes: A comparison of the siamang *Hylobates syndactylus* and the white-handed gibbon *Hylobates lar*. *Behaviour* **133**:321–356.

REICHARD, U. AND V. SOMMER (1997) Group encounters in wild gibbons (*Hylobates lar*): Agonism, affiliation, and the concept of infanticide. *Behaviour* **134**:1135–1174.

11

ORANGUTANS:

THE SHY APE

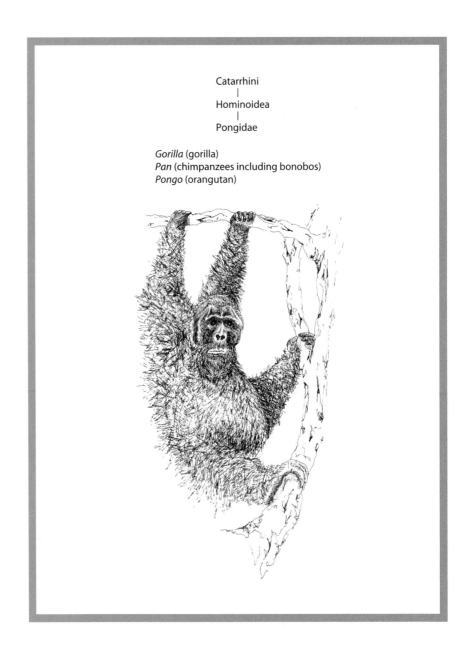

Catarrhini
|
Hominoidea
|
Pongidae

Gorilla (gorilla)
Pan (chimpanzees including bonobos)
Pongo (orangutan)

INTRODUCTION

Numerous molecular studies suggest that orangutans (*Pongo pyg-maeus*) were the first of the three great apes to split off from the lineage that eventually gave rise to hominids (see Fig. 10-1). This is thought to have occurred around 14 mya, and as noted in Chapter 1 orangutans are the only ape associated with any kind of appreciable fossil record (*Sivapithecus*). The orangutan is also the only great ape who lives in Asia instead of Africa. Sadly, although they were once widespread (living, for instance, in China), this endearing ape is today restricted to relict populations in the rain forests of Sumatra and Borneo (Fig. 11-1). This habitat reduction occurred partly because of shifts in Asia from subtropical to more temperate climates.

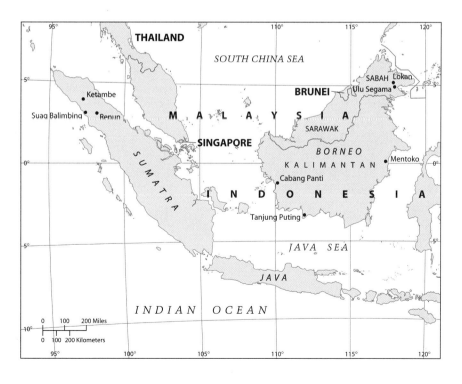

Figure 11-1 Field studies have been conducted on orangutans in Sumatra and Borneo at the locations shown here.

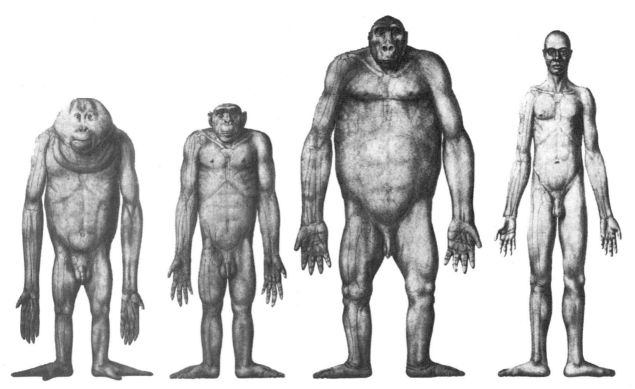

Figure 11-2 This classic illustration compares a male orangutan, chimpanzee, and gorilla with a man. For comparative purposes, the hair has been removed and the lower limbs straightened and turned out in the apes. Notice the long limb proportions and brachiating shoulder anatomy (see Chapter 10) that humans share with apes.

Figure 11-3 Unlike humans, great apes do not have chins, as can be seen in this lateral view of an orangutan skull.

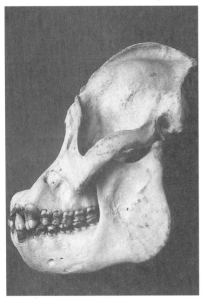

Although orangutans are distinctive from the other apes in a number of remarkable ways that are detailed in this chapter, they also share certain features with them. It goes without saying, for example, that orangutans lack tails and have the Y-5 lower molar pattern. They also exhibit the typical ape brachiating morphology, that is, mobile shoulder joints that are placed to the side, broad chests, and relatively long arms (Fig. 11-2). In addition, *Pongo* shares certain characteristics exclusively with the other two genera of great apes, *Pan* (chimpanzees including bonobos) and *Gorilla* (gorillas): Dentally, the great apes, particularly males, have large projecting canines that take up a good deal of space at the front of the jaw. This gives the sequence of teeth, or **dental arcade**, a U-shape and causes the whole jaw to stick out in front. Consequently, there is no chin in the human sense. Instead, the mandible slopes down and back from the front teeth (Fig. 11-3) and there is a thickening on the inside of the middle lower border known as a **simian shelf**. The huge prognathic jaws are accompanied by large cheek bones (**zygomatics**). As their name implies, the great apes are very large compared to the other nonhuman primates. They are also sexually dimorphic in body and (except for bonobos) canine size, and are comfortable assuming upright positions in trees and sometimes on the ground. Terrestrially, the African apes are prone to knuckle walking (see Chapter 12), while orangutans fist walk (see below). The great apes usually lack the ischial callosities of Old World monkeys and gibbons, which goes along with their proclivity for sleeping curled up in nests rather than in sitting positions. Great apes eat fruit, leaves, and other plant food. A taste for insects, nuts, honey, and meat varies with the species, population, and opportunity. These and other features are discussed below.

THE GREAT APE PATTERN

BONY FEATURES

1. Large canines that are sexually dimorphic in size
2. U-shaped dental arcades with 2.1.2.3. formula
3. Prognathic jaws with simian shelf (no chin)
4. Large zygomatics
5. Ischial tuberosities (callosities), rare
6. Big and sexually dimorphic (sagittal crests on some males)

OTHER FEATURES

7. Fist walk (orangutans) or knuckle walk (gorillas, chimpanzees) when on ground
8. Relatively long lives
9. Delayed maturity with long dependence on mothers
10. Capable of some symbolic behavior in the laboratory

What functional interpretation accounts for the great ape as opposed to the ape pattern? You will recall from Chapter 10 that most of the traits that characterize the ape pattern are related to brachiation in arboreal habitats. Five of the six bony features listed for the great apes, on the other hand, are related to the fact that males have bigger bodies (trait 6) and larger canines (trait 1) than females. Although big front teeth in both sexes are supported by an expanded and reinforced front jaw (traits 2 to 4), this package may have been driven by selection for dramatically larger canines in males. (In other words, because of the way genes mix and match from generation to generation, particular features that are selected for in one sex are likely to have some effect in the opposite sex.) Not only are great apes sexually dimorphic compared to hylobatids, but they also spend much more time on the ground (especially chimpanzees and gorillas), which is reflected in traits 5 and 7. As we will learn in this and subsequent chapters, large body size in the great apes is associated with longer lives (trait 8) and delayed maturation (trait 9) that results in prolonged nurturing of youngsters. Finally, some workers believe that the great apes have better cognitive abilities than other nonhuman primates (trait 10), an idea that is explored at the end of this chapter and in the remaining text.

Figure 11-4 Orangutans such as this female from Borneo occasionally stand upright, which is one of the reasons they were once viewed as wild people.

In addition to manifesting the general great ape pattern, orangutans have a number of highly interesting features that make them one of the most unusual and puzzling primates alive today. In fact, human explorers once believed that orangutans were wild people that lived in the woods because of their physical appearance and the fact that they sometimes stood upright (Fig. 11-4). As already noted, orangutans are restricted to relict populations living in the rain forests and swamps of Sumatra (*Pongo pygmaeus abelii*) and Borneo (*Pongo pygmaeus pygmaeus*). Although some workers classify the two forms as separate species because of recent genetic findings, this text keeps them as subspecies because of behavioral similarities and the fact that they interbreed easily in captivity. In general, the Sumatran subspecies is usually thinner and has a longer face than the species from Borneo.

Orangutan faces are turned upward (i.e., are dish shaped) and can appear very human-like, partly because of the narrow spacing between the eyes and the relatively modest browridge, compared to chimpanzees and gorillas. Among the apes, *Pongo* has the closest approximation of a forehead. In some individuals, this feature is set off by short bangs that almost appear to be trimmed. The remaining hair on the head and arms is long, although males, particularly those from Sumatra, have a variety of beards and mustaches. Hair color varies from a paler sandy or orange in the Sumatran population to a darker maroon in Bornean orangutans. Because of these hues, *Pongo* is sometimes called the "red ape." Skin color also varies from light to dark, and its pigmentation is frequently blotchy.

Secondary sexual characteristics that distinguish fully grown males from adult females are among the most striking features of orangutans. Adult males have huge laryngeal air sacs that spread out underneath the skin on their chests, arms, and shoulders, which gives them a generally flabby appearance. (Females have much smaller air sacs.) These throat pouches are especially large in *P. p. pygmaeus* males, and act as resonating chambers for loud calls that can be heard for at least half a mile (Fig. 11-5). (The other apes have less extensive air sacs.) Adult males also have huge flaps of skin on the sides of their faces, or cheek pads. Although it has been suggested that these flaps might act as a kind of funnel or megaphone for channeling loud calls, this explanation seems a little weak. Perhaps females simply prefer males with cheek flaps; in other words, these structures may be the result of sexual selection. Another possibility is that the face may just be riding along with the flabby air sacs; that is, cheek pads may be the result of genetic linkage, or the product of developmental events that influence more than one structure (these are known as **pleiotropic effects**). Because paddings of connective tissue cushion the sides of the sagittal crest in males, they have very large, high heads compared to females. Scent glands that are located in the middle of the chest are another male trait, but their function is mysterious.

Figure 11-5 Compared to males from Sumatra (left), male orangutans from Borneo (right) have especially large throat pouches.

Orangutans are also extremely sexually dimorphic in body weight, with adult females in the wild averaging about 37 kg, which is 50% to 60% of the average adult male weights listed for the two subspecies in Table 11-1 (Fig. 11-6). (Zoo animals that are provisioned sometimes reach twice these weights.) This amount of dimorphism is very large but not unheard of for primates, since gorillas (see Chapter 12) and baboons (see Chapter 9) are also very dimorphic. In general, smaller species can move more easily through dense canopies. On the other hand, terrestrial primates tend to be large, probably because firm ground offers better support for heavier individuals than do relatively flimsy branches. Although orangutan females are almost exclusively arboreal, adult males do much of their serious traveling on the ground, which is in keeping with their large sizes. (As we saw in the last chapter, marked sexual dimorphism is frequently associated with ground living.) The orangutan is extraordinary because it is not just

Figure 11-6 Orangutans are highly sexually dimorphic, with males commonly weighing twice as much as females.

TABLE 11-1 *Summary Statistics for Orangutans*		
	P. p. abelii	***P. p. pygmaeus***
Adult mean body weight F	37 kg (82 lb)	37 kg (82 lb)
M	66 kg (146 lb)	73 kg (161 lb)
Adult brain weight (g) F	(subspecies unspecified)	338
M		416
Mean		377
Maximum known life span (yr)	50	59
Habitat[a]	Sumatra: f, s, mf	Borneo: f, s
Locomotion[b]	s, qc, fw	s, qc, fw, b
Diet[c]	f, l, b, fl, a	f, l, sh, b, so, a
Mean no. of offspring	1	1
Seasonal breeders	No	No
Mean gestation (days)	260	244
Interbirth interval (yr)	8 (6–12)	7–8
Social organization	Semisolitary females/solitary males (roving male promiscuity)	
Territorial	Some males	Some males
Ratio of adult females to males		
Infanticide	0	0
Average group size	1–3	1–3
Density	5/km^2	1–3/km^2
Home range (females)	150–1000 ha	40–777 ha
Conservation status[d]	H	H

Note: Information in this table has been selected from numerous sources and should be viewed as tentative, i.e., subject to change in light of new information. Sources include but are not confined to: Tuttle, 1986 (body weights, brain weights); van Schaik and van Hooff, 1996 (ranges); Rodman and Mitani, 1987 (densities and other statistics); Rowe, 1996 (conservation status and other statistics). Abbreviations: +, yes; 0, no; var, variable. Blank entries indicate that no information is presently available. Dietary items are listed in approximate order of decreasing importance to the species.

[a]Habitat (Southeast Asia): f, forests; mf, montane forest; s, swamps.

[b]Locomotion: b, bipedal standing; fw, fist walking; qc, quadrumanous climbing; s, suspensory.

[c]Diet: a, animal prey; b, bark; f, fruit; fl, flowers; l, leaves; sh, shoots; so, soil.

[d]Conservation status: H, highly endangered.

the largest living arboreal primate, it is also the largest living arboreal mammal.

Orangutans move very carefully as they progress through the trees, which is adaptive since their large size makes them susceptible to falling and injuring themselves (healed fractures are fairly common findings in skeletal collections). Generally, the lighter females and young travel in the middle and upper levels of the canopy, while heavier males move through the lower parts. Orangutans do not leap through the trees as other primates including chimpanzees do. Instead they swing, sway, or clamber slowly from place to place while maintaining a strong grasp on a vine or branch with at least one extremity.

Orangutans have extremely flexible hip, knee, and ankle joints, which allow a much greater range of movement in the legs than is possible for humans or most other primates, as shown by dissection studies of Mary Ellen Morbeck of the University of Arizona and Adrienne Zihlman of the University of California, Santa Cruz. Because of this, and because the feet of orangutans are built for hanging on (and actually look like an extra pair of hands), *Pongo* is frequently described as **quadrumanous** (Fig. 11-7, color plate). Gisela Kaplan and Lesley Rogers of the University of New England, Australia, report that orangutans, despite having reduced thumbs that are located far down their wrists, can oppose the thumb to the fingers for precision gripping and manipulation.

Although they are superbly adapted for arboreality, orangutans come to the ground at times and, as noted, larger males are reported to do so much more often than lighter females and young. Terrestrial locomotion is by quadrupedal fist walking, in which the knuckles and palms of the balled-up hands (fists) contact the ground, while the weight of the hind limbs is carried on the outside edges (rather than the soles) of the feet. Orangutans may also "crutch walk" in this fashion by swinging both legs forward through the arms. This contrasts with the knuckle walking of the African apes, in which the entire weight of the forelimbs is placed on the knuckles and that of the hind limbs is distributed over the soles of the feet. One reason for descending to the ground is to seek and eat roots and vegetables that are found there.

Orangutans have unique dental features that are probably best explained by their diet. Unlike those of the African apes, the two upper central incisors are very large while the two lateral incisors are small. Orangutans also have thickened enamel on their molars, resembling the human condition and, again, differing from that of gorillas and chimpanzees. (Extra or supernumerary molars are also fairly common in *Pongo*.) One would expect large front teeth to be useful for biting into fruit (see Chapter 1), and thickened enamel on molars to aid in chewing hard or fibrous substances. Both expectations are borne out by the orangutan diet, which is surprisingly varied for an animal that is classified as basically frugivorous.

Kaplan and Rogers described in detail the enormous variety of foods eaten by orangutans: Favorite fruits include some that are difficult to harvest such, as the durian whose hard prickly shell is cracked open before it is ripe. Eating the fruit while it is green allows orangutans to favorably compete with other animals who prefer it ripe, such as elephants. Orangutans also eat leaves, shoots, bark, ants, termites, nuts, seeds, eggs, snails, worms, and honey. The central incisors are particularly good for peeling ribbons of bark from trees and gnawing the inside surfaces (discarding the inedible fibrous parts). Curiously, orangutans frequently eat soil (a behavior that has also been reported for other primates including gorillas), although it is not known whether this is for its mineral content, medicinal value, or as an aid to digestion.

Figure 11-8 The diets of orangutans are higher in calories and carbohydrates during mast fruitings (left) than during low-fruit periods when menus include less nutritious leaves, stems, figs, and bark (right).

Cheryl Knott of Harvard University recently studied the feeding ecology of *P. p. pygmaeus*, with fascinating results. Knott notes that periods called mast fruitings in which a large proportion of trees simultaneously bear fruit occur in Borneo about once every 4 to 7 years. During these periods, orangutans gain weight owing to diets that are high in calories and carbohydrates and low in fiber (Fig. 11-8). By analyzing urine samples, Knott determined that females have higher estrogen levels during mast fruitings, and this finding corresponds with her observation that they mate more frequently during these events. Knott's research may have implications for understanding the exceptionally long birth intervals (discussed below) that characterize *Pongo*. Her analyses of urine samples collected during periods of severe fruit shortages, on the other hand, show that orangutans (like humans) burn fat deposits when they cannot get enough to eat. Although mast fruitings are infrequent, smaller peaks in fruiting occur annually and are interspersed with longer low-fruit periods. The orangutan's proclivity for getting fat (Fig. 11-9) therefore may be an adaptation for coping with food shortages.

One of the most surprising features regarding diet is that orangutans have a taste for meat. If the opportunity arises, *Pongo* will occasionally eat birds, lizards, squirrels, or mice. There has even been a report of an adult female eating a gibbon that she found dead. Unlike chimpanzees, however, orangutans move too slowly to actively pursue most prey. One primate, however, the slow loris (*Nycticebus coucang*, see Chapter 3), moves so slowly that it is potential prey for even the ponderous orangutan. Sri Suci Utami of the Ketambe Research Center in Sumatra and Jan van Hooff of Utrecht University in The Netherlands report eight incidents in which four solitary orangutans captured and ate slow lorises. The orangutans were from two populations and, interestingly, all were females. One female named Getty was responsible for five of the incidents, and during two of them she allowed her 5-year-old son to take some meat.

The entire process involving discovery, capture, and consumption of the slow loris was observed for several of the incidents, which appeared to be very similar. All captures were made up in trees. In one case, Getty stared at the loris for about 15 minutes before slowly approaching it and grabbing it by the scruff of the neck. Another time, she slapped the loris until it fell to the ground, then pursued it while making high-pitched noises. Once the lorises were captured, they were instantly killed and eaten:

> The orangutan then sat down or made a small nest at a height of 10–15 m and started to eat the prey. First, she sucked out the brain and eyes and ate the palms of the hands. She ripped the neck and the chest apart with her teeth and ate the genitals. She picked out the entrails from the body cavity using a finger like a fish hook. She continued tearing apart the body using her teeth. She chewed all the bones and skin thoroughly before swallowing. After 2 h all body parts had been eaten. (Utami and van Hooff, 1997:161).

This report is especially interesting because another great ape, the chimpanzee, also has a taste for meat. However, as noted by Utami and van Hooff, orangutan hunting is a "stumble-upon-and-capture type" that differs from the more deliberate and strategic hunting done by (mostly) male chimpanzees (see Chapter 13).

Social Behavior

Inspired by pioneering fieldwork that was conducted on orangutans by a number of investigators during the 1960s, the late archeologist Louis Leakey encouraged Birute Galdikas (currently at the Orangutan Research and Conservation Project, Indonesia) to take up where they left off. As a result, in 1971 she began what would become the longest continuous field study of orangutans in the wild. According to Galdikas, the principal difficulty with studying orangutans is finding

REVIEW EXERCISES

1. What features set the great apes apart from gibbons and siamangs?
2. Discuss the functional interpretation that best explains the great ape pattern.
3. In what ways is the appearance of *Pongo* unique?
4. How do orangutans move through the trees? What is meant by the term *quadrumanous*?
5. What is *Pongo*'s form of locomotion on the ground? Which sex spends more time there?
6. Describe the dentition and diet of orangutans.
7. What are mast fruitings, and how might they explain why orangutans tend to get fat?
8. Does *Pongo* ever eat meat?

Figure 11-9 As is the case for people, orangutans tend to put on weight when food is plentiful.

them. Once located, an orangutan is followed until it makes and occupies a tree nest at night. The next morning, the observer arrives at the tree before dawn, and continues following the animal through the day. Based on the dedicated work of a number of primatologists who have studied orangutans since Galdikas began her long-term field observations, it is apparent that this ape is remarkably antisocial and somewhat mysterious.

Compared to other apes and monkeys, orangutans spend very little time in social interactions with nonrelatives. Instead, most of their waking hours are spent feeding, moving, and resting. The most common social unit for *Pongo* is a mother with one or two offspring, and adult females are thought to spend over 90% of their time alone with their young. However, more than one adult female with offspring may occasionally travel together for brief durations. For example, one of the biggest groups on record was reported by Galdikas, who observed three adult females, three infants, two juveniles, and one adolescent female traveling together for over an hour. Other workers have described numbers of orangutans peacefully feeding together in fruiting durian trees, but with little social interaction. Knott, for example, reports that she has seen as many as eight individuals feeding together in a fruiting tree during a mast season. Of all orangutans, adolescent females are the most gregarious, spending up to half of their time with other adolescent females or subadult males. However, despite these occasional female-centered groups, the most common occurrence by far (outside of mothers with dependents) is of solitary adults and subadults.

It would be a mistake to think that orangutans are always silent and uncommunicative, however. As summarized by Russell Tuttle, youngsters and, to a lesser extent, adults make a variety of squeaks, barks, screams, grunts, and groans during fearful or playful situations. Adults that are alarmed or annoyed may also produce sounds that have been characterized as kisses, gorkums, lorks, and grumphs, and they sometimes break branches or engage in other visual displays. The best known orangutan vocalization is the dramatic long call produced by adult males, in which

> the laryngeal sacs are voluminously inflated and contract heavily and irregularly during the call, which begins as a low soft grumble with vibrato, increases to a leonine roar, and terminates decrescendo with soft grumbles and sighs. During the roar, the subject protrudes and parts his lips and extends his neck. The lips are closed during the grumbling finale. (Tuttle, 1986:223)

Adult males are loners that move about in ranges overlapping with those of one or more females. On the occasions when they interact with other males, adults dominate subadults. Curiously, some but not all male subadults experience prolonged delays before reaching a fully adult appearance. This bimaturism among sexually mature males is not well understood. Although little is known about the actual sizes of male ranges, they are presumed to be larger than those of females (see Table 11-1). They also show a surprising degree of overlap with the ranges of other adult males, particularly during mast fruitings, as documented by Knott. Long calls probably serve as a spacing mechanism since they advertise not only the presence of adult males, but also their approximate position relative to other males (i.e., "stay away"). Of course, females are subject to hearing long male calls too, but whether or not these calls function in male-female communication is a matter of debate. Although John Mitani has shown that females usually do not respond directly to long male calls, other workers report that females occasionally respond to these calls by approaching their sources. Carel van Schaik of Duke University and van Hooff therefore speculate that tracking and responding differentially to long male calls may be one way that female orangutans express their preferences for certain partners. If so, long male calls serve not only as spacing mechanisms, but also as advertisements for receptive females ("Here I am").

Adult females travel within their ranges with their dependent offspring, often for many years. Adolescent females may establish ranges that are next to, or overlap with, those of their mothers. The interbirth interval can be longer than 8 years, at least partly because nursing (lactation) suppresses ovulation, menstruation, and sexual interest. The females that become sexually receptive may mate with a number of males at any time during their 22- to 30-day menstrual cycles. (Notably, sexual swellings that advertise estrus in many Old World primates are lacking in orangutans.) Despite the fact that, when circumstances permit, adults of both sexes mate promiscuously if infrequently (i.e., dispersed polygyny described in Chapter 3), van Schaik and van Hooff call the orangutan social organization "roving male promiscuity." In addition to promiscuous matings, a couple may form a consortship that lasts for several days. As a result of all this, ovulating females are few and far between and adult males seem to change their long-term ranging patterns to maximize their exposure to them, while minimizing their contacts with other competing males.

Subadult and adult males frequently force copulation with sexually unreceptive and actively resistant females. It is not surprising, then, that nonconsorting adult females go out of their way (e.g., by changing trees) to avoid subadult and adult males. Although adult males have an aversion for each other, Mitani has observed that most of the rare male-male interactions that take place do so in the presence of females. Furthermore, large or high-ranking males disrupt matings by other males, and adult males displace subadult males. As noted earlier, the degree of sexual dimorphism seen in orangutans is extreme and begs explanation. Peter Rodman of the University of California, Davis, and Mitani make a strong argument that the high degree of sexual dimorphism in orangutans is a consequence of male-male competition for sexual access to females.

On the other hand, Zihlman notes that male orangutans have fewer constraints on body size than females, and that small body sizes of the latter contribute significantly to the large amount of sexual dimorphism that typifies the species:

> Female body size (within the chimpanzee range) is probably constrained by the demands of gestation and lactation in a highly arboreal habitat and in associating with juvenile offspring. Females forage for seasonal foods and travel with infant and juvenile offspring through the forest canopy, which requires locomotor skill, time, and energy. Maintaining small body size might help reduce energy requirements and minimize potential feeding competition between females and juveniles. (Zihlman, 1997:101)

Mothering

Unlike most primates, infant orangutans (and gibbons) do not have much opportunity to play with peers. Instead, they enjoy a prolonged period of intense interaction with their mothers, and perhaps one sibling. Next to humans, orangutans have the slowest rate of development of any primate and have been observed nursing in captivity for as long as 6 years. The mother-infant relationship is therefore crucial for the learning, socialization, and survival of young orangutans.

Kaplan and Rogers, who have done research on wild Bornean orangutans for a number of years, beautifully described (and photographed) one mother-newborn interaction. The picture that emerges is quite different from the usual portrayal of orangutans as antisocial, solitary, and distant:

> We have observed a . . . mother with a six-day old infant. She fondled and touched the infant repeatedly over a two-hour observation period, investigating the infant's arms, legs, taking the small hands gently into hers and inspecting the fingers. She also took the infant's head and placed the face as far into her mouth as she could. She cradled and held it closely seeming entirely absorbed in her baby and, judging by her facial expression, she was smiling. (Kaplan and Rogers, 1994:84–86)

After the first week of life, the infant clings to its mother by hanging onto her hair. At first, it is carried on the mother's abdomen (ventrally). Later, the infant will perch on a hip or cling to the mother's back as the two travel. During the first 5 years of life, young orangutans learn from their mothers everything that they need to know to survive: what foods to eat, the dangers to avoid, ways to move safely through the canopy, and how to build sleeping nests. As Kaplan and Rogers point out, orangutans are not just schooled by their mothers, they also receive love, attention, and affection from them (Fig. 11-10). It is not until they reach the age of about 7 years that orangutans begin to wander off by themselves. If they become orphaned much before that, they are not able to fend for themselves and are likely to die unless they can get themselves adopted.

Kaplan and Rogers report that an orangutan named Abbie was no more than 3 years old when her mother died. Being resourceful, upon

Figure 11-10 Orangutan mothers are at times completely absorbed with their infants.

first meeting the researchers in the rain forest, Abbie attempted to adopt one of them (Kaplan) as a surrogate mother:

> She had no particular curiosity for any item that Gisela possessed, not even the tripod. Instead, after a few hesitant advances, she clambered up from behind, eventually settling on Gisela's hips not, as it turned out, for a brief cuddle but seemingly permanently. (Kaplan and Rogers, 1994:94)

Abbie was one of the lucky orphans who managed to survive. This is what happened when Kaplan and Rogers encountered her again 2 years later:

> When Abbie reached Gisela, she simply sat down opposite her. A few greeting silences passed, usually with eyes averted, but Abbie remained seated close. Inexplicably, she eventually took Gisela's hand, turned its palm upwards and ran her index finger over the lines on the hand. (Kaplan and Rogers, 1994:95)

One of the most important observations made by Kaplan and Rogers concerns the stereotype of orangutans as solitary primates. Females spend much of their adult life in the company of their young, and most orangutans spend their first 5 to 7 years interacting intensely with their mothers (and perhaps a sibling). These orangutans are neither alone nor antisocial. Although it's true that orangutans are not as gregarious as most other anthropoids, it is a mistake to think of them as totally isolated and lacking in social life. In fact, the relatively calm social life of orangutans is tied to their feeding ecology. Because orangutans are large-bodied primates that prefer fruit, they are usually unable to find enough food in one place to support group feeding. As noted earlier, however, orangutans do get together when infrequently occurring mast fruitings provide an exception to this ecological constraint. It is also a mistake to think of orangutans as less intelligent than chimpanzees or gorillas simply because they are, by nature, more subdued.

Orangutan Cognition

The average volume of the orangutan brain is about 375 cubic centimeters (cc), which is larger than the brains of monkeys, but considerably smaller than the rough average of 1350 cc for humans. (One gram of brain tissue equals approximately 1 cc of brain volume.) Great apes obviously have larger bodies than monkeys, and it naturally takes a somewhat bigger brain to run a larger body. Interestingly, when average brain volume is analyzed taking body size into account, that of *Pongo* (and indeed the other great apes) is no more impressive than the average brain sizes for a number of monkeys. Humans, on the other hand, have brains that are three times the size that would be expected for an ape or monkey that weighs as much as a person (see Neural Note 12). Nevertheless, many workers believe that great apes

REVIEW EXERCISES

1. Are orangutans very gregarious?
2. Describe the most common social unit for *Pongo*.
3. What kinds of sounds do orangutans make?
4. Describe the ranging patterns of male and female orangutans.
5. What is the typical age difference between orangutan siblings?
6. Why has the mating system of *Pongo* been described as "roving male promiscuity"?
7. Discuss the social ramifications that appear to be associated with a scarcity of ovulating (i.e., receptive) female orangutans.
8. What may account for the high degree of sexual dimorphism in *Pongo*?
9. How long do infant orangutans stay in close contact with their mothers?
10. What kind of playmates do young orangutans have?
11. Why is it incorrect to characterize *Pongo* as totally isolated and lacking in a social life?

are more cognitively sophisticated than monkeys, based on findings from apes that have been reared with humans, as well as results of laboratory experiments, and observations of apes in the wild. If apes do prove in some regards to be more intelligent than monkeys, it is likely to be due as much to the wiring and neurochemistry of their brains as to their sizes. Although they are relatively shy compared to the other apes, orangutans have exhibited a surprising amount of human-like cognitive capacities under some circumstances.

PROJECT CHANTEK Chantek is perhaps the world's most famous orangutan, having been intensively studied for years by Lyn Miles and her colleagues at the University of Tennessee at Chattanooga. When he was 9 months old, Chantek began to interact intensively with humans in a trailer on campus and to learn American Sign Language (ASL) for the deaf (Fig. 11-11). At first, he learned signs from teachers who molded his hands into the correct shapes. As time progressed, however, Chantek began to imitate his teachers' signs more spontaneously. Within a few years, he had acquired about 150 signs and was producing multisign sequences. He also invented a few signs (e.g., "eye-drink" for contact lens solution), and occasionally used his newfound language to deceive his human caretakers. By the time he was 4 years old, Chantek's cognitive development had paralleled but lagged somewhat behind that of similarly aged children. For example, unlike most humans, Chantek failed to recognize himself in a mirror by about 2 years of age, but could do so when he was about 4. The ability to recognize one's own image in a mirror is considered to be a sign of advanced cognition that is lacking in nonhuman primates other than the great apes (see Neural Note 11). Imitation is also regarded as a sign of higher cognition, and Chantek (and other orangutans) excelled here too by imitating faces, sounds, and actions; by deferring imitations until later; and by engaging in pretend play. As Miles points out, all of these behaviors occurred within social contexts; that is, Chantek had been encultured by humans. What about orangutans in more natural settings?

Figure 11-11 Chantek is well known for his ability to use American Sign Language. (Photo courtesy of Lyn Miles.)

Of Frontal Lobes and Mirrors

In a series of intriguing experiments with mirrors, psychologist Gordon Gallup of the University at Albany demonstrated that, unlike other mammals including monkeys, orangutans and chimpanzees have a sense of self or "me," much as humans do. To show this, Gallup painted red dye over the ears and eyebrows of a variety of anesthetized animals, which were then positioned so that they would awaken near mirrors. Upon encountering their images, most animals did not appear to recognize the reflections as their own. Some responded by attacking or threatening that other animal in the mirror, while others investigated its patches of dye by directly touching the mirror. Orangutans and chimpanzees, however, touched their own dye marks while looking in the mirror; that is, they recognized that the other animals in the mirror were themselves. Gallup's famous mark test continues to be widely used to investigate animal cognition. The evidence for gorillas is equivocal, although at least one gorilla (Koko) shows signs of self-recognition (see Chapter 12). A number of other gorillas, however, do not. It may be the case that all normal, adult great apes are capable of recognizing their own images under some circumstances.

What parts of the brain are involved with self-recognition? Frontal lobes are probably involved and, interestingly, these are steeper in orangutans and chimpanzees (and humans) than they are in gorillas, as pointed out by Karl Zilles and Gerd Rehkamper of the University of Cologne, Germany. Temporal lobes are known to be important in human recognition of specific people, their names, and related biographical information. Emotional reactions, on the other hand, including those to specific people, are processed by structures deep within the brain that are part of the limbic system. It is likely that all of these parts of the brain participate, to a greater or lesser degree, in self-recognition in the great apes and in humans. Primates with smaller brains apparently lack the precise neurological wiring in these areas that is a prerequisite for self-recognition in mirrors and, presumably, a sophisticated sense of self.

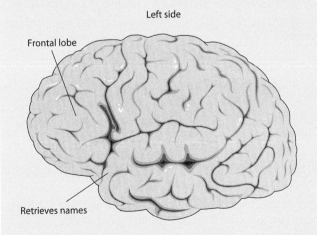

Left side

Frontal lobe

Retrieves names

Exterior of brain

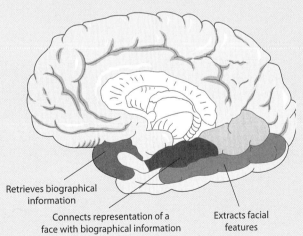

Right side

Retrieves biographical information

Connects representation of a face with biographical information

Extracts facial features

Inside surface

The tips of the temporal lobes (or temporal poles) are activated when humans recognize and name a familiar face.

Anne Russon of York University in Toronto studied imitation in orangutans that had been captured from the wild, orphaned, and then rescued and kept in a camp in Borneo by human caretakers until they were fit for release back into the forest. Most of these orangutan rehabilitants and their offspring were free-ranging; that is, they moved freely in the camp and its surrounding reserve. The orangutans varied in the extent of their day-to-day contact with humans, and Russon notes that their behaviors occurred freely in keeping with their own desires and interests. Thus, while not totally wild, these subjects seemed ideal for providing information about the extent to which *Pongo* learns new behaviors by observing the demonstrations of others (i.e., by true imitation).

Orangutan rehabilitants seem to enjoy imitating behaviors that they have observed in humans. In fact, Russon documented an extensive list of incidences in which orangutans engaged in both immediate and delayed imitations of human practices such as applying insect repellent, sawing wood, brushing teeth, sweeping with a broom, and putting on T-shirts. For example:

> Supinah watched a worker sharpen an axe blade: he dipped a stone in a pail of water then rubbed the stone back and forth along the blade's edge. After a minute he offered her the blade and stone and gestured to her to do the same. She accepted them and rubbed the stone against the blade's face. The worker retrieved his tools and resumed work; Supinah resumed watching. Five minutes later she motioned for the blade and stone, and the worker gave them to her. She took them, but this time retrieved the pail of water and dipped the stone into it before rubbing the stone against the blade's face. (Russon, 1996:168)

It is important that behaviors such as these be interpreted within social contexts. As Russon observes, orangutan rehabilitants selectively imitate demonstrations of people whom they prefer as social partners. Furthermore, because the complete individual histories of these orangutans are unknown, the extent to which previous training by human caretakers is responsible for instances of imitation is unknown.

Orangutans also engage in tool use that is less obviously social. In more natural settings, for example, they are known to drop branches on human observers, or wave them in a threatening manner. Vines are used to get from point A to point B, and large leaves are sometimes used as rain hats. Rogers and Kaplan recently discovered that *Pongo* occasionally uses sticks, with great concentration, to scratch the ground, but whether this is for play or some other purpose is not clear. Using tools is one thing; however, making them is another. When it comes to tool manufacture, wild orangutans have a reputation for lacking the keen abilities of their chimpanzee cousins (see Chapter 13).

That reputation is in the process of being modified thanks to a report by van Schaik and colleagues that wild Sumatran orangutans at Suaq Balimbing (see Fig. 11-1) of both sexes and all ages (except young infants) habitually make tools in trees and use them to extract termites, ants, bees, and honey from tree holes. Adults also make and use tools to remove irritating hairs and seeds from ripe fruits:

> Tools were made from live branches taken near the tree hole. Twigs and leaves were always removed. In some cases, the bark was also stripped with the teeth, and tips were sometimes chewed, thereby fraying them, or split, thereby flattening them. In most cases the tool was kept between the teeth when used. (van Schaik, Fox, and Sitompul, 1996:186)

Remarkably, tool dimensions vary by task (i.e., the orangutans appear to have a tool kit, Fig. 11-12), and a number of observations suggest that orangutans make instantaneous adjustments in the dimensions of their tools as they go along. As van Schaik and colleagues conclude, the natural cognitive abilities of wild orangutans are on a par with those of wild chimpanzees, but tool use only emerges under certain conditions such as the high rate of insectivory that characterizes the particular population they studied.

Rogers and Kaplan also reported a new form of tool use by three orangutan females at a rehabilitation center in East Malaysia. One 15-year-old mother was recorded engaging in this behavior on videotape, which permitted detailed analysis. She left a feeding table with a mouthful of bananas, and then picked 6 to 12 large, oblong leaves that she held in her right hand as she climbed high into the canopy. She then fashioned the leaves into a fan and spat chewed-up bananas into it. Both she and her infant alternately fed from the leaf plate. During this 22-minute sequence, the mother took all of the food into her mouth, rearranged the leaves, and then spat the food onto the plate again, six times. Rogers and Kaplan speculated that this process may aid digestion by increasing the exposure of food to saliva, or it may provide a means for orangutans to eat a large mouthful of food in private. Whatever the case, this observation is extremely interesting because it provides another rare example of orangutans actually making and using tools in a manner that does not appear to be imitating the actions of people.

Although they are shy compared to the other great apes, orangutan rehabilitants are surprisingly social for a species that has long been classified as solitary because of its behavior in the wild. Orangutans are also clearly intelligent, demonstrating a capability for recognizing themselves in mirrors, true imitation, and a certain amount of rudimentary tool use and production. The differences between orangutans and the other great apes in their dispositions, propensity for tool use, and curiosity are delightfully captured by Benjamin Beck of the National Zoological Park in Washington, D.C.:

> There is an anecdote that circulates among zoo folk describing the results of placing a screwdriver in the cages of an adult gorilla, chimpanzee, and orangutan. The gorilla would not discover the screwdriver for an hour and then would do so only by stepping on it. Upon discovery, the ape would shrink in fear and only after a considerable interval would it approach the

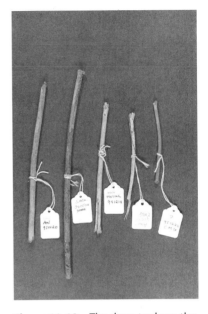

Figure 11-12 The three tools on the left were made by orangutans to extract insects from tree holes. The two on the right were used to process fruit. (Photo courtesy of Carel P. van Schalk.)

In the Think Tank

Daniel Shillito is a doctoral student in the Department of Psychology at the University at Albany, who collected data for his dissertation while working in a unique facility that enables scientists from around the world to examine the behavior and cognitive abilities of orangutans. In a sense, rather than visiting the field, the field visited Shillito and the other researchers that frequent the Think Tank of the National Zoological Park in Washington, D.C. In his own words:

DANIEL SHILLITO (PHOTO BY JESSIE COHEN, NATIONAL ZOOLOGICAL PARK.)

One of the best things about the Think Tank is that participation in its various activities is completely voluntary on the part of the orangutans. Some choose to reside there full-time, while others commute to the exhibit from the nearby Great Ape House, which is located across the zoo. The animals travel between the two exhibits using a series of towers linked together by steel cables, 40 feet above the heads of zoo visitors. This 500-foot-long system, called the O-Line, permits visitors to observe the orangutans overhead engaging in their natural form of locomotion while providing the animals with a variety of social choices. Animals decide when and where to travel, associating with some orangutans while distancing themselves from others, much as they would in the wild. The O-Line thus allows zoo scientists to examine the social and biological factors influencing individuals' social-distancing decisions.

Once at the Think Tank, the orangutans are free to interact with each other, or to participate in a variety of ongoing experiments that explore their cognitive abilities, which are not as well un-

WHEN THEY ARE IN THE MOOD, ORANGUTANS AT THE NATIONAL ZOOLOGICAL PARK USE THE O-LINE TO COMMUTE TO THE THINK TANK. (PHOTOS BY JESSIE COHEN, NATIONAL ZOOLOGICAL PARK.)

derstood as those of the other great apes and many monkeys. Because a common feature of all of the studies at the Think Tank is that the orangutans choose to participate in them, scientists need to design studies that challenge the animals enough so that they are motivated to participate, yet are not so cognitively taxing or ambiguous that the animals become frustrated or lose interest in the task. For experiments to be successful, researchers must also develop relationships with orangutans that are built on mutual respect, trust, and cooperation.

Most of my colleagues' and my ideas for experiments originate from our day-to-day interactions with the animals. For example, the idea for one of our experiments came rather serendipitously when one of the animal keepers accidentally left a bottle of fruit-flavored vitamins outside of the orangutans' enclosure. When my colleague and I arrived on the scene, one of the orangutans, having not been trained to do so, reached through the bars of her enclosure, gently grasped my colleague, and moved him along her enclosure to a location directly in line with the vitamins. She then manipulated and turned him so that he was oriented toward the vitamins. The orangutan waited until the experimenter was facing the vitamins before pushing him toward them.

Using an elaboration of the animal's natural behavior, Rob Shumaker, Ben Beck, Gordon Gallup, and I have shown that some orangutans are able to understand what people currently do and do not see, a task that human children are unable to solve until they are between 2 and 3 years old. In our study, an orangutan watched us hide food in a location outside of her enclosure and beyond her reach. We then presented the orangutan with two naive experimenters: One experimenter could see, while the other experimenter was unable to see because he was wearing a bucket over his head. We figured that if the orangutan understood the different visual perspectives of the two experimenters, she would direct only the sighted experimenter to the hidden food, while ignoring the experimenter wearing the bucket. As we predicted, the orangutan directed the sighted experimenter to the food. However, on some of the trials she chose to manipulate the experimenter wearing the bucket. What was so surprising was that she removed the bucket from the visually impaired experimenter (so he could see) before she showed him where the food was hidden. While completely unexpected, both strategies demonstrated that the orangutan was able to understand (and even alter) the two experimenters' points of view.

In another experiment Rob Shumaker, Ben Beck, Gordon Gallup, and I examined the ability of orangutans to cooperate with an experimenter on a tool-use task. To do this, we put an experimenter in one of the animal enclosures and then placed some cookies beyond his reach. From an adjacent enclosure, an orangutan watched the human trying unsuccessfully to obtain the cookies. Previous testing demonstrated that an orangutan in the human's situation could use a tool (e.g., a stick) to rake in the cookies; however, on these trials the experimenter had no tools in his enclosure. All of the potential tools were located in the orangutan's enclosure. In order to solve the task then, the human needed the orangutan's assistance. After observing the human's dilemma, the orangutan found a tool within its own enclosure and voluntarily gave it to the experimenter. The experimenter was then able to use the tool to rake the cookies within his reach and share them with his orangutan partner. We found similar results with another orangutan.

Because an important goal of the Think Tank is to demonstrate how scientists assess the cognitive abilities of orangutans and other animals to zoo visitors, all experiments are conducted in full view of the public. (Interestingly, on one trial of the cooperative tool-use task, a zoo visitor, unaware of the ongoing experiment, asked the experimenter if he needed the visitor's help in reaching the cookies!) The Think Tank is an incredibly exciting place to visit because it is rewarding for all parties concerned: Scientists are able to examine orangutans' impressive cognitive abilities and better understand how they view and organize their world. Zoo visitors share not only the researchers' findings, but also their sense of excitement about doing the experiments. Last but not least, the orangutans have free access to a stimulating environment that enriches their lives.

tool. The next contact would be a cautious, tentative touch with the back of the hand. Thus finding it harmless, the gorilla would smell the screwdriver and try to eat it. Upon discovering that the screwdriver was inedible, the gorilla would discard it and ignore it indefinitely.

The chimpanzee would notice the tool at once and seize it immediately. Then the ape would use it as a club, a spear, a lever, a hammer, a probe, a missile, a toothpick, and practically every other possible implement except as a screwdriver. The tool would be guarded jealously, manipulated incessantly, and discarded from boredom only after several days.

The orangutan would notice the tool at once but ignore it lest a keeper discover the oversight. If a keeper did notice, the ape would rush to the tool and surrender it only in trade for a quantity of preferred food. If a keeper did not notice, the ape would wait until night and then proceed to use the screwdriver to pick the locks or dismantle the cage and escape. (Beck, 1980:68–69)

Although this sketch is merely anecdotal, researchers who work with great apes (see In the Field) believe that it accurately portrays their personalities and cognitive styles.

THROUGH A GLASS DARKLY

Because they resemble us in so many ways, the great apes hold a special fascination for people. Unfortunately, this has contributed to their undoing and orangutans have disappeared from many of their former haunts. The decimation of populations of *Pongo* is due not only to deforestation by humans for agricultural and logging purposes (an old story), but also to the fact that the species have been intensively hunted and killed either because they are pests that raid crops or because parts of their bodies are used for trophies (Fig. 11-13) or medicine. Additionally, untold numbers of orangutan mothers have been killed and their young captured for the pet and zoo trade and, incredibly, this great ape is still sometimes hunted for sport. As you know, efforts are currently underway to rehabilitate domesticated orangutans and return them back to the wild. Despite the fact that orangutans are protected in a number of reserves and national parks in Borneo and Sumatra, however, *Pongo* continues to be highly endangered.

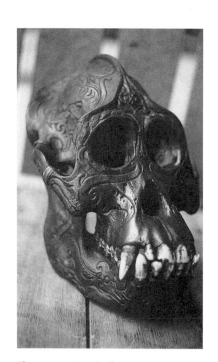

Figure 11-13 Sadly, orangutans are hunted for body parts such as this skull that was carved to sell to people.

SUMMARY

The great apes share a suite of features including large bodies and a high degree of sexual dimorphism in body and (usually) canine size. They also build platforms, or nests in which to sleep at night. Because the ancestors of orangutans were the first to branch off of the common stock that led to the three extant apes, *Pongo* provides the central focus of this chapter. Orangutans are unusual-looking because of their upturned faces, high foreheads, red hair, and somewhat flabby bodies. They are also highly sexually dimorphic, with adult males frequently being twice the size of adult females. *Pongo* is extremely careful and almost double jointed when moving through the canopy, but fist walks (or crutch walks) when on the ground. Compared to other apes and monkeys, orangutans spend very little time in social interactions with nonrelatives. Instead, most of their waking hours are spent moving, feeding, and resting. A mother with one or two offspring is the most common social unit, while adult males are usually loners who move about in ranges that overlap with those of one or more females. Interbirth intervals are extremely long, and receptive females are scarce. This raises the possibility that the high degree of sexual dimorphism that characterizes orangutans may be the result of male-male competition for sexual access to females. Although they are relatively subdued socially, various lines of evidence suggest that orangutans are highly intelligent. The famous orangutan Chantek was able to learn about 150 signs and to use them in multisign sequences. He also developed the ability to recognize himself in a mirror and, like other orangutans, excelled at imitating human behaviors. In more natural settings, orangutans use branches, vines, and leaves as rudimentary tools. Recently, orangutans have even been observed fashioning tools including a kind of plate from leaves that is used for holding chewed bananas. Clearly, although they are somewhat retiring compared to their lively chimpanzee cousins, orangutans appear to exhibit human-like cognitive capacities under some circumstances.

FURTHER READING

GILBERT, B. (1996) New ideas in the air at the National Zoo. *Smithsonian*, **27**(June):32–43.

KAPLAN, G. AND L. ROGERS (1994) *Orang-utans in Borneo*. Armidale, NSW, Australia: University of New England Press.

KNOTT, C. (1998) Orangutans in the wild. *National Geographic* **194**:30–57.

VAN SCHAIK, C. P. AND J. A. R. A. M. VAN HOOFF (1996) Toward an understanding of the orangutan's social system. In: W. C. McGrew, L. F. Marchant and T. Nishida (eds) *Great Ape Societies*. Cambridge: Cambridge University Press, pages 3–15.

REVIEW EXERCISES

1. How large are orangutan brains compared to those of humans?
2. Who is Chantek, and what are some of the signs of higher cognition that he manifests?
3. What evidence is there that orangutans have a sense of self or "me"?
4. How has Gordon Gallup demonstrated this experimentally?
5. What parts of the brain are involved in self-recognition?
6. How does cognitive development compare in children and young orangutans?
7. Name some of the human activities that orangutans have been observed to imitate.
8. What kinds of tools do orangutans use in more natural settings?
9. Describe the new form of tool use recently reported for *Pongo* by Rogers and Kaplan.
10. What is the Think Tank and what goes on there?
11. Do you think that orangutans have a bright future? If not, why are they endangered?

12 GORILLAS: THE LARGEST PRIMATES OF ALL

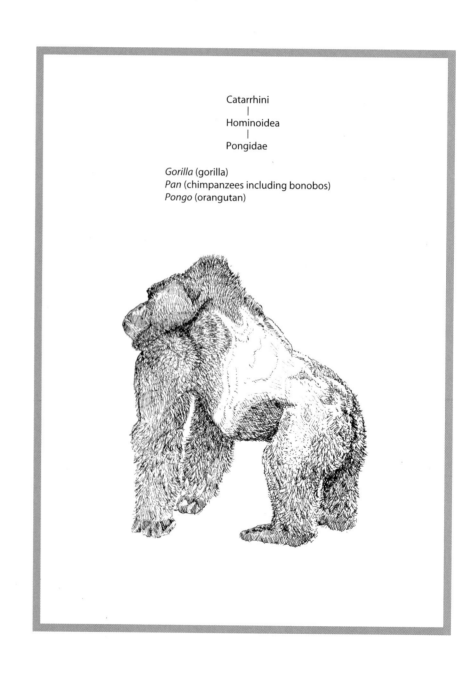

Catarrhini
|
Hominoidea
|
Pongidae

Gorilla (gorilla)
Pan (chimpanzees including bonobos)
Pongo (orangutan)

INTRODUCTION

We now leave the Asian apes and move to equatorial Africa, which is home to the few relict populations of gorillas and chimpanzees that have managed to survive—at least so far—in the face of *Homo sapiens*'s domination of the planet (Fig. 12-1). Comparative genetic studies indicate that gorillas, the subject of this chapter, share a common ancestor that lived about 7 mya, with chimpanzees and hominids. Gorillas are therefore a step closer to humans than orangutans are, although their fossil record has yet to be discovered. Since the first gorilla was described in the 1870s, this species (*Gorilla gorilla*) has been woefully misunderstood. Are they vicious King Kongs that wreak terror on human beings (Fig. 12-2), or are they gentle and sweet? Is that old response to the question, Where does a gorilla sleep? (anywhere he wants) really true? Or are adult males that weigh several hundred pounds just too big to spend much time, sleeping or otherwise, in trees? One also wonders if gorillas are big dim-witted giants that cannot hold a candle to the sparkly intelligence of the other great apes or

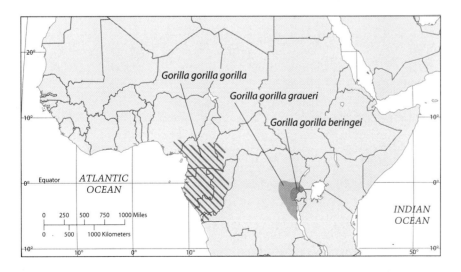

Figure 12-1 Populations of gorillas are confined to small refuges in tropical Africa.

Figure 12-2 Gorillas have been the focus of much negative press, as shown by this illustration of King Kong.

if, on the other hand, they are so smart that captive animals are even able to use sign language to tell about their past experiences. Inquiring minds will find discussion of these and other gorilla topics below.

Although gorillas conform, of course, to the great ape pattern described in Chapter 11, they share certain features of the skull with chimpanzees that differ from those of orangutans. For example, gorillas and chimpanzees lack rounded foreheads and have dramatic browridges over the eyes (**supraorbital tori**). Their skulls are relatively narrow and long compared to *Pongo* skulls, and their faces are inclined in a more forward than upward direction. The African apes, like humans, also have hollow spaces called **frontal sinuses** within the frontal bone above the eyes. Unfortunately, the exact functions of the browridges (which appear in some fossils of early human ancestors) and frontal sinuses (also present in humans) remain somewhat mysterious. Finally, gorilla and chimpanzee molars lack the thick enamel found on those of orangutans and humans.

The word that best describes the appearance of living gorillas is majestic. For one thing, they are the largest of the contemporary primates, yet seem to conduct many of their day-to-day affairs with an air of quiet serenity. The species is highly sexually dimorphic, with many adult males reaching at least twice the size of adult females (Table 12-1). Males also have pronounced sagittal crests along the lengths of their skulls that merge with another crest that runs across the back of the skull (Fig. 12-3). These bony crests support huge chewing and neck muscles that, when padded with connective tissue in the living animal, give the top of the head a startling contoured appearance reminiscent of an exaggerated 1950s beehive hairdo (Fig. 12-4). Deep-set eyes stare out from underneath dramatic browridges, and the ears, when visible at all, seem small and delicate in comparison to other features (Fig. 12-5). Gorilla noses are big, and look almost as if they have been squashed onto the middle of their dark, bare faces. Because of their brachiating anatomy (see Chapter 11), gorillas appear as if they are all arms and shoulders (see Fig. 11-2). They (and chimpanzees) have well-developed laryngeal air sacs that are internal,

TABLE 12-1 Representative Statistics for Gorillas

		G .g. gorilla (western lowland)	G. g. graueri (eastern lowland)	G. g. beringei (mountain)
Average adult body weights	F	71.5 kg (157 lb)	80.0 kg (176 lb)	97.7 kg (215 lb)
	M	169.5 kg (374 lb)	175.2 kg (386 lb)	159.2 kg (351 lb)
Average adult brain weight (g)	F	443 (subspecies unspecified)		
	M	535 (subspecies unspecified)		
	Mean	489 (subspecies unspecified)		
Maximum known life span (yr)		50		50
Habitat[a]		f, s, mf	f, bf, ma	mf, bf
Locomotion[b]		kw, ac, bs, s	kw, ac, bs	kw, ac, bs
Diet[c]		f , l, sps	l, sps, f	l, sps, f
Mean no. of offspring		1	1	1
Seasonal breeders		0	0	0
Mean gestation (days)		256		256
Interbirth interval (yr)		4		3.9
Social organization[d]		U, M	U, M	U, M
Territorial		0	0	0
Mean ratio of adult females to males		3:1		3:1
Infanticide				+
Group size		10 (4–16)	14 (5–20)	9 (≤ 37)
Density		1.0/km^2	0.27–0.47/km^2	
Home range		700–1400 ha	270–6500 ha	400–800 ha
Conservation status[e]		H	H	H

Note: Information in this table has been selected from numerous sources and should be viewed as tentative, i.e., subject to change in light of new information. Sources include but are not confined to Jungers and Susman, 1984 (body weights); Tuttle, 1986 (brain weights); Watts, 1996 (diet), 1991 (*G. g. beringei*); Tutin, 1996, Tutin and Fernandez, 1993 (*G. g. gorilla*); Yamagiwa et al., 1996 (*G. g. graueri*); Rowe, 1996 (conservation status and other statistics). Abbreviations: +, yes; 0, no. Blank entries indicate that no information is presently available. Dietary items are listed in approximate order of decreasing importance to the species.

[a]Habitat (Africa): bf, bamboo forest; f, forests; ma, marshes; mf, montane forest; s, swamps.

[b]Locomotion: ac, arboreal climbing; bs, bipedal standing; kw, knuckle walking; s, suspensory.

[c]Diet: f, fruit; l, leaves; sps, stems/pith/shoots.

[d]Social organization: M, multimale, multifemale; U, one-male, multifemale.

[e]Conservation status: H, highly endangered.

rather than mostly external as is the case for orangutans. Compared to the smaller chimpanzees, gorillas have relatively larger teeth. Their bodies are also generally stockier, with huge stomachs (guts) and broad chests. They have short, broad hands and feet that are rather human-like (Fig. 12-6). Their coat color runs from black to brownish grays, but varies with subspecies. Mature males of all three subspecies are apt to lose hair from their chests, while that on their backs turns a handsome, silvery gray. For this reason, fully mature males are called **silverbacks**.

12 GORILLAS: THE LARGEST
PRIMATES OF ALL

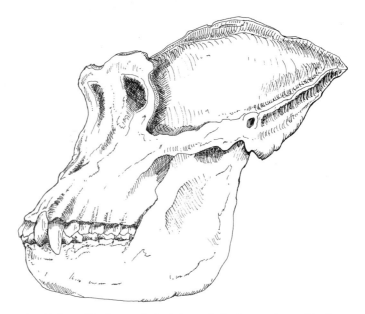

Figure 12-3 Male gorillas have relatively long skulls that are topped by large sagittal crests.

Figure 12-4 When fleshed out, the sagittal crests of male gorillas make the tops of their heads appear enormous.

THREE KINDS OF GORILLAS

The one species of gorilla (*G. gorilla*) is divided into three subspecies that live in distinct regions of tropical Africa (see Fig. 12-1). The most widespread and numerous subspecies is the western lowland gorilla (*G. g. gorilla*) that spans from parts of Nigeria to the western edge of the Democratic Republic of the Congo (formerly Zaire). The eastern lowland gorilla (*G. g. graueri*) is a distant cousin (geographically

Figure 12-5 Gorillas have deep-set eyes that stare out from underneath their browridges, as shown by this photograph of an adult mountain gorilla.

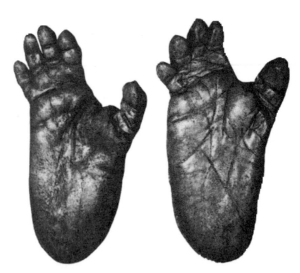

Figure 12-6 The feet of gorillas (western lowland on the left, mountain gorilla on the right) look somewhat like those of humans, except for their divergent big toes.

speaking) that lives in the eastern part of the Democratic Republic of the Congo, while the rare and highly endangered mountain gorilla (*G. g. beringei*) lives in tiny enclaves at high altitudes in Uganda, Rwanda, and the Democratic Republic of the Congo. More is known about the rare mountain gorilla than the other two subspecies because of the path-breaking research, first of George Schaller, and more recently of Dian Fossey, who was murdered in 1985 at the Karisoke Research Center in Rwanda. (Fossey's fieldwork is described in her book *Gorillas in the Mist*, which has been made into a movie.) Although the habitats of the three subspecies of gorillas do not overlap, all three are sympatric with chimpanzees in parts of their ranges.

The western lowland gorilla tends to be somewhat smaller than its eastern cousins (but there are exceptions, see Table 12-1). Instead of the luxuriant jet-black hair found in mountain gorillas, the coat of the western lowland gorilla is shorter and brownish gray to slightly auburn. Unlike the eastern subspecies, the top of the head is brown or reddish, and the silvery color on the backs of mature males extends farther down from their saddle region to the rump and thighs. Another distinction is that the top of the nose of *G. g. gorilla* has a distinct little knob of flesh that overhangs the division between the nostrils (Fig. 12-7). The coloring of the eastern lowland gorilla resembles that of the mountain gorilla, while its face is longer and its chest is wider than those of the western lowland gorilla. Curiously, *G. g. beringei* sometimes has webbed toes. The subspecies of gorilla also vary a good deal genetically, in fact so much so that Maryellen Ruvolo of Harvard University calculates that the western group may have been separated for some 3 million years from the two eastern groups. (If so, it is possible that two rather than one species of gorilla should be recognized, but this has not yet become widely accepted.)

Gorillas live in forests that are found from sea level in the west, to

1. Based on comparative genetic studies, how long have gorillas been around?
2. In what ways do the skulls of the African apes differ from that of the orangutan?
3. Everyone knows that gorillas are big, but how heavy can they grow to be?
4. How do the tops of silverbacks' skulls look, and why?
5. Describe the gorilla face.
6. Name the three subspecies of *Gorilla* and show where each lives on a map.
7. You see a gorilla in a zoo. What features do you observe to determine its subspecies?

Figure 12-7 The nose of the western lowland gorilla has a little knob of flesh at its tip, as seen in this newborn.

altitudes of up to 3900 meters (13,000 feet) in the mountains of the Virunga Volcanoes in the most eastern part of their range. Western lowland gorillas inhabit a variety of terrain including primary and secondary forests, as well as the edges of open areas. They have been observed crossing open savannas, and feeding in marshes and streams. Despite their name, lowland gorillas have also been reported living in montane forests up to 3000 meters (10,000 feet). The eastern lowland gorilla inhabits bamboo forests in addition to marshes and primary and secondary forests. The mountain gorilla, on the other hand, is restricted to montane and bamboo forests that are above 2700 meters (9000 feet). In general, gorillas are said to prefer open-canopy areas, which filter more light than the denser canopies found in primary forests.

LOCOMOTION—ON THE GROUND AND IN THE TREES

Because they spend most of their waking hours moving about quadrupedally on the ground, gorillas are regarded as terrestrial. Like chimpanzees, gorillas are knuckle walkers that place their weight on the soles of their feet, but not on the palms of their hands. When **knuckle walking,** the weight of the arms is transmitted across the tops of the flexed fingers and knuckles of the hands (Fig. 12-8). Although young gorillas are known to climb and swing in trees, it has often been argued that adult gorillas (particularly males) spend very little time in trees. Embedded in this notion are the ideas that the weight of large gorillas cannot be supported physically by trees, and that larger animals may also lack agility. These assumptions, which resulted from studies of mountain gorillas, need modification in light of research on western lowland gorillas by Caroline Tutin of the Station for the Study of Gorillas and Chimpanzees in Gabon and Melissa Remis of Purdue University (see In the Field).

Remis discovered that, despite their great size, western gorillas of all sizes regularly use trees for feeding and nesting (Fig. 12-9). However, she also found that, unlike smaller females, silverbacks rarely suspend themselves from trees or venture onto smaller branches; that is, large males remain nearer the sturdy cores of trees while other individuals forage at the periphery. Not surprisingly, Remis discovered that the distribution of gorillas in a tree at any given time depends partly on the number, size, and sex of the other gorilla occupants. Availability of fruit also influences arboreal behavior, with silverbacks becoming more terrestrial when fruit is scarce. An important conclusion reached by Remis and Tutin from their research at different West African sites is that one should not generalize the behavior of mountain gorillas to that of all gorillas, as has frequently been done in the past.

Gorillas do not swim but, again, the general belief that they hate water and avoid entering it at all times is in need of modification (Fig.

Figure 12-8 Gorillas (and chimpanzees) walk on the knuckles rather than the palms of their hands.

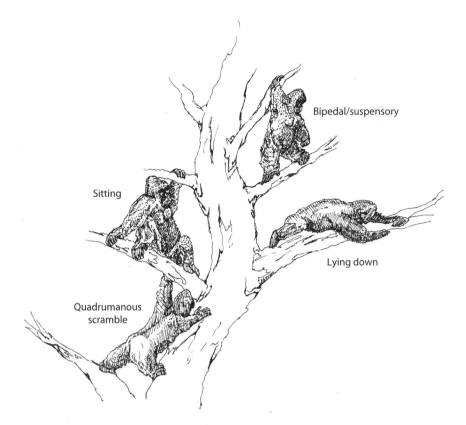

Bipedal/suspensory

Sitting

Lying down

Quadrumanous scramble

Figure 12-9 Contrary to generalizations based largely on studies of mountain gorillas, western lowland gorillas of all sizes use trees regularly.

Figure 12-10 This photograph belies the notion that gorillas hate water! (Photo courtesy of Claudia Olejniczak.)

Figure 12-11 This male gorilla is engaging in an elaborate chest-beating display.

12-10), according to Liz Williamson of the University of Stirling in Scotland and her colleagues. Lowland gorillas have been observed bathing and feeding in streams, and mountain gorillas have been reported wading through brooks. Although there are reports of gorillas drowning in captivity, some zoo animals have been observed happily splashing in moats and pools.

All three subspecies of gorillas engage in a limited amount of bipedal standing. This behavior appears, for example, in the famous chest-beating display that was first described over a hundred years ago (Fig. 12-11). Despite the fact that gorilla chest beating captured the public imagination, the entire sequence of behaviors involved in this display was not fully described until George Schaller did his classic research during the late 1950s. Imagine yourself to be an unsuspecting primatologist who inadvertently threatens a 157.5-kg (350-lb) silverback, and it responds with the full chest-beating display, so dramatically described by Shaller:

1. Hooting. The display begins as the animal emits a series of soft, clear hoots, which start slowly but grow faster and faster until the individual sounds merge into one another at or near the climax of the display.
2. "Symbolic feeding." The hooting is sometimes interrupted at one point as the animal plucks a leaf from nearby vegetation and places it between the lips.

Gorilla Letters

Melissa Remis, a primatologist at Purdue University, has spent over a decade studying the feeding ecology and positional behavior of western lowland gorillas. When she began her dissertation research in 1988 in the Central African Republic, Remis lived in the field with a team of Bayaka (Pygmy) assistants. What follows are excerpts of letters that she wrote to her advisor (Adrienne Zihlman) over a period of 4 years. Notice how the gorillas became more habituated to Remis's presence over time, and the positive effect that this had on her research.

MELISSA REMIS EXAMINES A GORILLA'S NIGHT NEST. (PHOTO COURTESY OF M. REMIS.)

21 October 1988 Tonight at dusk we heard gorilla screams not 500 meters from camp. We followed the tracks and fear scent of a male gorilla who had a leopard on his trail. We saw the tracks of a skirmish. Evidently, the very large adult male gorilla displayed at and wrestled with the leopard, pulled a sapling out of the ground, rolled over a log, and then turned and fled.

23 October 1988 The night before last we heard chest beating all night from different locations. Yesterday we returned to the site and tracked forward, observing that the lone male had joined up with a juvenile and the pair had an uneasy night, constructing and moving between six consecutive nest sites. At each site, the gorilla trail was superimposed by leopard pug marks, but we found no sign of serious injury to any of the participants. So much for gorillas having no nonhuman predators!

January 1991 We have been working hard and found the group three times last week. Each time the gorillas were in the trees (including the silverbacks), all eating bark and pith of vines. I am busy naming gorillas. By counting nests, I think that Combetti's group (two silverbacks) has grown from 12 to 15 plus one infant in the mother's nest. I think this is a combination of infants maturing to make nests of their own and a female immigrating into the group. It is puzzling because on occasion I've tracked the group to find only a subgroup (?) of five nests in the area where there should be more. Sometimes the nest counts vary from 10 to 15 on consecutive nights. So maybe, just maybe, I am seeing fission-fusion behavior in a gorilla group. Or maybe a splinter group is beginning to break off. I feel like I am in training for a career as a private detective!

3. Rising. Just before the climax the gorilla rises onto its hind legs and stands bipedally.
4. Throwing. As the animal rises, it often grabs a handful of vegetation and throws it into the air.
5. Chest beating. The chest beat occurs at the climax of the display, and usually consists of a rapid, alternate slapping of the chest.

February 1992   We are lucky to have found a new tracker who will hold his ground in the face of hysterical charging silverbacks that still seem to be incredibly annoyed by our presence. Nevertheless, sometimes they simply knuckle slowly out of view and disappear into the dense vegetation. On other occasions, they tolerate us. Recently I sat within 5 meters of a silverback who was so busy stuffing his face with wild ginger that he seemed oblivious to my presence. I tried to compose myself as he walked straight toward me. Worried does not even begin to describe my feelings when I thought that at any moment he would realize we were practically in his lap and startle so badly that he would leap on top of us in a rage. Wonga said my hand was shaking but I continued to take notes at a rapid clip. A female gorilla, Jasmine, was the first to become uncomfortable about our proximity. Her mouth dropped aghast, and she quickly turned on her heels and, with the silverback following, slowly eased away from us.

April 1992   I'm still shaken from the charge of a lifetime I got yesterday. I didn't even know that we were close to gorillas. We were in a thick vine tangle when all of a sudden this snarling scream-ing beast of a gorilla charged directly at us. This was very different from the more typical side "mock rush" behind vegetation that I have gotten somewhat accustomed to. This gorilla came hurling straight at us so that all I really saw was his canines (big and black). He didn't stop until Mokedi was literally in my lap, clinging on for dear life. The silverback stood up bipedally and flung a rotten log at us, but missed. He didn't turn around until he was within 1 meter of us. It was quite a shock that he left without pummeling us into the ground. Now I really question whether Carl Akeley knew what he was talking about when he insisted that silverback charges are merely bluffs!

September 1992   It is rainy season and the gorillas are on mad dashes through the forest, traveling farther out of the study area and into the unknown upper reaches north of camp in rampant pursuit of ever sweeter water-vine fruits. I have no idea what is wrong with the formerly favored fruits that are now ignored and rotting off the trees. This means I spend from 6 A.M. to 6 P.M. hiking just to pick up their trails. I am exhausted. After 8 hours of tracking, the gorillas mysteriously appeared, climbing out of a tree to see me. It was a terrific hour of gorillas approaching me one by one, and then each scampering away. This work is 99.9% sweat, exhaustion, and frustration, but the remaining 0.1% is amazing. These animals are radically transforming our notions of what it means to be a gorilla. Despite their body size, they are climbing 40 meters in trees, consuming fruit, traveling far each day and over each year, and they may even be changing their grouping patterns to accommodate the search for fruit. So much for gorillas as terrestrial cowlike herbivores!

Melissa Remis tracked and studied gorillas with the help of Bayaka (Pygmy) assistants. (Photo courtesy of M. Remis.)

6. Leg kicking. While beating its chest, the gorilla may kick one leg into the air.
7. Running. Immediately after the climax and sometimes during the latter part of it, the gorilla runs sideways for several feet.
8. Slapping and tearing. While running or immediately afterwards, the animal slaps the vegetation and tears off branches by hand.

REVIEW EXERCISES

1. Why are gorillas regarded as terrestrial?
2. How do gorillas move quadrupedally on the round? (Try knuckle walking yourself.)
3. Do large male gorillas always avoid climbing in trees?
4. What lesson did Melissa Remis and Caroline Tutin learn about the assumption that behaviors observed in mountain gorillas are the norm for all gorillas?
5. Do gorillas swim? Do they hate water?
6. Try doing the chest-beating display, using all nine steps.
7. When is a silverback likely to display?

9. Ground thumping. The final gesture in the sequence is a single loud thump of the ground with the palm of the hand.

This impressive display may require as long as thirty seconds to complete, although all but the first two acts follow each other in one continuous, violent motion which is usually finished in five seconds or less. (Schaller, 1963:222)

Gorillas often beat their chests when they become excited, for instance by the presence of another gorilla group. According to Schaller, although gorillas of all ages engage in aspects of the chest-beating display, only silverbacks perform variations of the entire sequence. As they beat their chests with cupped hands, silverbacks produce hollow "pok-pok-pok" sounds, rather than the mere slaps of black-backed males, females, and youngsters. Interestingly, this distinctive sound may be due to the inflated air sacs of silverbacks functioning as sound chambers (resonators).

DIET AND OTHER DAY-TO-DAY ACTIVITIES

Diet varies with subspecies, although gorillas in general tend toward vegetarianism. Mountain gorillas are basically folivorous, eating mostly leaves, shoots, and stems of about 150 different plants enumerated by David Watts of the University of Michigan, who has spent more time than anyone doing research on mountain gorillas in the wild. These plants include bamboo, wild celery, nettles, and thistles. *G. g. beringei* very infrequently eats ripe fruit and invertebrates. The eastern lowland gorilla is also largely folivorous, and eats a wide variety of trees, vines, grasses, herbs, and so on. Fruit, ants, and bark are also consumed in small quantities, and *G. g. graueri* is known to raid crops of taro, peas, and maize. The balance of fruit and other plants that are consumed is considerably different in the western lowland gorilla than in the two eastern subspecies. In addition to other types of plants, *G. g. gorilla* eats a high percentage of ripe fruit, especially during the wet season. Western lowland gorillas also consume a small amount of invertebrates on a regular basis, including caterpillars and termites.

According to Kelly Stewart and Alexander Harcourt of the University of California, Davis (who focused mainly on mountain gorillas), a typical day in the life of a gorilla involves peaks of feeding in early morning and late afternoon, interspersed with relatively rapid travel. At midday, gorillas take a 1- to 2-hour siesta. Sunbathing is also a frequent activity. During the rest of the day, animals spread out and move slowly, feeding as they go. A total of about 45% of an adult's day is spent feeding. Aggression within gorilla groups is usually mild. However, Stewart and Harcourt note that, despite an abundance of food, rates of aggression are higher during feeding periods than during rest.

As noted in earlier chapters, larger-bodied primates lean toward folivory and do not have to go as far afield to feed as do smaller frugivorous primates. This trend appears to hold for gorillas, as body size tends to be smaller, and day range larger, in the more frugivorous western lowland gorilla than in the eastern gorillas (see Table 12-1). Another interesting difference associated with the presence or absence of appreciable amounts of fruit in the diet has to do with nesting behavior. All gorillas make nests in which to sleep at night (and sometimes during the day). These cushions of leaves and branches may be on the ground (all subspecies), or in the case of the western lowland gorilla, sometimes in trees. Harcourt observed that mountain gorillas defecate in their nests at night, but that because of the lack of fruit in their diet, their dung is dry and does not foul their coats. Western lowland gorillas who eat a good deal of ripe fruit, on the other hand, rarely defecate in their nests.

Gorillas live in relatively stable social groups. Most of the time, these groups contain only one adult silverback male and several adult females with their young. According to Martha Robbins of the University of Wisconsin, however, approximately 40% of mountain gorilla groups contain more than one silverback, and these have fewer adult females per adult male than one-male groups do. When multiple silverbacks live in one group, they are likely to be related to each other (e.g., father and son) and to form coalitions during intergroup encounters. Multiple adult males also form a linear hierarchy that is based on body size and probably age. Gorillas are atypical for group-living primates in that, before they breed, both males and females usually emigrate away from their natal groups. The females transfer directly to another group or to a solitary male. Males, on the other hand, usually do not transfer to other bisexual groups. Instead, they may link up with all-male groups, or be joined by emigrating females. Males may wander alone for several years, however, before they are joined by a female and settle down to begin their own group.

Although gorillas are generally peaceful souls when they are not threatened, severe tensions erupt when males attempt to attract females away from established silverbacks. The majority of such encounters include vigorous displays such as chest beating, as well as physical violence between adult males. Although males are sometimes successful at attracting individual females, takeover of an entire group by an outsider rarely if ever occurs. Instead, when a dominant silverback dies, his group may be taken over by a (probably related) silverback within the group, or the group's females may disband and join other groups. According to Watts, infants of mothers that are unaccompanied by silverbacks are especially susceptible to infanticidal attacks from males, and females presumably join groups to seek protection of a resident silverback against such aggression. Watts also notes, however, that the frequency of infanticide in mountain gorillas is relatively low compared to the number of aggressive encounters that occur.

Once a one-male, multifemale group is established, its silverback defends it for as many years as possible, sometimes for the rest of his life. (Gorillas' ranges overlap and they are not territorial, however.) During encounters in which outside males attempt to attract females,

away, resident silverbacks are likely to direct aggression toward their females to prevent them from leaving. Although herding of females is well known for some other primates such as hamadryas baboons (see Chapter 8), it has only recently been described in detail for mountain gorilla by Pascale Sicotte at the University of Calgary:

> Typically, it involves strut and display towards the female, who may nevertheless try to circumvent the male and approach the opposing group. The male can also position himself between his female and the other group, and drive her back to his group by strutting towards her. In a particular case that I witnessed, the male repeatedly grabbed and bit the female as she was trying to slip away from him. Interestingly, other females that were present during the episode intervened on several occasions on behalf of the herded female, the most active being her relatives. . . . They approached, screaming and "pig grunting" . . . at the male as he was biting the female and holding her to the ground. It is worth mentioning that in this particular case the male kept a close watch on the same female not only for the whole duration of the encounter, but also the following day, as the groups were moving apart from each other. (Sicotte, 1993:25)

It is clear that male-male competition is extreme in gorillas, in fact so much so that Stewart and Harcourt believe it accounts largely for the high degree of sexual dimorphism that characterizes this ape.

Emigrating females may try out more than one male before they settle down with a particular one, or join a small group. Watts has shown that female mountain gorillas do not form clear agonistic dominance hierarchies, although they can sometimes be ranked based on their nonaggressive interactions. Adult females also do not form strong alliances with each other, which is consistent with their usual origins from different natal groups. Social grooming is less frequent in gorillas than in many primates, and is particularly rare between adults of the same sex. When grooming does occur, it is often between mother and infant, or directed by females toward the silverback, which is the dominant animal in a group, not to mention the center of its social life (Fig. 12-12). Indeed, during midday rest periods, adult females and youngsters seek to be near the protective silverback.

Figure 12-12 Silverback mountain gorillas are the social centers of their groups.

REPRODUCTION

As is often the case for primates that live in one-male groups, the sex lives of gorillas are somewhat staid. Gorillas are basically polygynous. Watts has shown that silverback mountain gorillas are generally intolerant of copulation attempts by subordinate males, which therefore engage in relatively few copulations. Females become sexually mature at approximately 7 years old, but do not conceive their first offspring for about 2 years. Because it is tough for males to establish their own groups, they are usually at least 15 years old before becoming fathers. The menstrual cycle averages about 28 days, while estrus, during which females copulate, lasts for only 1 to 4 days. Receptive females do not have sexual swellings, although there may be slight labial swelling in estrous females that have never given birth. Females solicit more copulations than males. They do so by approaching and presenting to males, and mating may be in either a face-to-face or back-to-front posture. According to Watts, the average duration of copulation in mountain gorillas is 80 seconds. After a gestation of around 8 months, the offspring is born at night. (Twins are rarely born.) Interbirth intervals are around 4 years, but because of high infant mortality the spacing between individuals that survive until breeding age is nearly double this length of time. Infants sleep in their mother's nest until they are about 3 to 4 years old, and gorilla mothers are described as highly caring and fiercely protective.

GORILLA COGNITION

For several reasons, gorillas have the reputation of being less intelligent than the other great apes: They have smaller brains than expected for primates of their size; that is, they are less **encephalized** than orangutans and chimpanzees. In the wild, gorillas have not been observed using tools to obtain food such as termites or nuts, as have chimpanzees (see Chapter 13) and to a lesser degree, orangutans (see Chapter 11). Gorillas simply use their strength to break open the hills and nests of ants and termites. With the exception of one controversial gorilla, Koko (see below), and again unlike the other great apes, gorillas have also been unsuccessful at the mirror self-recognition test. All in all, this is not a very good report card. However, as is sometimes the case for humans, these indicators of cognition may fail to reflect the true abilities of gorillas. For one thing, because of a quirk in how encephalization is quantified, it is usually underestimated for bigger-bodied mammals including primates, and this fact has not been appreciated for gorillas. In other words, it is not so much that gorillas have small brains—it is that they have large bodies. In fact, gorilla brains average about 490 g (see Table 12-1), which is 115 g larger than the average brain weights of both orangutans and chimpanzees (see Neural Note 12).

REVIEW EXERCISES

1. What is the main dietary difference between western lowland and other gorillas?
2. Describe the association between frugivory versus folivory, and body size and range size. How does this relate to gorillas?
3. Where do gorillas sleep—anywhere they want, in nests that they build, or sitting in trees?
4. How many silverbacks usually reside in a gorilla group? Does this vary?
5. Which sex of gorilla emigrates away from the natal group before breeding? Is this generally the norm for group-living primates?
6. How are new gorilla groups formed?
7. What behavior may account for the high degree of sexual dimorphism in gorillas?
8. Describe the social relationships between adult females and between the silverback and other group members.
9. How old are gorillas when they first breed? Describe their mating and breeding patterns.

Brain Size in Primates

The volume of the braincase, or **cranial capacity**, is easy to measure from skulls (including those of fossils) and approximates actual brain weight closely enough to be considered, for all practical purposes, equivalent. Thus, a cranial capacity of 100 cubic centimeters (cc) is taken to represent a brain weight of 100 g. The graph shows the ranges of cranial capacity for various groups of primates, as well as the average capacities for male and female apes and humans. The first thing to notice is that brain sizes do not overlap between great apes and humans, or between great apes and the other primates, including gibbons. Next, observe that females, with smaller bodies on average than males, also have smaller average brain sizes. Further, the apes that are especially sexually dimorphic in body size—the orangutan and the gorilla—are also more dimorphic in their brain sizes.

As you may have noticed, body sizes are not provided in the graph, and this might be an important omission. After all, a New World monkey is smaller than a gorilla, so one would not expect it to have as large of a brain. While this is indeed true, brain size scales differently than body size. In other words, brain size generally does not keep up with increased body size; bigger-bodied animals don't need that much more in the way of brains than their smaller-bodied relatives. Take, for example, the gorilla. Its body is several times larger than those of chimpanzees and orangutans, yet its average cranial capacity of roughly 490 cc represents only about a 30% increase over the average of approximately 375 cc for both *Pan troglodytes* and *Pongo*. And gorillas do just fine.

However, this does not mean that we can ignore body size, because some species do have larger brains than one would expect for their body sizes. The prime example of this is our own species. Human adults are usually larger bodied than chimpanzees and orangutans, but much smaller than gorillas. Thus if human brain size were scaled accordingly (referring to the previous example), the average would be between 375 and 490 cc. But human brains average about 1340 cc; they are approximately three times the size that would be expected in an ape with a human body size! This is what is meant when humans are described as highly encephalized.

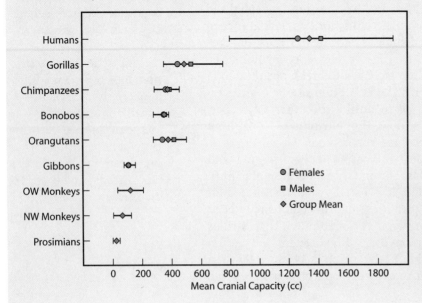

The volume of the braincase, or cranial capacity, is a good approximation of brain size.
NW = New World; OW = Old World.

Figure 12-13 Koko has been trained to use American Sign Language by psychologist Francine Patterson.

As far as the lack of tool use in the wild, one should remember that "necessity is the mother of invention." Or, as Daniel Shillito (a graduate student at the University at Albany) said, "Gorillas live in a great big salad bowl—they don't *have* to use tools to find food." With respect to the western lowland gorilla that must search outside of the salad bowl for the ripe fruit that it consumes, S. E. MacDonald of York University in Canada recently tested spatial memory of this subspecies in a foraging task in which two subjects were required to remember various locations that had been baited with highly preferred food the day before. MacDonald found spatial memory to be excellent in the adult, which may even have used a counting strategy to minimize the number of sites visited.

To date, at least two gorillas have been schooled in sign language. Of the two, the western lowland gorilla named Koko is the most proficient, and the most famous. Koko is 1.5 meters (5 feet ¾ inch) tall and weighs 126 kg (280 lb). Trained by Francine Patterson of The Gorilla Foundation, Koko was able to use over 500 signs by the time she was 26 years old (Fig. 12-13). She likes to play with dolls, kittens, and humans (e.g., playing chase), enjoys drawing, and is fond of the color red. According to Patterson, Koko refers to herself as "Fine Animal Gorilla," and recognizes herself in the mirror. In reviewing Koko's achievements, however, it is important to keep in mind that she has been encultured by humans. In other words, Koko has acquired an advanced education—an opportunity that most gorillas never have. Much of what is known about Koko comes from the popular media, rather than the scientific literature where details of primatological research are traditionally made available for evaluation and replication by other investigators. For this reason, some primatologists are skeptical about many of the claims that are made for Koko (e.g., that she has thought about where gorillas go when they die—"comfortable hole bye"). This does not, however, mean that the descriptions of Koko's accomplishments are necessarily inaccurate. Rather, it means that the jury is still out on Koko and the other signing gorilla, Michael (who reportedly has signed about his mother's death, caused by poachers). It also underscores the importance of presenting discoveries in scientific, as well as popular publications.

One can also look to zoo animals for indications about cognition. This must be done with great caution, however, since zoos have long been notorious for crushing the spirits of highly intelligent animals. Although many of today's zoos have been modernized, they still lack the complexity and social conditions found in natural habitats. On the upside, however, zoos also lack some of the challenges that frequently bring species to the brink of extinction in the wild. With these caveats in mind, it is interesting to revisit two little-known western lowland gorillas, Kongo and Lulu, that grew up in the Central Park Zoo in New York City over 25 years ago. Because they spent most of their lives in zoos, it seems reasonable to assume that their intelligence, curiosity, and proclivity for manipulating objects reflect the *minimum* potential for their species:

Lulu had somehow grabbed the spoon which had held her vitamins and was quietly playing with it. He [Kongo] eyed in covetously; . . . headed for Lulu, head lowered, eyes threatening. Promptly she handed it over to him and strolled away.

He sat down with his new toy and looked it over very carefully. He scraped the floor with it, listening to the metallic scrape that it made on the harsh cement. . . . He pounded the spoon and rattled it and clacked it against the floor. . . . The long metal handle was bending. He looked at it very closely, turning it around in his hand. He gave another whack on the floor, and it changed again. He held it up and scrutinized it carefully. Then, deliberately, he bent it further out of shape. . . . He looked at it again, and straightened it to its original shape. He did not continue to bend the stem but stopped precisely when it was in its correct form. . . . [He] came to the front of the cage, where he deliberately fitted the edge of the spoon into the slot in a metal bolt in the cage door. He tried to turn it as if it were a screwdriver! . . . A few days later he again tried to unscrew some screws that held the supports to the Plexiglas windows before his cage. And he succeeded enough so that the panel which they held wobbled around in the air.

Lulu, too, continually demonstrated the use of materials as tools, for she often dipped lettuce leaves or paper towels (when she could get them) into some water and then, using them like a sponge, squeezed the water out into her mouth. (Green, 1978:72–73)

These observations of a quarter of a century ago do more than illuminate the tool-using potential of gorillas. They also captured something of their curiosity and playfulness:

In midsummer as I sat beneath the trees outside their cages, a parakeet appeared as if from nowhere. Leaf-green with the delicacy of yellow softening his feathers, he fluttered at the bars for a moment, then perched there. Lulu's eyes fastened on the lovely creature. She came slowly toward the little bird and then sat down. . . . Lulu did not take her eyes off the bird. She . . . climbed carefully toward it. Slowly, gently, she reached out her middle finger and brushed its lovely tail. She watched as the bird fluttered lightly around the bars and once more landed above her. Slowly she began to climb toward it again; it took off. It flew into the waving branches of the tree above me. I lost sight of it. But Lulu stayed poised at the bars, spellbound, her eyes gazing on the bird until the leaves stirred and it flew away. (Green, 1978:74)

"Aha! No, you don't." Raoul or Eddie would cry out triumphantly as Kongo made a swipe toward the nozzle and missed. But when he did not miss, the fun began. Hand over hand he would pull the hose into the cage, gathering it up in the far corner, piling it around and around on top of itself the way a sailor does. Four men, at first, stood outside, trying to pull it out after the first call for help. Kongo would let it slacken and then, with a slight twist of a wrist, jerk the hose back into the cage, winding it around the area like a huge snake. Now six men strained, pulling on it, their bodies tensed against the floor. And Kongo sat relaxed, still holding it easily in one hand. He gave a little tug, barely visible, and the men could not contain it, lurching forward or tumbling backward as Kongo played with them. It would take the six men fifteen minutes of tremendous labor, bribes, and unabashed trickery to get the hose away from the gorilla. It was usually the bribes that finally did the trick. It was one game that I never saw Kongo tire of. (Green, 1978:75)

The bottom line from reviewing the reports about Koko, Michael, Kongo, and Lulu is that it is time to reexamine, and perhaps revise, the negative assessments in the scientific literature about gorilla intelligence.

THE BIGGER THEY COME,
THE HARDER THEY FALL

Gorillas appear to have few enemies other than humans, although they are occasionally preyed on by leopards. Unfortunately, this does not mean that all is well with these magnificent animals. Far from it. The western lowland gorilla is vulnerable to extinction, the eastern lowland gorilla is endangered, and tragically, the mountain gorilla is almost gone. Some of the reasons for this sad state of affairs are all too familiar by now: deforestation associated with exploding human populations, and lack of enforcement of existing conservation laws. Gorillas are fiercely protective of their young and, for every baby captured, a number of group members are killed. There are other threats to gorillas as well. Perhaps because of their awesome size, gorilla body parts hold a special mystique for collectors of animal trophies. As a result, they are still hunted in western Africa for their skulls, heads, teeth, and hands. In western and central Africa, gorillas are also hunted for bushmeat, which is thought not only to be tasty, but also to have spiritual value. The mountain gorilla, which at last count numbered only a few hundred individuals, is currently under added pressure from humans—this time in the form of war, which broke out in Rwanda in 1990. Thanks to a dedicated group of primatologists, visits to the mountain gorilla groups at Karisoke have continued (even though it has been impossible since 1994 to return to the camp itself) and research is ongoing on another population of mountain gorillas at Bwindi. In sum, it is both horrible and ironic that gorillas have been branded with an unfair reputation for brutality toward humans (remember King Kong), while at the same time being sub-

REVIEW EXERCISES

1. Give three reasons why gorillas have a reputation for not being very bright. Elaborate why these observations may not be sound.
2. What does it mean to say that a species is highly encephalized?
3. Who are Koko and Michael? What claims have been made about their cognitive abilities?
4. Why are scientists hesitant about believing these claims?
5. What effect do you think living in a zoo has on the cognitive life of highly intelligent animals such as gorillas?
6. Is it true that gorillas have no inclination to use tools? Discuss both the pros and cons.
7. What lessons do we learn from Kongo and Lulu?
8. What factors threaten gorilla survival?
9. Review and answer the questions posed in the first paragraph of this chapter.
10. What do you think will become of gorillas, and why?

jected to various acts that, if directed at humans, would be recognized for what they are—atrocities, through and through.

SUMMARY

Partly because of their awesome size and famous chest-beating display, gorillas have been woefully misunderstood for more than a century. This chapter explores some of the conflicting assertions about their nature, locomotor patterns, and intelligence. Gorillas are a highly sexually dimorphic species (*Gorilla gorilla*), divided into three subspecies that live in distinct regions of tropical Africa. The most widespread and numerous subspecies, the western lowland gorilla (*G. g. gorilla*), has only recently been studied to any degree in the wild. Emerging research on this subspecies requires revision of certain earlier generalizations, for example, that gorillas do not eat much fruit, or that adult males avoid spending much time in trees. The two eastern subspecies include a second lowland gorilla (*G. g. graueri*) and the famous mountain gorilla (*G. g. beringei*). All gorillas are terrestrial knuckle walkers that tend toward vegetarianism. They feed, travel, and rest during the day, and build nests in which to sleep at night. Gorillas are polygynous, and live in relatively stable social groups that usually contain one silverback male, several adult females, and their young. Both males and females emigrate away from their natal groups before they breed, but only females transfer to other units. This pattern is highly unusual for a group-living primate. The ranges of different groups may overlap, but gorillas are not territorial. However, severe male-male tension erupts over females, and the physical combat that is frequently associated with such competition may account for gorillas' high degree of sexual dimorphism. The silverback is dominant to all other animals in a group. Adult females exercise female choice when they migrate into a unit where they establish and maintain a relationship with a silverback.

It is a commonly held belief that gorillas are less intelligent than the other great apes. However, upon inspection, it appears that certain measures that have traditionally been used to assess cognitive abilities in primates may not reflect the true abilities of gorillas. Reports about the mental lives of four western lowland gorillas (Koko, Michael, Kongo, and Lulu) suggest that the stereotype of gorillas as being relatively mentally slow compared to the other great apes needs to be reexamined.

FURTHER READING

FOSSEY, D. (1983) *Gorillas in the Mist.* Boston: Houghton Mifflin.
GREEN, S. K. (1978) *Gentle Gorilla: The Story of Patty Cake.* New York: Richard Marek Publishers.

SICOTTE, P. (1993) Inter-group encounters and female transfer in mountain gorillas: Influence of group composition on male behavior. *American Journal of Primatology* **30**:21–36.

WATTS, D. P. (1996) Comparative socio-ecology of gorillas. In: W. C. McGrew, L. F. Marchant and T. Nishida (eds) *Great Ape Societies*. Cambridge: Cambridge University Press, pages 16–28.

13 OUR COUSINS:
THE CHIMPANZEES

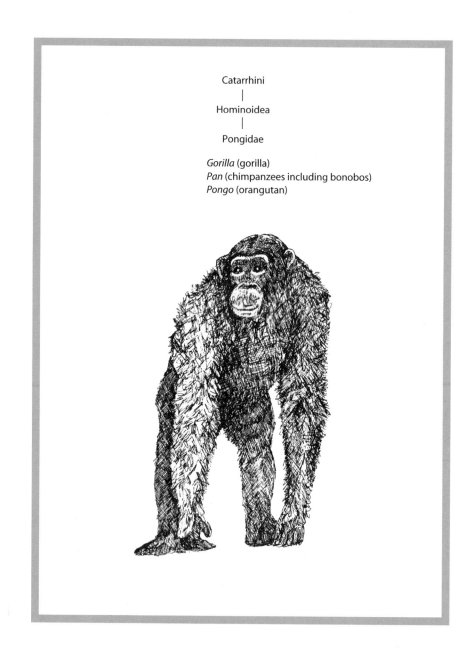

Catarrhini
|
Hominoidea
|
Pongidae

Gorilla (gorilla)
Pan (chimpanzees including bonobos)
Pongo (orangutan)

INTRODUCTION

A moment's reflection will confirm that, although you and your first cousin have different parents, you share grandparents who are potential great-grandparents to your combined children (i.e., to first-cousins-once-removed). Similarly, humans and chimpanzees are cousins of sorts (but many, many times removed) because they are descended from a common ancestor that lived in Africa approximately 5 to 6 mya. Because humans and chimpanzees are such close relatives, our genes are nearly identical. Compared to the other nonhuman primates, chimpanzees look like us, act like us, and even catch many of our diseases. They are not, however, quite human. Both the similarities and differences between chimpanzees and people have fascinating implications for understanding human evolution, which are discussed in Chapter 14.

Two species of chimpanzees are currently recognized. The best known is *Pan troglodytes* (the so-called common chimpanzee), which has received much attention because of ongoing research that began in Tanzania in the 1960s with the fieldwork of Jane Goodall at Gombe and Toshisada Nishida (of Kyoto University) in the Mahale Mountains. The second species, *Pan paniscus*, is sometimes called the pygmy chimpanzee. Because it is not much smaller than *P. troglodytes*, however, the alternative common name of bonobo will be used in this text. Bonobos have not been nearly as well studied as their sister species, with the first systematic field research beginning in 1974 by Takayoshi Kano of Kyoto University. Bonobos comprise one species that is not further divided into subspecies (i.e., they are monotypic) and they live exclusively in parts of the Democratic Republic of the Congo (formerly Zaire). The remaining chimpanzees, on the other hand, are traditionally viewed as comprising three more widely distributed subspecies: *P. t. verus* in West Africa, *P. t. troglodytes* more centrally, and *P. t. schweinfurthii* at the Gombe National Park and other eastern sites (Fig. 13-1).

However, this traditional view of chimpanzee taxonomy is currently in a state of flux because of two recent mitochondrial DNA

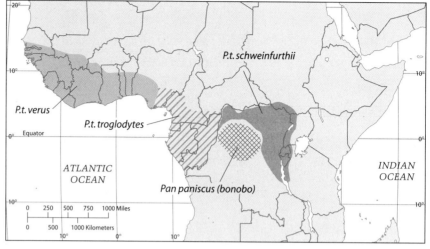

A

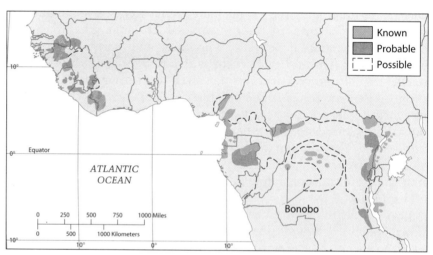

Figure 13-1 Although bonobos and the other chimpanzees once ranged throughout much of Central Africa (**A**), today they are restricted to much smaller pockets within their former ranges (**B**).

B

(mtDNA) studies. The first, by Phillip Morin of the University of California, San Diego, and his colleagues, suggests that *P. t. verus* should perhaps be elevated to a third chimpanzee species. Another study, by M. Katherine Gonder at Hunter College and others, finds the mtDNA of Nigerian chimpanzees distinct enough to warrant their being placed in a fourth subspecies, *P. t. vellerosus*. It remains to be seen whether or not these revisions will be generally accepted. For the time being, this text will adhere to the traditional division of *P. troglodytes* into three subspecies representing relict populations of one species that was once more widespread and geographically continuous.

Chimpanzees are smart, social, sexual, and political animals. They fight, reconcile, make and use tools, hunt, make war, and sometimes practice infanticide and even cannibalism. Some populations engage in unique behaviors that appear to be transmitted in a way that is similar to human cultural traditions. Chimpanzees have also proved to be extremely clever in laboratory and language experiments. It is not surprising, then, that they have long been used as models for speculating about the evolution of human behavior. There is a wrinkle in this model, however—namely bonobos, which differ dramatically from other chimpanzees in a number of respects (see below). Because bonobos share a common ancestor with chimpanzees that lived somewhere around 2.5 mya, their morphology and behavior are just as potentially informative for forming hypotheses about the earliest ancestors of humans. As Adrienne Zihlman (1996a:36), puts it, "our two closest living relatives together express a wider range of behaviours than either taken alone and offer a broader platform for speculating about the early history of our own multiplex genus and species."

TWO SPECIES, TWO LOOKS

Despite descriptions of *P. t. verus* as masked or pale-faced, *P. t. troglodytes* as black-faced, and *P. t. schweinfurthii* as long-haired, the three subspecies of *Pan* appear so similar that most zoos are not sure which subspecies they are exhibiting. The coat is generally dark, although older individuals may have gray on their backs. Both males and females sometimes have short white beards and may become bald (Fig. 13-2). Chimpanzees have big ears, prominent browridges, and projecting muzzles. Their feet and hands are black, while the faces vary from dark to blotchy combinations of black, brown, and pink. Infants have white tail tufts (but no tails, of course) and lighter faces; both are darker in adults. Chimpanzees generally weigh less than humans and are sexually dimorphic, with males and females of the largest subspecies for which there are data (*P. t. troglodytes*) averaging about 60 kg (132 lb) and 47 kg (104 lb), respectively (Table 13-1).

Bonobos are another story. Although their body sizes are about the same as those of *P. t. schweinfurthii* that lives to their north in the

Figure 13-2 Male and female *Pan troglodytes* are sexually dimorphic, but not to the degree that gorillas and orangutans are.

TABLE 13-1　Summary Statistics for Chimpanzees

		P. t. verus (far western)	P. t. troglodytes (central western)	P. t. schweinfurthii (eastern)	P. paniscus (bonobo)
Average body weights (in wild)	F		47.4 kg (104.5 lb)	33.2 kg (73.2 lb)	33.2 kg (73.2 lb)
	M		60.0 kg (132.3 lb)	43.0 kg (94.8 lb)	45.0 kg (99.2 lb)
Adult brain weight (g)	F		360 (P. t. subspecies unspecified)		345
	M		389 (P. t. subspecies unspecified)		353
	Mean		375 (P. t. subspecies unspecified)		349
Maximum known life span (yr)			53 (P. t. subspecies unspecified)		40
Habitat[a]			ws, g, lrf, mrf (P. t. subspecies unspecified)		lrf, s
Locomotion[b]			kw, ac, s, bs, l (P. t. v. and P. t. s.)		kw, ac, s, bs, l
Diet[c]		+n	f, l, fl, s, i, a (P. t. subspecies unspecified)		f, l, sps, h, i, a
Mean no. of offspring		1	1	1	1
Seasonal breeders		0	0	0	0
Gestation (days)			230–240 (P. t. subspecies unspecified)		230–240
Interbirth interval (yr)		4.6–5.1		5.5–6.0	4.5–>5.0
Social organization[d]		F-F, P-M	F-F, P-M	F-F, P-M	F-F, P-M
Territorial		+	+	+	0
Approximate mean ratio of adult females to males				4:1	1:1
Infanticide				+	0
Community size		19–20		20–105	10–58
Density			0.08–7.00/km² (P. t. subspecies unspecified)		2–3/km²
Home range　Forest		8–27 km²		6.8–19.4 km²	58 km²
Savanna		275 km²		560 km²	
Conservation status[e]		H	H	H	H

Note: Information in this table has been selected from numerous sources and should be viewed as tentative, i.e., subject to change in light of new information. Sources include but are not confined to Jungers and Susman, 1984 (body weights); Tuttle, 1986 (brain weights); Doran, 1996 (locomotion, home ranges); Wrangham, 1984 (gestations); Takahata et al., 1996 and Sugiyama, 1994 (interbirth intervals, adult sex ratios); Nishida and Hiraiwa-Hasegawa, 1987 (social organization, density P. paniscus); Sakura, 1994, Nishida, 1990, and Boesch, 1996 (community sizes); Moore, 1992 (density P. troglodytes, savanna home ranges); Rowe, 1996 (habitats, diets, conservation status). Abbreviations: +, present; 0, not present. Blank entries indicate that no information is presently available. Dietary items are listed in approximate order of decreasing importance to the species.

[a]Habitat (Africa): g, grassland; lrf, lowland rain forest; mrf, montane rain forest; s, swamps; ws, savanna woodland.

[b]Locomotion: ac, arboreal climbing; bs, bipedal standing; kw, knuckle walking; l, leaping; s, suspensory.

[c]Diet: a, animal prey; f, fruit; fl, flowers; h, honey; i, insects; l, leaves; +n, nuts in addition to food eaten by other P. troglodytes; s, seeds; sps, stems/pith/shoots.

[d]Social organization: F-F, fission-fusion; P-M, patrilineal multimale, multifemale.

[e]Conservation status: H, highly endangered.

Democratic Republic of the Congo, their general appearance differs significantly from that of *P. troglodytes*. Bonobos are relatively slender, with longer hind limbs and narrower chests that are topped by shorter clavicles. The molars are not as large as those of other chimpanzees. Nor does coloration change as infants mature. The expressive face remains black, and adults retain the white tail tuft. Bonobos have relatively small heads and distinctive hairstyles with parts down the middle of the forehead, which minimize the apparent size of their ears. Perhaps the most startling difference in the appearance of the

Figure 13-3 Male and female bonobos are more muscular than the other chimpanzees.

two species has to do with muscle definition which is much better in bonobos of both sexes than in *P. troglodytes* (Fig. 13-3). (Humans with comparable "pumped-up" looks are frequently body builders.) It is not just that bonobos have better muscle definition, however. By dissecting and weighing their muscles, Zihlman demonstrated that bonobos actually have more muscle mass than other chimpanzees of comparable body weights. Like well-muscled ballet dancers, *P. paniscus* has relatively long, graceful limbs (Fig. 13-4).

It should be kept in mind, however, that despite their differences, both *P. troglodytes* and *P. paniscus* are chimpanzees and, as such, share certain features. As summarized by Richard Wrangham of Harvard University:

> Both species have ... relatively long arms (reaching just below the knee when standing erect). The top of the head is rounded or flattened (there is no

Figure 13-4 Bonobos have long graceful limbs and smaller heads than other chimpanzees.

sagittal crest), and the neck appears short. Their ears are large and projecting, while the nostrils are small and lie above jaws that project beyond the upper part of the face (prognathous muzzle). . . . Males are larger and stronger than females, . . . both sexes have prominent genitals. (Wrangham, 1984:126)

OF HABITATS AND HABITS

Although chimpanzees are primarily forest animals, they occupy a variety of habitats in both West and East Africa. For example, populations of well known nut-cracking chimpanzees (*P. t. verus*) live in the dense tropical rain forest of the Taï National Park in the Ivory Coast of West Africa. Other members of *P. t. verus* at Mt. Assirik, Senegal, live under very different conditions. Their habitat is much drier and includes a mosaic of open grasslands and forest (i.e., traditional savanna), according to James Moore. Some populations of *P. t. schweinfurthii* occupy another kind of savanna, which Moore calls miombo woodland, at some sites in Tanzania. Unlike more traditional savanna, miombo is savanna woodland; that is, it contains forested patches that are separated by regularly spaced trees rather than open grasslands. As discussed in Chapter 14, Moore's suggestion that savanna woodland might have provided greater protection from predators for early hominids than more open savanna may have important implications for understanding the evolution of early hominids.

Other Tanzanian chimpanzees have mistakenly been identified as savanna dwellers, according to Craig Stanford of the University of Southern California. For example, *P. t. schweinfurthii* at Gombe is primarily a forest chimpanzee, since the open grassland part of its range is rarely used. Although they occur in West and East Africa and provide important models for studying early hominids (see Chapter 14), savanna chimpanzees have not been nearly as well studied as forest chimpanzees. It is therefore difficult to generalize about population density and range sizes in *P. troglodytes*. However, Moore points out that population densities appear to be much lower in savanna than in forest chimpanzees (i.e., about one-fiftieth the density), and that savanna populations are estimated to be associated with extraordinarily large home ranges (see Table 13-1). Compared to the diversity of other chimpanzee habitats, those of bonobos are more restricted, being limited to tropical rain forests and swamps in the low, southern basin of the Zaire River.

Despite the variety of habitats occupied by chimpanzees, their locomotor and positional behaviors are remarkably similar across chimpanzee (and gorilla) species. All of the African apes travel terrestrially between feeding and resting sites by knuckle walking, and Diane Doran of the State University of New York at Stony Brook estimates that 85% of their locomotor activity is therefore quadrupedal. Climbing in trees is another important activity. Chimpanzees including bonobos engage in a small amount of suspensory locomotion through trees, and also build night nests for sleeping in the canopy (gorillas sometimes, and chimpanzees rarely, build nests on the ground). Both

REVIEW EXERCISES

1. Why does this text refer to chimpanzees as our cousins?
2. Name the two species of *Pan* and give their common names.
3. Name the three recognized subspecies of *Pan troglodytes* and show on a map where they live.
4. Why are bonobos described as monotypic?
5. Show on a map where bonobos live.
6. Why have chimpanzees been favored as models for studying human evolution?
7. Explain why there is a wrinkle in using *Pan troglodytes* as a model for early hominids.
8. Compare and contrast the physical appearances of the two species of chimpanzees.

species of chimpanzees engage in a small amount of bipedalism that is more common in trees than on the ground. Doran notes that there are consistent sex differences in the degree of arboreality across chimpanzees, with females spending more time in trees. Interestingly, bonobos are reported to be the most arboreal of the African apes (which is consistent with their relatively curved finger bones), although the extent to which this is an artifact of the tendency for unhabituated animals to stay in trees when humans are around remains to be determined. According to Doran, bonobos do more leaping (and diving) than the other chimpanzees and use this behavior along with bridging, suspensory movement, and quadrupedal locomotion on branches to travel over "arboreal highways."

Chimpanzees including bonobos spend much of their day feeding and are extremely eclectic omnivores that have large components of fruit in their diets (see Table 13-1). In addition to the usual fruits, leaves, and different plant parts, both species eat a number of insects including ants, bees, and termites (see Box 13). Each also partakes of a

---

Box 13

Eating Bugs

Students frequently cringe at the thought of chimpanzees scarfing down juicy caterpillars, termites, and worms. But they shouldn't. Not only are these and other bugs good for chimpanzees, as described by columnist Sandy Bauers, but also they are good for people, who in fact eat them in many parts of the world. Insects are high in protein and good (unsaturated) fat. While a lean quarter-pounder has about 250 calories in its meat, an equivalent 4 oz of termites has a whopping 700 calories plus more thiamine and riboflavin than the hamburger. No wonder people in many undeveloped nations eat termites, lo-custs, mealworms, crickets, some species of ants, grasshoppers (which supposedly have a shrimplike taste), and wax worms (reminiscent of bacon). In Mexico, caterpillars are an expensive delicacy. Chimpanzees eat their bugs alive, but many people prefer theirs boiled, roasted, or fried. As Bauers observes, the official term for bug eating is **entomophagy**, and it is possible to find an occasional bug tasting, bug banquet, or bug-containing trail mix in the United States—if you look hard enough. So don't be faint-hearted. Next time you prepare salad, think twice about washing those insects off the greens!

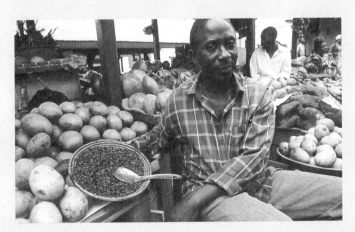

Various bugs including termites are highly nutritious and are eaten by many people. Here, a vendor sells dried termites in the Central Market of Kampala, Uganda.

Figure 13-5 William McGrew collects data on social grooming of two adult male chimpanzees at Gombe. (Photo courtesy of W. C. McGrew.)

Figure 13-6 This adult female chimpanzee from Gombe is modifying an ant dipping wand during ant dipping season. (Photo courtesy of W. C. McGrew.)

variety of small mammals including rodents and ungulates. There are also differences in the diets of the two species. Bonobos eat more fibrous plants such as shoots, stems, and herbs and ingest a wider range of invertebrates like worms. Unlike bonobos, other chimpanzees consume monkeys, and some populations of *P. t. verus* crack open and eat nuts. In fact, as detailed later, some of the most dramatic behavioral differences between bonobos and the other chimpanzees are related to how they obtain food.

Getting Fed: The Role of Tools and Hunting

One must be careful about comparing bonobos with the other chimpanzees because the former have been well studied at only two sites in the Democratic Republic of the Congo—Wamba and Lomako. To add to this, the bonobos at Wamba have been artificially fed by researchers, which is a practice that has been avoided or abandoned elsewhere in Africa because it is known to distort group behavior. These caveats aside, it appears that bonobos differ significantly from the other chimpanzees in how they get food: *P. troglodytes* systematically hunts for meat and habitually uses a variety of tools to obtain other food; *P. paniscus* appears to do neither.

Goodall's initial observation that Gombe chimpanzees (*P. t. schweinfurthii*) fished for termites by poking prepared blades of grass into their mounds was nothing short of stunning. As she put it,

> on a never-to-be-forgotten day in October, 1960, I had watched David Greybeard, along with his close friend Goliath, fishing for termites with stems of grass. Thinking back to that far-off time I re-lived the thrill I had felt when I saw David reach out, pick a wide blade of grass and trim it carefully so that it could more easily be poked into the narrow passage in the termite mound. Not only was he using the grass as a tool—he was, by modifying it to suit a special purpose, actually showing the crude beginnings of tool-*making*. What excited telegrams I had sent off to Louis Leakey, that far-sighted genius who had instigated the research at Gombe. Humans were not, after all, the *only* tool-making animals. (Goodall, 1990:5)

Since Goodall's discovery 40 years ago, 34 populations of free-ranging chimpanzees have been observed using tools, as summarized by primatologist William McGrew of Miami University in Ohio (Fig. 13-5). McGrew notes that these observations fall into 19 patterns of habitual tool use (i.e., practices observed at least 10 times), and that 11 (58%) of these are related to obtaining food: Plant probes (e.g., the blade of grass employed by David Greybeard) are used by numerous populations to fish, dip, gouge, or dig for termites, ants, bees, gum, and honey. Remarkably, chimpanzees not only are capable of modifying their tools (Fig. 13-6), but also sometimes use a set of different tools to accomplish a task (Fig. 13-7). Some chimpanzees use picks to extract marrow from bones, and leaves as drinking sponges. Others have been observed eating medicinal plants, possibly as a means of self-treatment for illness!

Figure 13-7 In the presence of other chimpanzees, this adolescent female from Gambia uses a set of different tools to get honey from a tree hole. The tools include a stout chisel (top left), a fine chisel (bottom left), a puncturing tool (top right), and a dipstick to extract the honey (bottom right). (Photos courtesy of W. C. McGrew.)

Figure 13-8 This adult female chimpanzee in Liberia uses a stone hammer to crack oil palm nuts on a stone anvil. (Photo courtesy of W. C. McGrew.)

Chimpanzees (*P. t. verus*) at various sites in western Africa use rocks to crack open nuts (Fig. 13-8). Interestingly, this activity is engaged in by females more than males in the Taï Forest, Ivory Coast. In fact, the most complex form of tool use yet reported for wild chimpanzees involves manipulation of three separate stones to open palm oil nuts at Bossou, Guinea, as detailed by Tetsuro Matsuzawa of Kyoto University: One stone serves as the surface or anvil upon which the nut is placed, a second stone is used as a wedge to secure the anvil, and the third is the hammer (Fig. 13-9). Matsuzawa also reports

Figure 13-9 Chimpanzees at Bossou, Guinea, use a third stone to secure the anvil on which nuts are cracked. (Photos courtesy of Tetsuro Matsuzawa.)

Figure 13-10   This captive adolescent female is using a twig to remove a loose tooth from the mouth of an adolescent male, while an adult male watches over her shoulder. (Photo courtesy of W. C. McGrew.)

that chimpanzees at Bossou may show a "rudimentary form of possession" because they carry around their favorite nut-cracking tools.

Tools are also used in contexts that do not involve food (Fig. 13-10), and these incidences are frequently related to emotional displays and weaponry (e.g., sticks serving as missiles, flails, or clubs). It is important to note that none of the food-related or other patterns have been observed at all of the sites, and that different populations engage in tool use to varying degrees. For example, opening nuts, ant dipping, and missile throwing have each been reported from 5 different sites. Of the 14 tool sites listed by McGrew, the highest number of patterns (11) was recorded at Gombe. This is not surprising, however, given the length of time that chimpanzees have been studied at Gombe, and their degree of comfort with researchers. In contrast, only one habitual pattern has been observed (so far) at 7 of the 13 other sites where tool use has been observed. In addition to the 14 tool-use sites, there are eight populations of *P. troglodytes* in which tool use has not been recorded. However, the studies of these populations represent mostly short-term surveys of unhabituated animals. Thus, to quote McGrew (1994b:28), "After more than 30 years of field study of chimpanzees in Africa, the accumulated data suggest that tool use is likely to be typical of the species."

As noted earlier, similar tool use is not typical of bonobos. In fact, during 900 hours of direct observation of *P. paniscus* at Wamba, Ellen Ingmanson of Dickinson College failed to observe *any* instances of the two main categories of tool use seen in *P. troglodytes*: i.e., employment of objects to obtain food or as weapons. She did, however, record other patterns of tool use. For example, adult males habitually drag leafy branches on the ground in association with directing group movement (see In the Field), and young bonobos frequently use sticks or parts of plants in the context of play. Bonobos' use of tools in social rather than agonistic or foraging contexts is in keeping with their highly social natures (see below). Interestingly, plants are also infrequently used as wipes or napkins, toothpicks, fly swatters, and back scratchers. Unlike other chimpanzees or gorillas, but like orangutans (see Chapter 14), bonobos habitually make and use rain hats (a practice that is apparently passed on through mothers):

> One of the best examples I observed was when a young adult male (Tawashi) moved about 10m, pulled up a large leafy plant, broke off the bottom of the stem, carried the main part of the plant back to where he had been sitting, and held it over his head in exactly the same manner as we would hold an umbrella. In fact, many of the leaves *P. paniscus* utilized as rain hats closely resemble those used by the village women and children during rain. (Ingmanson, 1996:197)

CHIMPANZEE THE HUNTER

Goodall's first report of hunting by chimpanzees, in the late 1960s, was every bit as startling as her initial report of tool use. Since then, wild chimpanzees have been observed hunting some 25 species of vertebrates in a variety of habitats that span from West to East Africa. Predation seems to occur during hunting seasons, and red colobus monkeys are, by far, the major prey at a number of sites including Taï, Gombe, Mahale, and Kibale. However, similar to the situation described for tool use, different populations of chimpanzees focus on distinct combinations of animals, and have their own methods for hunting them.

Take for example the chimpanzees of the Taï Forest, Ivory Coast, which have one of the strictest diets when it comes to meat, perhaps because they are very adept at using rocks to obtain fats from at least five species of nuts. Unlike other chimpanzees, those at Taï hunt only primates (especially red and black-and-white colobus monkeys), and do so in a highly coordinated manner. According to Christophe Boesch and Hedwige Boesch-Achermann of the Max Planck Institute in Leipzig, a hunt is usually undertaken by a group of males, each of which assumes a separate role of driver, pursuer, blocker, or encircler. Once the hunt is on, the chimpanzees give loud barks, and some remain on the ground while others climb into trees. Such cooperation makes it possible to capture prey, despite the fact that the tall forest provides escape routes for the fleeing monkeys. Not surprisingly, Boesch and Boesch-Achermann have found that the chance of success increases with the number of hunters and how well they are organized. The Taï chimpanzees kill a relatively high percentage of mature compared to young monkeys, and successful hunters are likely to share their kills with others.

Hunting at Gombe is quite different, as detailed by Stanford. Although red colobus monkeys account for about 80% of their prey, unlike the chimpanzees at Taï, those at Gombe kill mostly young monkeys, and also hunt bushpig, bushbuck, and baboons. Again, it is males who do most of the hunting, but those at Gombe do not assume specific roles. Instead, they appear to hunt in a disorganized but simultaneous manner. Thus, a pair of chimpanzees might drive a colobus monkey into the top of a tree, or onto the ground, where it is then grabbed and flailed to death by one of them. Despite the apparent lack of specific roles by Gombe hunters, however, Stanford reports a positive correlation between the number of males in a hunting party and success rate. Nor are Gombe hunters reported to be as willing to share their kills as are Taï hunters (but see below). Interestingly, the entire carcass of a monkey is usually eaten (flesh, viscera, and bone).

It would be a mistake to think that hunting is less of a social activity at Gombe than at Taï, however. As summarized by Stanford, Gombe males are more likely to hunt if one or more estrous females are present, and successful hunters tend to share their kills preferentially with these receptive females. Further, McGrew has shown that, compared to other females, those that receive generous helpings of meat produce more offspring that survive past infancy. At Mahale, the alpha male, Ntilogi, withholds meat from his rivals but shares it with his allies, as described by Nishida and his colleagues. The chimpanzees at Gombe tend to go on hunting binges from time to time that involve large parties with a relatively high number of estrous fe-

REVIEW EXERCISES

1. What range of habitats does *Pan troglodytes* live in? What about *Pan paniscus*?
2. Describe the locomotor patterns of bonobos and other chimpanzees. How do they differ?
3. Compare and contrast the diets of bonobos and other chimpanzees.
4. When was tool use first discovered in wild chimpanzees, and by whom?
5. What are the majority of chimpanzee tool-use patterns for? Give examples.
6. Do any patterns of tool use appear in all wild chimpanzees?
7. Do bonobos use tools the same way that other chimpanzees do? Explain.

Why Am I Doing This?

Like many primatologists, Ellen Ingmanson of Dickinson College has discovered that, although fieldwork can be tedious, the anticipation of discovery makes it all worthwhile. Ingmanson's research focuses on bonobos, and she remembers one particularly rewarding day in the field that brought new insights into their behavior. In her own words:

BINOCULARS IN HAND, ELLEN INGMANSON SEARCHES FOR BONOBOS. (PHOTO COURTESY OF E. INGMANSON.)

Field observations are often difficult and frustrating. One must get up before dawn; hike through jungle, swamps, and streams; crawl through bushes while keeping an eye out for snakes; swat at insects; and swelter in the heat and humidity—all of this to get a better view of your subjects. And that is on a good day! On bad days, you may not even find them. Even well-habituated groups of bonobos (*Pan paniscus*) have an uncanny ability to melt into the forest and disappear, leaving few signs that can be followed. I am often asked why I bother with fieldwork when it can be so uncomfortable. But for those of us who study primates in the field, there is an irresistible draw that keeps bringing us back, and getting us out of bed before the crack of dawn. It is the anticipation of potential discovery that is associated with each moment of observation. The mere possibility that today might be the day when something new comes to light is incredibly addictive. Bonobos are wonderful creatures, and I never tire of observing them (I do tire of slogging through swamps, however!). Each animal has a unique personality, and I find it endlessly fasci-

males. No wonder Stanford concludes that "meat is thus a social, political, and even reproductive tool"! Because Stanford has thoroughly studied both chimpanzees and red colobus monkeys, he is in the unique position of knowing both the hunters and the hunted as individuals, which

> makes my research intriguing but a bit heart wrenching. In October 1992, for example, a party of thirty-three chimpanzees encountered my main study group, J. . . . The result was devastating from the monkeys' viewpoint. During the hour-long hunt, seven were killed; three were caught and torn apart right in front of me. Nearly four hours later, the hunters were still sharing and eating the meat they had caught, while I sat staring in disbelief at the remains of many of my study subjects. (Stanford, 1995b:52)

nating to piece together the various daily activities that make up their lives. Some days are more rewarding than others, though.

While great new discoveries are few and far between, there are moments of new understanding and realization that make all the work worthwhile. One of these moments occurred for me when I first began to grasp the meaning of branch dragging by the bonobos. I had been in the field for about 2 months, and was beginning to adjust to the routine and to become familiar with the group of bonobos I was observing at Wamba. It was midmorning, and the group had been sitting around resting and grooming after their early-morning feeding activity. Many were sitting in groups on the ground; some were in nearby trees. I was leaning against a tree with my day pack next to me and my notebook in my hand, surrounded by bonobos. Everything had been fairly quiet and relaxed, even a bit drowsy. Then I began collecting all of my paraphernalia and stuffing it in my pack, preparing to move on. Suddenly I stopped myself and asked, "Why am I doing this?" The group was mostly still sitting around, not yet showing many signs of moving. But a few of the males had been up and had begun pulling long, leafy branches behind them through the forest underbrush, making a distinctive noise. This is when I realized that I was keying in on a bonobo signal—branch dragging—without even being conscious of it. As I collected data on branch dragging over the next several months, I

THIS BONOBO'S BRANCH DRAGGING TELLS HIS GROUP WHICH WAY TO GO. (PHOTO COURTESY OF E. INGMANSON.)

found that it was more than just a signal telling the group to get ready to move. It also told them which direction to move, as the males would drag the branch away from the group in a particular direction, then back again along the same line. This was generally the same direction in which the group then traveled, often directly to another food source. While it took many, many hours of observation and data analysis to work out the details of branch dragging, I will never forget the sense of joy, wonder, and comprehension that I felt that first morning when I found myself responding to a bonobo communication signal.

Bonobos also eat animal foods, especially invertebrates such as ants and termites. Although they have not been observed hunting monkeys, there are now several reports of predation on small ungulates called duikers, at unprovisioned sites in the Lomako Forest. In one incident, reported by Noel Badrian and Richard Malenky of the State University of New York at Stony Brook, a large group of excited bonobos of all ages surrounded a small adult male that had killed an infant duiker. Although the bonobos constantly made begging gestures accompanied by vocalizations, the individual tried to avoid his conspecifics and was not observed sharing the meat. Because they have not been well studied in the wild, the full extent to which bonobos hunt is probably not known. As outlined earlier, despite differences between populations, hunting is a highly social activity for *P.*

troglodytes. In addition to differences in their apparent hunting practices, when it comes to social life in general, *P. troglodytes* and *P. paniscus* are very different.

SOCIAL LIVES OF BONOBOS AND OTHER CHIMPANZEES: LIKE NIGHT AND DAY

The broad outlines of social organization are similar in all chimpanzees. Both species are extremely outgoing and gregarious compared to orangutans and gorillas. All populations of *Pan* live in multimale, multifemale **fission-fusion** social groups that frequently divide into a number of transitory subgroups. The size and composition of these temporary parties are fluid, that is, subject to change related to food acquisition or social endeavors, and lasting from a few minutes to days. Thus, members of the genus *Pan* can truly be described as "party animals." As befits their gregariousness and need to keep in touch when apart, each species has a repertoire of vocalizations ranging from noisy to soft, and human-like greetings and appeasements, including touching, hugging, and kissing. The daily rhythms of chimpanzees include feeding, resting, grooming, traveling, and building (and occupying) night nests at the end of the day. Another important similarity of the two species, and one that is not very common in higher primates, is that males remain with their natal groups for life, while females disperse to other groups around the age of sexual maturity.

Despite all of these similarities, *P. troglodytes* and *P. paniscus* lead markedly different social lives tied to fundamental differences in the sexual behavior of females that profoundly affect the basic nature of relationships between individuals. Although sexual swellings associated with estrus are conspicuous and of approximately equal duration in females of both species, female bonobos remain partially swollen for a longer part of their cycles, according to Yukio Takahata of Naruto University of Education and colleagues. Furthermore, bonobos become sexually receptive much sooner after giving birth, and are therefore sexually receptive over longer periods throughout their interbirth intervals than are other chimpanzee females. Not surprisingly, then, bonobos engage in more-frequent and more-varied sex than their relatively staid cousins—to such an extent, in fact, that Takahata and colleagues (1996:153) suggest that "female bonobos may have developed 'continual receptivity and attractivity'" compared to other chimpanzees. As put by one of the leading authorities on *Pan* politics, Frans de Waal of Emory University,

> bonobos have more ways of inviting each other sexually, more ways of engaging in sex, and a greater variety of facial expressions and vocalizations associated with sexual intercourse than do chimpanzees. The chimpanzee's sex life is rather plain and boring; bonobos act as if they have read the *Kama Sutra*. (de Waal, 1997a:25)

Figure 13-11 Two male bonobos reassure each other by rubbing their rumps together. (Photo courtesy of Takayoshi Kano.)

Bonobos of all ages and combinations interact sexually. Males rub rumps as an apparent gesture of reassurance (Fig. 13-11), and mount each other to express dominance or to release tension. Females also frequently engage in homosexual behavior, known as genitogenital (or GG) rubbing (Fig. 13-12) as a means of greeting or, again, to dispel social tension. Even youngsters engage in sexual interactions with each other and with adults. Kano notes that whenever bonobos first encounter an excellent feeding site, they become excited and engage in much heterosexual and homosexual behavior, while noisily calling. Face-to-face sex with eye contact is also practiced regularly, and copulations generally take longer in bonobos than other chimpanzees. Given these behaviors, it is not surprising that males and females spend more time together in bonobo than in other chimpanzee societies, or that female bonobos groom each other more and generally have stronger bonds with each other than do other female chimpanzees. Male bonobos, on the other hand, do not form as many all-male groups as male *P. troglodytes* (and they maintain closer ties with

Figure 13-12 Two bonobo females rub genitals as a means of relieving tension.

their mothers), perhaps because the wide availability of sexual partners is an incentive to remain in mixed groups.

Why do bonobos and other chimpanzees differ so much socially? Ordinarily, one might seek the answer to this in ecological variables related to habitat or food supply. However, Taï chimpanzees live in rain forests similar to the habitat of bonobos, and chimpanzees and bonobos have access to basically the same foods (see Table 13-1). The key to social differences between the two species may well be female behavioral differences (which are likely to be related to different reproductive physiologies) because, although males of both species are usually ready and willing to mate, sexually receptive *P. troglodytes* females are fewer and farther between than is the case for bonobos. Amy Parish of University College London agrees that female sexuality is a crucial factor for understanding social differences between the two species of *Pan*:

> Bonobo females are remarkably skillful in establishing and maintaining strong affiliative bonds with each other despite being unrelated. Moreover, they control access to highly desirable food, share it with each other more than with males, engage in same-sex sexual interactions in order to reduce tension, and form alliances in which they cooperatively attack males and inflict injuries.... The ultimate advantage of friendly relationships among females is an earlier age at first reproduction, which results in a large increase in lifetime reproductive success. (Parish, 1996:61)

It appears that the relatively staid sex life of *P. troglodytes* chimpanzees leaves them with some time on their hands (compared to bonobos who are happily partying), which might help explain why they developed relatively greater use of tools and, in the case of males, a predilection for hunting and intergroup conflicts.

Dominance hierarchies exist in both sexes of both species of *Pan*. In the case of *P. troglodytes*, the lowest-ranking male is dominant to the highest-ranking female. The situation is murkier for bonobos, however, because males are generally more tolerant and rarely threaten or attack females. Interestingly, high-ranking *P. troglodytes* females appear to have significantly higher infant survival, faster maturing daughters, and more rapid production of young than do low-ranking females because of better access to good foraging areas, according to Anne Pusey of the University of Minnesota and her colleagues. Pascal Gagneux of the University of California, San Diego, and his colleagues, on the other hand, have dispelled the long-held assumption that chimpanzee infants are all sired within their communities, at least for *P. t. verus* in which genetic tests revealed that 7 of 13 infants from the Taï Forest could not have been fathered by any of the group's males. It thus appears that these infants were conceived when estrous females left the group for 1 or 2 days. (As discussed in Chapter 10, furtive matings also occur in presumably monogamous primates such as gibbons.)

All-male parties of *P. troglodytes* occur much more frequently compared to *P. paniscus*, and these groups often patrol the borders of their territories (occasionally accompanied by adolescent males or mature females) in search of strangers from other groups. Chimpanzee patrols may end in mere vocal confrontations, or in physical combat that can result in the death of males, females, or infants. (Infanticide has also been reported in other contexts.) Indeed, *P. troglodytes* chimpanzees at Gombe were so aggressive that one community eventually exterminated another. Fortunately, chimpanzees are also skilled at peacemaking within their own social groups, as thoroughly documented by de Waal. By all reports, bonobos do not patrol their borders, kill each other, or commit infanticide; they seem generally more passive and gregarious than chimpanzees.

CULTURE AND COGNITION IN *PAN*

As we have seen, populations of *Pan* vary tremendously in their patterns of tool use, preferred prey, hunting practices, relationships with conspecifics, and levels of aggression. Different populations also have unique dialects when they vocalize, and McGrew has observed a form of grooming (called the "grooming-hand-clasp") at Mahale (it has now also been reported at Kibale and Taï) that does not occur at Gombe. What accounts for the fact that chimpanzees in one population habitually fish for termites, while those in another simply dig them out of their mounds? Why do the meat-loving chimpanzees of Taï ignore the bushbuck in their habitats, unlike the Gombe chimpanzees which add this species to the red colobus monkeys so avidly hunted by both populations? For what reason do bonobos make rain hats, while chimpanzees at Gombe sit miserably in the rain? A possible explanation is that different groups accidentally or deliberately invent certain behaviors that are then picked up by other individuals. If the practices become widespread enough, they may be passed to future generations. As discussed for Japanese macaques (see Chapter 9), some primatologists refer to such traditions as precultural. Certain primatologists go even further by claiming that chimpanzees, like humans, have unique cultural traditions. This assertion has been debated by others on the grounds that there is little evidence for the ability of chimpanzees to deliberately teach each other new activities, or to imitate them. However, Matsuzawa notes that chimpanzees at Bossou (which, as described earlier, produced the most complex form of tool use in wild chimpanzees to date) acquire their nut-cracking skills from direct experience and observational learning. Furthermore, even some skilled human behaviors may be acquired and passed on without being consciously taught or imitated, as McGrew points out. One thing is for sure—if bonobos and other chimpanzees do have cultural traditions, then culture may no longer be considered a hallmark of humanity.

Chimpanzees are extremely smart as evidenced not only from the wild (as described throughout this chapter), but also from the laboratory. Matsuzawa is a comparative cognitive scientist who has studied *P. troglodytes* under both conditions. He reports that an ape named Ai

REVIEW EXERCISES

1. Which animal does *P. troglodytes* hunt the most?
2. Compare hunting practices of chimpanzees at Taï and at Gombe.
3. Why does Craig Stanford refer to meat as "a social, political, and even reproductive tool"?
4. Do bonobos ever hunt for meat? If so, how frequently and what kind?
5. In what ways are the social organizations of bonobos and other chimpanzees similar?
6. Compare and contrast sexual behavior in the two species of *Pan*.
7. How does the near-constant sexual receptivity of female bonobos affect male-male, male-female, and female-female relationships?
8. Compare dominance hierarchies and aggression in bonobos and other chimpanzees.

(Japanese for "love") and other chimpanzees at the Primate Research Center in Kyoto have form and color perception that are similar to those of humans. Interestingly, chimpanzees are quicker than humans at recognizing upside-down pictures and forms. They can also recognize and name 11 colors (using both abstract pictures that stand for words [lexigrams] and Japanese characters). Sarah Boysen of Ohio State University has established that chimpanzees are capable of at least remedial counting; for example, they differentiate between piles that contain small but different numbers of candies, and Ai learned to count from 1 to 9. Laboratory chimpanzees are also adept at doing simple drawings, puzzling out how to obtain out-of-reach baits with objects, being deceptive, and recognizing themselves in mirrors.

No discussion of chimpanzee cognition would be complete without reference to the modern era of ape-language studies, which began in the 1960s when two psychologists from the University of Nevada, Beatrice and Allen Gardner, began teaching American Sign Language (ASL) to a young female chimpanzee named Washoe. Washoe ultimately acquired nearly 200 signs that she strung together in short, childish sentences (e.g., "cookie more"). Her first signs included "come-gimme," "sweet," "up," and "more." Washoe babbled to herself in signs before falling asleep, used signs for the chimpanzee equivalent of swearing, and sometimes produced almost poetically creative signs. For example, she identified two swans as "water birds," and other chimpanzees as mere "black bugs." Further, Washoe eventually taught a number of signs to her adopted son, Loulis.

Equally important, Washoe stimulated a number of researchers to undertake their own language studies on other chimpanzees. One chimpanzee named Sarah used plastic pieces as substitutes for words and another, Lana, learned to manipulate a series of lexigrams in a computer language known as Yerkish. The investigators who studied Lana, Duane Rumbaugh and Sue Savage-Rumbaugh of the Yerkes Regional Primate Research Center in Georgia, eventually studied two male chimpanzees, Austin and Sherman, that learned to use Yerkish to request tools from each other in order to obtain and share food.

The all-time star pupil at Yerkes, however, is a bonobo named Kanzi (Fig. 13-13). Unlike some of the apes used in earlier studies, Kanzi not only has acquired a large number of signs, but also produces spontaneous utterances that reflect what is on his mind, rather than merely labeling objects in response to the question, What's that, Kanzi? Kanzi understands spoken English and even attempts to say a few words including "carrot," "onion," "raisin," "snake," "open," and "right now." The effort to actually speak is surprising because apes lack the laryngeal anatomy needed for human-like speech. Are bonobos generally better at ape language than chimpanzees, or is Kanzi simply a bonobo genius? Only time and study of more bonobos will tell.

Meanwhile, chimpologists have a privileged perspective from which to compare the cognitive abilities of humans and apes, and can tell us certain ways in which humans are unique. As Savage-Rum-

Figure 13-13 When it comes to ape language, Kanzi (a bonobo shown here with Sue Savage-Rumbaugh) is a star.

baugh is the first to acknowledge, there is an enormous difference between chimpanzees and humans. People can attend to multiple tasks, plan ahead, cooperate more fully, and easily transmit inventions to others. Both chimpanzees and young children use gestures combined with two-word utterances (pressing computer keys or saying words) to ask for cookies, tickles, games, and favorite people. However, the chimpanzee never grows beyond this stage, while normal children always do. By the time a child is 5 years old, he or she begins asking profound questions like, Where did I come from? In short, humans are just plain smarter than chimpanzees. The reason for this is that humans have bigger brains that are more specialized for conscious thought based partly on language (see Neural Note 13).

WILL THEY SURVIVE?

It would be a terrible shame if these fascinating and complicated apes, our closest cousins, became extinct. But this could happen despite the fact that bonobos and other chimpanzees are legally protected in some parks and reserves. In fact, populations of *Pan* have already been eliminated entirely in a number of African countries, owing to deforestation as well as ravages associated with pet, zoo, trophy, and bushmeat trades. Compared to these threats from humans, the dangers that chimpanzees experience from cats (e.g., leopards and panthers) or hostile conspecifics is minimal. Unfortunately, all chimpanzees are currently vulnerable to extinction at the hands of *Homo sapiens*, while the western *P. t. verus* is even more highly endangered.

SUMMARY

Two species of *Pan* are currently recognized: *P. troglodytes*, which is divided into three subspecies (*P. t. verus*, *P. t. troglodytes*, and *P. t. schweinfurthii*), and *P. paniscus*, the bonobo. Compared to other

REVIEW EXERCISES

1. What is meant by the term *culture*?
2. Do you think that chimpanzees have cultural traditions? Why or why not?
3. What kinds of nonlinguistic cognitive skills have chimpanzees demonstrated in laboratory experiments?
4. Who is Washoe and what have her contributions been to primatology? Answer the same question for Kanzi.
5. What do language-trained chimpanzees have to say?
6. Can bonobos and other chimpanzees make the same speech sounds that people can? Why or why not?
7. Why are people more intelligent than apes?
8. How do the frontal lobes of great apes and humans differ?
9. What is the outlook for chimpanzee survival? Who are their worst enemies?

Ape Language, Ape Brains

Verbal language has both sensory (understanding) and motor (speech) components. One reason why apes do not speak is because their larynges and throats are anatomically incapable of producing the full range of human-like speech sounds. Nevertheless, because they vocalize a good deal in social situations, one must look elsewhere for a satisfactory explanation of why great apes lack human-like speech. An important limitation on the ability to speak is due to the size and organization of pongid frontal lobes. Not only do great apes have brains that are approximately one-third the size of those of humans (see Neural Note 12), but also their cerebral cortices are configured differently in exactly those parts of the frontal lobe that facilitate speech in humans. Specifically, the frontal lobes of humans have a triangular patch of gray matter, which is called Broca's speech area on the left side. Although other nearby regions are also important for symbolic speech, humans require an intact Broca's area for normal speech. Great apes have a completely different pattern of convolutions in their frontal lobes and, of course, they do not speak. This difference in the sulcal patterns of ape and human frontal lobes is one of the few distinctions, in addition to overall size of the brain, that permit an observer to identify a brain or endocast as either human-like or apelike. As such, it has important implications for studying brain evolution in our early hominid ancestors (see Neural Note 14).

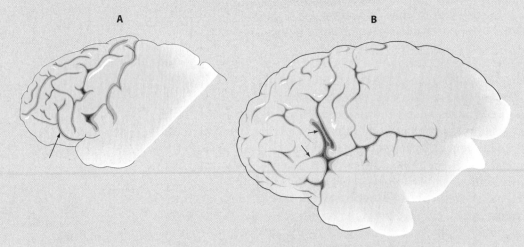

The left frontal lobes of chimpanzees (**A**) and humans (**B**) are noticeably different in their pattern of convolutions. In place of the simple bulge shown by the chimpanzee brain (**arrow**), human frontal lobes have a triangular patch of cortex (two sides indicated here by **arrows**) known as Broca's speech area.

chimpanzees, bonobos are relatively slender, with smaller heads and well-muscled, yet graceful bodies. The other chimpanzees are found in a wide variety of habitats, contrary to bonobos that live only in lowland rain forests. Both species knuckle walk on the ground, and each eats a wide variety of plant and animal foods. However, *P. troglodytes* and *P. paniscus* differ dramatically in how food is obtained, with tool use and hunting common in the former, but rare in bonobos. Furthermore, different populations of chimpanzees use different tools, have distinct methods of hunting, and focus on select prey in a manner that is suggestive of human cultural traditions. All members of *Pan* live in fission-fusion social groups, with females emigrating upon reaching sexual maturity. In contrast to other chimpanzees, however, female bonobos are much more sexually receptive, which profoundly affects social relationships, not only between males and females, but also between members of the same sex. Compared to *P. troglodytes*, bonobos do not patrol their borders, kill each other, or commit infanticide; they seem generally less aggressive. Chimpanzees perform well on laboratory tests, and both species have demonstrated an ability to learn human-like languages that are nonverbal, albeit at remedial levels compared to people. Great apes do not speak because their brains are smaller and organized differently in the frontal lobes than are human brains. Sadly, our closest cousin *Pan* is vulnerable to extinction because of the activities of humans.

FURTHER READING

BOESCH, C. AND H. BOESCH-ACHERMANN (1991) Dim forest, bright chimps. *Natural History* **100**(9):50–57.

McGREW, W. C., MARCHANT, L. F. AND T. NISHIDA (1996) *Great Ape Societies.* Cambridge: Cambridge University Press.

SAVAGE-RUMBAUGH, S. AND R. LEWIN (1994) *Kanzi: The Ape at the Brink of the Human Mind.* New York: John Wiley and Sons.

STANFORD, C. B. (1995b) To catch a colobus. *Natural History* **104**(1):48–55.

DE WAAL, F. (1997) Bonobo dialogues. *Natural History* **106**(5):22–25.

14 THE EARLIEST HOMINIDS

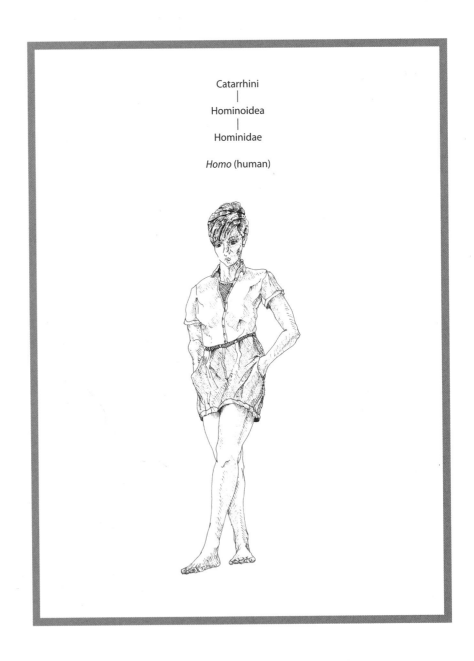

Catarrhini
|
Hominoidea
|
Hominidae

Homo (human)

INTRODUCTION

The world was startled 75 years ago by the discovery of a fossilized skull and endocast from a "man-ape" at a site called Taung in South Africa. Described in the scientific journal *Nature* by an anatomist named Raymond Dart, the combination of apelike and human-like features in this little fossil had never before been seen in a primate—living or dead (Fig. 14-1). Dart therefore assigned the specimen to a new genus and species, *Australopithecus africanus* ("southern ape from Africa"), and claimed that it represented a very early member of the human family, now known as the **Hominidae**. Although Dart's announcement that this fossil represented an extinct population of hominids that was ancestral to humans (a so-called missing link) met with controversy from both the public and scientists alike, his general view of human evolution has since been vindicated by the discovery of hundreds of other fossils at numerous sites in southern, eastern, and (recently) central Africa (Fig. 14-2). In fact, today paleoanthropologists accept that a population of australopithecines (as they are commonly called) branched off somewhere in Africa before 2.0 mya and gave rise to the genus *Homo*.

The devil, however, is in the details. A variety of australopithecine species are now known from African sites that range from 4.4 to around 1.0 mya, and paleoanthropologists argue passionately about which ones were directly ancestral to *Homo* and which were not. Predictably, there is a strong tendency for scientists to favor the fossils that they discovered (or intensively studied) as the legitimate ancestors of *Homo*. Consequently, a certain amount of wishful thinking often goes into the reconstructions and interpretations of early hominids, a trend that is all too often facilitated by the fragmentary nature of the fossils themselves (see Chapter 1). Nevertheless, there remains an intense curiosity about the nature of our earliest relatives that first ceased to be apes, and the initial circumstances that started them on the evolutionary journey that eventually led to ourselves. What did the very first hominids look like? How did they behave, and why did

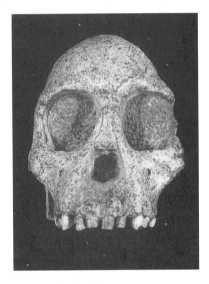

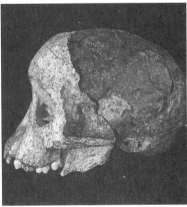

Figure 14-1 This fossilized facial portion of the skull and associated endocast, known as the Taung child, was the first australopithecine ever discovered. Note the combination of a human-like rounded forehead, lack of browridges, and small canines with an apelike projection of the upper jaw and small endocast. (Photos courtesy of Glenn Conroy.)

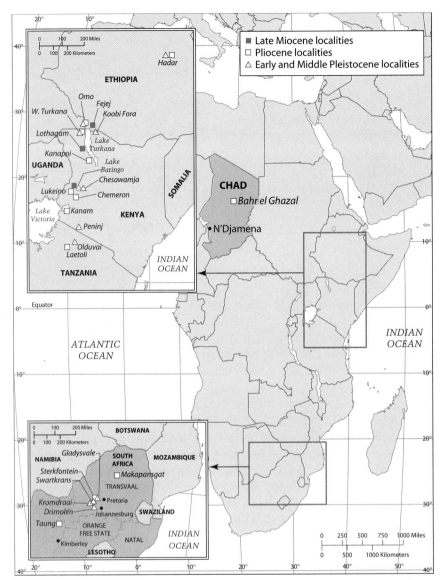

Figure 14-2 Although fossils of australopithecines have been found in various southern and eastern African sites for decades, they have only recently been discovered at Chad in central Africa.

some of their descendants eventually become human? Although one might think that the answers to these questions can best be sought directly in the fossil record, it is often too murky and its interpretations too controversial to provide any but the broadest of answers. Instead, many of the most reasonable discussions regarding these and other questions concerning the earliest hominids are based on living primates.

THE AUSTRALOPITHECINES

The many fossils that have been discovered over the years and grouped with the little skull from Taung in the species *A. africanus* are called **gracile australopithecines** because of their comparatively round, smooth skulls that have foreheads and (usually) lack sagittal crests. These specimens have relatively small molars and large incisors that cause the upper jaw to protrude, giving a slight upward curve to the face, and their cranial capacities are in the ape range. Recent research on a fairly complete *A. africanus* skeleton (a rare find) by Lee Berger of the University of Witwatersrand in South Africa and Henry McHenry of the University of California, Davis, reveals that, from the neck down, *A. africanus* appeared apelike with long arms and short legs. Gracile australopithecines are recognized only in South Africa, where they are known from slightly before 3.0 mya to around 1.5 mya.

Over a decade after Dart made his announcement, a second kind of australopithecine was discovered in South Africa and assigned to a new taxon, *Paranthropus robustus* (today, some paleoanthropologists substitute *Australopithecus* for *Paranthropus*). The skulls of **robust australopithecines**, as they are commonly called, are much more rugged-looking than those of their gracile cousins. For one thing, they lack true foreheads. They also have large sagittal crests along the tops of the skull (especially males) and massive cheekbones (zygomatics), both of which anchored chewing muscles that must have been impressive. The molars of robust australopithecines are huge, while the front teeth are comparatively small. This entire configuration results in an enormous flat-looking face.

Despite the dramatically different appearances of their skulls, the cranial capacities are virtually indistinguishable in the two kinds of australopithecines (see Neural Note 14), both of which seem to have been quite sexually dimorphic. Although they were once believed to have larger, heavier bodies than gracile australopithecines, robust australopithecines, in fact, may have been roughly the same size, based on estimates from parts of the skeleton by contemporary paleontologists such as McHenry. In South Africa, females of both types appear to have weighed around 30 kg (66 lb) while males were approximately 40 kg (88 lb). The East African robusts are believed to have been somewhat heavier, with females weighing around 35 kg (77 lb) and males estimated to weigh about 50 kg (110 lb). It should be noted, however, that these estimates are extremely tentative and could change with the discovery of even one more fossil. Because relatively complete portions of a skeleton and associated skull have yet to be identified for a robust australopithecine, many of the details about its body, including limb proportions, can only be reconstructed with a good deal of guesswork (Fig. 14-3). Today, robust australopithecines are known from East as well as South Africa. The oldest specimen discovered (so

Figure 14-3   A good deal of imagination is needed to flesh out the bones of early hominids, as seen in this illustration of gracile and robust australopithecines.

far) lived in Kenya about 2.6 mya, and the genus appears to have become extinct by about 1.0 mya.

The environments associated with australopithecines were extremely variable and included forests, woodland savanna, open savanna, floodplains, rivers, streams, and lakes. At times, there were probably marked seasons, and the fossil record of pollen and animal bones shows that climatic fluctuations caused shifts in the overall balance between wetter wooded conditions and drier more-open grasslands. Because environments were so varied, the exact combination of forests, grasslands, and so on that were frequented by gracile versus robust australopithecines is subject to speculation. From their teeth, however (see Chapter 1), we may infer that the diets of gracile australopithecines included a good deal of leaves and fleshy fruits, while robust australopithecines relied more on harder foods like seeds and nuts.

What exactly is it about australopithecines that removes them from the apes and places them in our own hominid family? The answer to this pertains to their form of locomotion. Fossilized parts of the pelvis, hips, knees, and feet of australopithecines show that they were able to walk bipedally for sustained periods of time. This suggests that unlike any living nonhuman primates including the apes, australopithecines were habitual bipeds. Bipedalism, in fact, is recognized by most paleoanthropologists as *the* hallmark of hominids. Does this mean that australopithecines had fully developed fluid, striding bipedal gaits like those of living people? Because their bones, although basically human-like, retained some features that are intermediate between those of apes and humans (Fig. 14-4), the answer to this question is a source of ongoing controversy (discussed below).

A number of different australopithecine species from various locations in East Africa have now been named. Robust species include *Paranthropus (Australopithecus) boisei* and *P. (A.) aethiopicus* in addition to the South African *P. (A.) robustus*. Besides *A. africanus* from South Africa, other nonrobust species (in this text the term *gracile* refers only to South African specimens) include *A. afarensis*, *A. anamensis*, and *A. garhi*. The oldest supposed early hominids described so far are dated to a little less than 4.5 mya and have been placed in a new genus and species, *Ardipithecus ramidus*. The problem with these remains is that their published descriptions suggest that they had extremely chimpanzee-like dentitions in many respects, and the evidence presented for bipedalism seems unconvincing. You will recall that there are no recognized fossils of chimpanzees (or gorillas, for that matter—see Chapter 1), and one wonders if, rather than being a hominid, *Ardipithecus ramidus* might not be an early member of the lineage that led to *Pan*. This exciting possibility awaits clarification from further descriptions of *Ardipithecus ramidus* fossils.

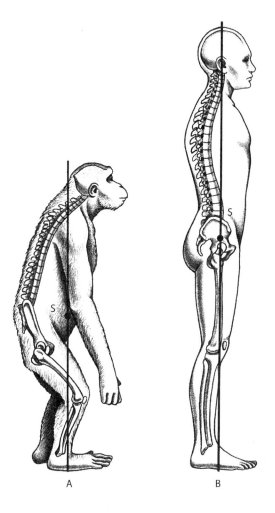

A   B

Figure 14-4 Chimpanzees (**A**) and humans (**B**) stand differently when they are in bipedal postures. Note that the line representing the center of gravity (**s**) falls very close to the hip, knee, and ankle joints in humans, but is far forward of the hip joint in chimpanzees. Differences in the anatomy of the hip region are one reason why chimpanzees waddle when they walk, while humans do not.

Brain Evolution in Early Hominids

Although endocasts provide information about both the size and the morphology of the brains of early hominids (see Neural Note 1), the quality of the revelations acquired from them depends on how complete they are. Skulls of early hominids are notorious for being fragmentary, and missing parts of endocasts traditionally have been reconstructed by using other endocasts as models. In particular, a complete endocast from a gracile australopithecine named "Mrs. Ples" (Sts 5) has been used extensively to model the missing parts on other endocasts from both gracile and robust australopithecines. A number of endocasts that reproduce previously missing parts in East African robust australopithecines have become available in the past 15 years, enabling colleagues and I to undertake a comparative analysis of endocasts of gracile and robust australopithecines.

Contrary to earlier studies, we found that the brains of robust australopithecines are no larger than those of gracile australopithecines. Rather, both hominids appear to have brains that averaged around 450 cc, which is well within the range for great apes (see Neural Note 12). However, the similarity between the endocasts of gracile and robust australopithecines stops there because of previously unrecognized differences in the forms and sizes of both their frontal and temporal lobes. In robust australopithecines, the bottom surfaces of the frontal lobes are rela-tively small and form a beaked shape when seen from the side, and their temporal lobes are relatively short with rounded front ends (called temporal poles). Both of these features resemble the appearance of endocasts from great apes. The frontal lobes in gracile australopithecines, on the other hand, have greatly expanded bottom surfaces, particularly at the front end, which gives the bottom part of the frontal lobe a large, blunt appearance from the side. Their temporal lobes, on the other hand, appear relatively long and pointed, as they are expanded forward near the outside part of the temporal poles. With respect to these two features, the endocasts of gracile australopithecines are more similar to those of humans than are the endocasts of either robust australopithecines or apes. These findings are consistent with recent research on ape and human brains by Katerina Semendeferi of the University of California, San Diego, who earlier hypothesized that many reorganizational events including expansion of the most anterior end of the frontal lobe took place in early hominid brains.

So what does it all mean in terms of hominid brain evolution? To answer this question, we can only make reasonable inferences based on what is known about some of the functions of the relevant areas in human and ape brains. For example, the most anterior ends of the frontal lobes in both apes and humans are involved in

The controversial nature of paleoanthropological interpretations is best exemplified by an approximately 3.2-mya fossil called Lucy that was discovered in the Hadar region of Ethiopia and attributed to *A. afarensis*. Much has been made of this specimen, partly because approximately 40% of its skeleton was recovered, which is astonishing for such an old fossil. Although Lucy was only about 105 cm (3 feet 5 inches) tall (Fig. 14-5), her teeth indicate that she was a young adult.

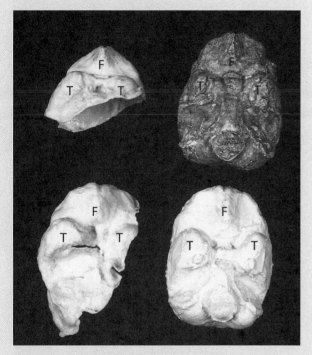

These photographs show the underneath surfaces of endocasts. Compared to endocasts of robust australopithecines (above), those from gracile australopithecines (below) have frontal lobes with relatively expanded bottom surfaces (F) and temporal lobes with more pointed and expanded front ends (T).

abstract thinking, planning of future actions, and undertaking initiatives, and this was probably also true of australopithecines. If so, gracile australopithecines may have been more advanced in these behaviors than robust australopithecines. The temporal lobes are known to be important in humans for recognition of specific people, their names, and related biographical information (see Neural Note 11). Interestingly, the lateral part of the temporal pole that is expanded in gracile australopithecines becomes ac-

tivated in people when they recognize and put a name with a familiar face! Did gracile australopithecines have rudimentary protolanguage in which they named people and objects in their environment? We do not know, but it is fascinating to contemplate the newly discovered features of their brains with respect to such questions. Our findings also lend support to the controversial suggestion that, among early hominid species, gracile australopithecines may have been the ones that gave rise to our own genus.

As so often happens in paleoanthropology, Lucy's discoverers proclaimed that she represented the species that gave rise to *Homo*, while sidelining the other australopithecines to a branch of hominids that became extinct without issue. Although *A. afarensis* was named and interpreted more than 20 years ago (in 1978), many workers continue to believe that Lucy was the mother of us all. Others, however, do not.

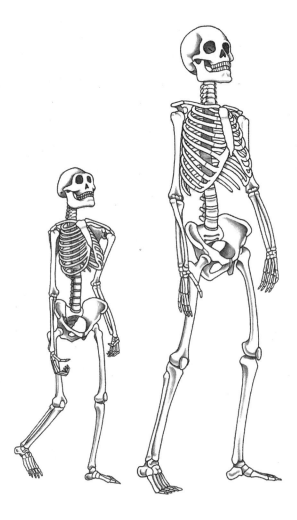

Figure 14-5 The reconstructed skeleton of Lucy (*A. afarensis*) on the left is much smaller and has relatively longer arms than the skeleton of a modern human female of average height.

Adrienne Zihlman was one of the first to raise questions about whether or not all of the specimens attributed to *A. afarensis* actually belonged to only one species. If so, this species exhibited at least as much (if not more) sexual dimorphism in body weight than that seen in any living species of primate. This would be remarkable, especially because canine size did not appear to be sexually dimorphic in the specimens attributed to *A. afarensis*, which contradicts the expectation for a species that is highly dimorphic in body size. Glenn Conroy of Washington University in St. Louis and I think that the pattern of cranial blood flow seen in the *A. afarensis* crania from Ethiopia rules *A. afarensis* out as an ancestor for *Homo*. However, the specimens attributed to this species from another country, Tanzania, appear to be different from the Ethiopian fossils—both in their cranial blood flow and in their feet, as discussed by Russell Tuttle who analyzed fossilized footprints and foot bones attributed to *A. afarensis*. After studying fossil arm bones, Bridgitte Senut of the National Museum of Natural History in Paris also suggested that *A. afarensis* is a conglomeration of more than one species. Given that so much of Lucy's skeleton is available for study, we might hope that paleoanthropologists would at least agree about how to interpret her morphology. Lucy's discoverers, however, portray her as a biped that was fully adapted to terrestrial living, while other workers interpret her curved finger and toe bones as reflecting a good deal of arboreal activity and her pelvis and hip morphology as indicative of bipedalism that was less fully developed than our own. In fact, the debate about the fine details of bipedalism in australopithecines in general (not just Lucy) continues to this day.

If interpretations of the exact nature of australopithecine bipedalism are controversial, assessments of its causes are even more so (Fig. 14-6). Did early hominids first become bipedal in conjunction with freeing of the hands to carry food, water, or babies, or did they use their hands primarily to make the primitive stone tools that are asso-

(a) Freeing hands

(b) Running after game

(c) Hunting (looking over tall grass)

(d) Efficient for staying cool

(e) Tool production

(f) Sexual signalling

REVIEW EXERCISES

1. What specimen represents the first australopithecine discovered? When and where did it live? Who described and interpreted this specimen?
2. Name the genera and species of gracile and robust australopithecines that are recognized today. In which countries has each been found?
3. How do gracile and robust australopithecines differ? How are they the same?
4. Why are australopithecines classified as hominids rather than pongids?
5. Who was Lucy and why is she controversial?
6. What selective pressures have been hypothesized as prime movers for the evolution of bipedalism?

ciated with their general paleoenvironments? Were bipedal hominids better able to see predators and game over tall grass? Did bipedalism help them not only to see but also to manipulate tools used to hunt small animals? Were bipedal walking and running associated with more stamina for following and keeping up with migrating game? Did bipedalism allow early hominids to reach higher and therefore feed more efficiently from bushes and trees? Was the recognition of sexual signals enhanced by bipedalism? Were upright hominids more efficient at keeping cool because they had reduced areas of skin exposed to the noonday sun (as suggested by Pete Wheeler)? Although all of these imaginative hypotheses have been, or are, entertained as explanations for the selective pressures that first led protohominids down the path to bipedalism, they are based as much (or more) on speculation as on scientific evidence.

At this point, it is still not even clear exactly when bipedalism occurred. A famous trail of fossilized footprints discovered in Laetoli, Tanzania, by the late archaeologist Mary Leakey shows that hominids had already achieved bipedalism by about 3.6 mya. What is not known, however, is how soon bipedalism developed after the 5.0- to 6.0-mya split between the ancestors of chimpanzees and hominids. Did selection for bipedalism occur at the time of the divergence, as many workers hypothesize, or was there a period after the split during which non-ape protohominids had not yet evolved bipedal locomotion? What much of this boils down to is the central question of what the common ancestor (CA) of *Pan* and *Homo* was like. Although one might expect that the best way to approach this problem is directly from the fossil record, as we have seen it is too fragmentary and its interpretations too controversial to yield much in the way of persuasive answers. For this reason, we can learn as much if not more about the nature of the CA by focusing on living primates.

MODELING THE COMMON ANCESTOR

Inferences about early hominids have a long tradition of being formulated, at least in part, by comparing them with living animals. Such referential models, as pointed out by Jim Moore, may be based on analogies due to common adaptations, for example, terrestrial living in savanna baboons (the model or referent) and early hominids (see Chapter 9). Social carnivores such as lions and wolves have also been used as analogous referential models for interpreting early hominids, especially some years ago when hunting was viewed as *the* primary adaptation that drove hominid evolution (the hunting hypothesis is now out of favor). A second kind of referential model discussed by Moore, and the one favored here, is based on common descent or homology rather than analogy. For example, Richard Wrangham compared the African apes with humans to identify behavioral traits in both that were presumably inherited from a CA.

But how does one select the best primate models for interpreting the CA? For assessing general topics such as the probable level of basic cognitive abilities, we can refer to collective evidence from the great apes, which are generally regarded as more cognitively advanced than monkeys. This approach is taken at the end of this chapter. As shown in this section, however, it is also possible to investigate more specific details of the behavior and physical appearance in the CA. This is done by selecting a homologous model based on the most recent molecular studies which strongly suggest that *Homo* and *Pan* share a most recent CA that lived 5.0 to 6.0 mya, while gorillas split off a bit earlier (see Chapter 10). In this case, *Pan* is used as the referential model instead of both *Pan* and *Gorilla* or all of the great apes together. A very important method advocated by Moore may also be incorporated, namely, that of focusing on variation within *Pan*. Thus, both *P. troglodytes* and *P. paniscus* are fair game as models. (Although the present analysis is at the generic level, it should be noted that Moore focuses on variation within *P. troglodytes*.) Finally, as Zihlman does, the fossil record may also be considered, at least to judge the best chimpanzee model for filling in the details about what the CA might have looked like, if not how it behaved.

Although no living primate is likely to be identical in all respects to the CA, Zihlman details why she slightly favors bonobos (Fig. 14-7)

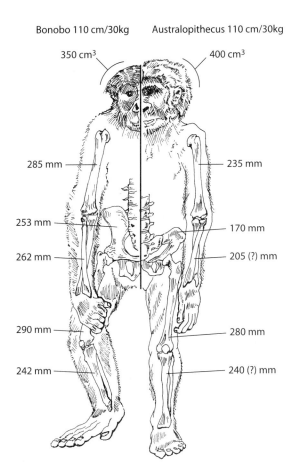

Bonobo 110 cm/30kg   Australopithecus 110 cm/30kg

350 cm³   400 cm³

285 mm   235 mm

253 mm   170 mm

262 mm   205 (?) mm

290 mm   280 mm

242 mm   240 (?) mm

Figure 14-7 According to Adrienne Zihlman, bonobos are the best referential model for the general physical appearance of the common ancestor.

over *P. troglodytes* as models, not only for assessing the physical appearance of the CA, but also for visualizing the possible transition from quadrupedalism to bipedalism:

> Comparing humans, *Pan troglodytes* and *Pan paniscus*, the latter appears to be the most "generalized" anatomically in its smaller body size, smaller facial and canine dimensions, and particularly in limb proportions. In addition, like humans, *Pan paniscus* is less sexually dimorphic in dentition. The existing reports of their tendency for bipedal behavior seem to follow from their having body proportions with more equal upper and lower limb lengths, and longer and heavier lower limbs. This predisposition for bipedal behavior and an inferred lower center of gravity suggest an easier transition from a quadrupedal ape to a bipedal hominid than previously imagined. (Zihlman, 1996:297)

Over 20 years ago, Zihlman predicted that newly discovered fossil crania, postcrania, and teeth, dated between 3.0 to 4.0 mya, would converge on bonobo-like morphology. The somewhat older fossils that have been discovered and attributed to *Ardipithecus ramidus* are compatible with this prediction. Although the final word is not yet in on whether or not this species is actually hominid (i.e., bipedal) rather than pongid, Zihlman's analysis that incorporates both *Pan* as a referential model and information from the fossil record gives us a pretty good idea of the CA's probable general physical appearance. The fossil record is obviously not much help, however, when it comes to sorting out the behaviors of the CA.

One way to begin to assess early hominid social life is to apply a slightly modified version of Wrangham's technique of examining the behavioral variation among African apes and living people in order to identify traits that are likely to have been present in the CA. Wrangham reasoned that social behaviors that are found in gorillas, bonobos, other chimpanzees, and across cultures of people that live in tribes or bands (rather than larger groups such as states) are likely to have been present in the CA:

> Thus the CA is implied to have commonly had closed social networks, hostile and male-dominated intergroup relationships with stalk-and-attack interactions, female exogamy and no alliance bonds between females, and males having sexual relationships with more than one female. (Wrangham, 1987:68)

Since Wrangham's study, molecular evidence has accumulated which indicates that chimpanzees share a more recent CA with humans than either does with gorillas. It is therefore reasonable to adjust Wrangham's findings accordingly by focusing on *Pan* but not *Gorilla*. Furthermore, rather than identifying only conservative traits, one can speculate that traits found in *P. troglodytes* and *P. paniscus* but not in living people may have been present in the CA, and that the human condition may be derived relative to the ancestral condition. Applying these modifications (along with some recent findings reported in Chapter 13) to Wrangham's data results in a somewhat more detailed hypothesis regarding the social nature of the CA: The CA probably lived in a fission-fusion society that was mostly closed to outsiders (estrous females are a likely exception). Males and occasionally females may have sometimes foraged alone. Females presumably migrated from their natal groups to breed, although the nature of their relationships with other females is not clear. Males, on the other hand, probably remained in their natal groups, where they may or may not have formed especially strong bonds with other adult males. Mating systems are likely to have been promiscuous and short term in the CA (with females possibly sneaking copulations with extra-group males), and the bewildering variety of systems that occur in living people (most notably long-term polygyny) may have been derived from such an ancestral state. Relationships between groups were at least tense, and may have been openly hostile, with stalk-and-attack interactions carried out mostly by males.

Did the Common Ancestor Hunt?

As discussed in Chapter 13, the extent to which bonobos hunt has yet to be determined. Nevertheless, several observations of wild bonobos preying on small ungulates as well as a report of an excited group of bonobos begging meat from an adult that had killed an infant duiker show that they share at least some hunting ability as well as a taste for meat with other chimpanzees and humans. It therefore seems reasonable to model the hunting ecology of australopithecines by using the chimpanzees in which hunting has been the most thoroughly documented (i.e., *P. troglodytes*) as referential models. This is just what Craig Stanford has done, and with fascinating results.

Stanford, who is an expert on the predator-prey ecology of chimpanzees and red colobus monkeys (see Chapter 13), focuses his comparative analysis of the behavioral ecology in early hominids and chimpanzees on the hunting and capture of medium-sized mammals weighing more than 5 kg (about 11 lb). Scavenging, on the other hand, refers to foraging for and feeding on carcasses that the foragers themselves have not killed. Although scavenging may or may not involve actively taking a carcass away from the predator that killed it, active pirating characterizes most of the observed cases of chimpanzee scavenging. Nevertheless, contrary to a number of workers who have suggested that the earliest hominids were scavengers rather than hunters, Stanford points out that scavenging among chimpanzees is noteworthy mainly by its rarity. Instead, chimpanzees like humans hunt and eat mammals on a regular basis and in a variety of environments (Fig. 14-8). Although dental evidence suggests that their diets consisted primarily of plant foods (as is true for chimpanzees), Stanford reasons that hunting also occurred in the CA, with scavenging eventually providing an additional small but regular part of the australopithecine diet (e.g., as indicated by cut marks on fossilized animal bones).

Classically, major adaptations in early hominids (e.g., bipedalism) have been viewed as correlates of a shift from more closed habitats to open-country savannas, an assumption shared by Stanford, Moore,

Figure 14-8 Chimpanzees have been observed hunting and eating meat at every field site where *P. troglodytes* has been studied on a long-term basis. (Photo courtesy of Curt Busse.)

and many others. Recent fossil evidence, however, suggests that early hominids were adapted to life in the trees as well as on the ground, and that at least some groups lived in wooded habitats. This, together with the fact that a wide range of patchy habitats and seasonal fluctuations characterized the relevant parts of Africa during the time when early hominids were evolving (as they do now), raises the possibility that the earliest hominids occupied a variety of habitats. So does the fact that different populations of chimpanzees are adapted to habitats that range from forests to savanna (see Chapter 13).

By using chimpanzees as a referential model, Stanford fleshes out a number of other details that may have characterized the behavioral ecology of australopithecines. In different chimpanzee populations, hunting reaches its highest intensity during either the dry (e.g., Gombe) or the wet (e.g., Taï) season, and it is likely that part of the reason for this is that chimpanzees supplement seasonal plant food shortages in their respective habitats with meat. Chimpanzees tend to favor certain parts of their larger home ranges at different times of the year, probably in response to seasonal variations in the distributions of plant foods. Stanford's research shows that chimpanzees in the Kasakela community at Gombe hunt more frequently within these core areas than in the peripheral parts of their home ranges. Put more simply, these chimpanzees go where the plants are, and they hunt where they go. The CA and its australopithecine descendants may have too.

But hunting in chimpanzees is not merely related to nutrition. It is a highly social activity engaged in mostly (but not entirely) by groups of adult and adolescent males. Hunting at Gombe is more frequent in larger than in smaller parties, and in the presence of one or more sexually receptive females (see Chapter 13). Furthermore, because of the manner in which meat is shared among the hunting party and with females that are present, Stanford concludes that male chimpanzees use meat to gain political and reproductive benefits and that this may also have been the case with australopithecines. Despite this analysis, however, Stanford is not a proponent of the suggestion that hunting in males was *the* driving force behind early hominid evolution, that is, the "man the hunter" hypothesis:

> It is important to stress that the evidence from recent chimpanzee research, while showing that hunting is primarily a male activity and that males may use meat in an effort to control female behavior, does not necessarily imply a "man the hunter" model. In those early interpretations, much weight was given to the cooperative and communicative abilities of males that derived from hunting together, which then became the driving force in human evolution. The chimpanzee evidence from Gombe shows that while communal hunting enhances success rates, cooperation and communal selfish behavior must be distinguished, although doing so can be difficult in field situations. Gombe males often appear to behave more selfishly than cooperatively during the hunt, perhaps because of the potential reproductive and political payoffs that come to meat possessors. Indeed, because female reproductive cycles influence hunting decisions,

females might be considered the driving force behind chimpanzee hunting. (Stanford, 1996:106)

Although Stanford believes that chimpanzees provide a good referential model for describing the probable behavioral ecology of the earliest hominids, he does not advocate extending the model to the earliest members of our own genus, *Homo*. Rather, early *Homo* may have been characterized by a markedly different socioecology in which meat procurement, butchery, and eating were assisted by the use of stone tools. At that time (around 2.5 mya according to Stanford), prey size increased and meat became a more important part of the diet.

The Evolution of Cognition in Early Hominids

Discussions about the cognitive abilities of the earliest hominids are extremely speculative and controversial, partly because there is so little evidence to go on. Because cranial capacities and body sizes were within the ranges of those for living apes, there is no reason to think that australopithecines were particularly encephalized. Although the archeological record is nearly mute on the nature of australopithecine material culture, the vast majority of which perished long ago, it does reveal that early hominids made and used primitive stone tools (Fig. 14-9). Wild chimpanzees sometimes use stone tools (e.g., to crack

Figure 14-9 Simple stone tools such as these represent all that is left of the material culture of some of the earliest known hominids: **1**, hammerstone; **2**, subspheroid; **3**, bifacial chopper; **4**, polyhedron; **5**, discoid; **6**, flake scraper; **7**, flake; **8**, core scraper.

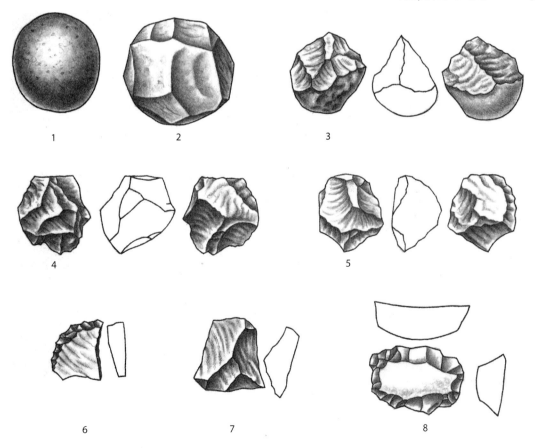

Thinking in the Rain

Peter Rodman from the University of California, Davis, is a leading expert on orangutan social systems. Below, he describes an observation that he made and photographed in the field. Although this incident occurred decades ago, it made such a vivid impression on Rodman that it has been on his mind ever since. Rodman's commentary beautifully illustrates the use of research on great apes for formulating hypotheses about human evolution. In his own words:

PETER RODMAN DOESN'T JUST OBSERVE ORANGUTANS IN THE FIELD. HERE HE IS HOLDING A GIANT WORM. (PHOTO COURTESY OF P. RODMAN.)

This orphaned female orangutan is sitting under my house in September 1970, in the sixth month of my first field study in East Kalimantan, Indonesia. She had just arrived in a large wooden crate brought by two forest department officers. My assistants, Sahar, Ali, Simin, and Darham, named her "Bontjel."

Bontjel emerged from the crate and immediately moved under my house. Shortly afterwards, there was a heavy afternoon rainstorm. Water poured off the eaves of the house in front of Bontjel, as she sat there, apparently quite forlorn, but also very dry. Her next action startled some of my unexamined presumptions about the behavior of orangutans, apes, and even humans. This is the first time I have attempted to explain why the behavior shown in this picture of Bontjel under her big leaf has been constantly on my mind ever since.

Wild orangutans often respond to rain by covering their heads with leaves, and the leaves are not usually as big as the one Bontjel is holding. This looks like a "rational" act, and we humans do the same thing. In the forest, Sahar and the other assistants—or I—often sat in a downpour holding a large leaf overhead "like an umbrella," looking more or less like Bontjel (forlorn and all). The rationale for our behavior is straightforward. It is more "comfortable" to stay dry. But why is it comfortable? Why do we "mind" being cold, and why, particularly, does it "make sense" to cover the head even if nothing else is dry? Briefly, the head is a highly vascularized body part, which means that chilling the head will chill the whole body profoundly even to the point of hypothermia in a tropical forest! Hypothermia is dangerous, and can be life-threatening directly or by weakening the system and allowing respiratory infections

nuts), but they do not modify them. It therefore appears that at least some early hominids were cognitively advanced compared to living apes.

Many workers who study the evolution of culture by analyzing the material archeological record believe that australopithecines did not have language, and that this capacity evolved only relatively recently

to take hold. In other words, it is not only sensible, but truly adaptive to keep at least the head dry in the rain if one has the anatomy to pick a leaf and hold it over one's head.

What in this picture of Bontjel is revealing? After all, it is raining, and she is sitting under a big leaf like a good hominoid keeping her head dry. The lesson is in the details. She was sitting completely dry under the house when the rain came. As the rain poured down, she moved out of her dry spot to the big macaranga leaves she could see on the ground behind the house. She was completely drenched and probably chilled as she picked up the leaf and dragged it back to the dry spot under the eaves of the house. In this case, her otherwise "sensible" behavior was inappropriate. It was much as if I were to wake up in the morning, run out in the rain to bring the newspaper into the house from the driveway, put a section of it on my head, and sit under it to read the rest of the paper at breakfast in a dry house—or if I were to mistake my wife for my hat, as Oliver Sacks has described in his fascinating account of neural pathologies of human behavior. If Bontjel had been in a tree outside in the rain, her action would have looked sensible in a way that we might never think about again (except to congratulate our pongid cousins on their rationality). Instead, her behavior suggested that this complex set of actions is organized very differently than a series of steps rationally assembled to solve the problem of staying dry. It is as if the pattern of behavior is laid down whole, to be triggered in certain circumstances when it not only makes sense, but also is adaptive. That it is an adaptive organized pattern is probably why it has evolved to be laid down whole somewhere in the brain. While sheltered under a leaf in the forest, my train of thought made the set of actions "sensible" to me. But the set of actions would probably occur without any "train of thought" at all, as it no doubt has among our hominoid ancestors long before we would invoke human rationality to explain the behavior. Bontjel's error taught me that much seemingly "learned" behavior could simply be a post hoc rationalization of behavioral patterns that evolved without the complex cultural context. To put it succinctly, incidents like this lead to the insight that the human mind makes sense of what we do at least as much as it guides us to do what makes sense.

BONTJEL SITS UNDER A BIG LEAF, CONTEMPLATING THE RAIN OUTSIDE OF HER SHELTER. (PHOTO COURTESY OF P. RODMAN.)

in the genus *Homo*. On the other hand, biological anthropologists like myself that specialize in the study of endocasts think that language had already begun to evolve by the time of early *Homo*. Furthermore, as discussed in Neural Note 14, new evidence from endocasts suggests that protolanguage may have begun to develop even earlier, that is, in the gracile australopithecine ancestors of early *Homo*. This de-

REVIEW EXERCISES

1. What is a referential model? Compare and give examples of analogous and homologous models.
2. What are the arguments for selecting *Pan* as a referential model for assessing the probable specific social behaviors of the CA, rather than using *Pan* and *Gorilla* or all three great apes?
3. Why does Zihlman favor bonobos as referential models?
4. Provide a hypothetical sketch of the possible social nature of the CA.
5. Do you think that the CA scavenged and hunted? Why or why not?
6. What does research on the great apes suggest about the general level of cognition in the CA?

bate, however, will not be resolved without a time machine. Until one is invented, discussions about the first occurrence of articulate speech and its associated sharpening of conscious thought are bound to remain highly speculative.

We can, however, refer to nonhuman primates—particularly the great apes—as general referential models for the probable lower boundaries regarding the cognitive abilities of the CA. Thus, we might expect the CA to have had a sense of self as indicated for great apes by mirror tests, and to have had cultural traditions that were at least as complex as those documented for chimpanzees. As discussed, tool production and presumably tool use were probably more sophisticated in the CA than in any of the apes. We are also safe in assuming that the CA had the potential for at least the level of conscious thought demonstrated by language experiments on the famous bonobo Kanzi.

As noted earlier, great apes are generally regarded as more cognitively advanced than monkeys, and the primatologists who have the privilege of working with great apes (see Chapters 11 to 13) appear keenly aware of their value as referential models for formulating hypotheses about the earliest hominids. In fact, some investigators have devoted decades to thinking about the evolutionary implications of their research on apes. For example, as detailed in In the Field, a remarkable chance observation of an orangutan caused Peter Rodman to spend 30 years pondering the evolution of the human predilection for learning patterns of adaptive behavior and for rationalizing those behaviors.

SUMMARY

Since the first australopithecine was discovered in South Africa 75 years ago, hundreds of additional specimens have been recovered. Gracile australopithecines (*A. africanus*) are recognized only in South Africa, from sometime before 3.0 mya to around 1.5 mya. Robust australopithecines (*Paranthropus*) have been found in both South and East Africa, dating from approximately 2.6 to 1.0 mya. Robust australopithecines have more rugged skulls than gracile australopithecines, and both share certain features with living apes and humans. Because *Australopithecus* and *Paranthropus* were bipedal, they are grouped with hominids rather than pongids, although the degree to which their bipedalism was refined is a matter of controversy, as is the topic of why hominids became bipedal in the first place. A famous australopithecine partial skeleton called Lucy is discussed along with the problems surrounding efforts to interpret her species (*A. afarensis*). Paleoanthropology is a very controversial field, which makes it difficult to infer the nature of the CA from fossil evidence alone. Fortunately, primates (and other animals) may be used as referential models for assessing the behaviors of the CA. A homologous referential model that focuses on *Pan* is used to speculate about the

CA's physical appearance as well as a number of its social behaviors including hunting and scavenging. A more general referential model (the great apes) is used to discuss the probable general cognitive abilities of the CA.

URTHER READING

FURTHER READING

BERGER, L. (1998) Redrawing our family tree? *National Geographic* **194**:90–99.

FALK, D., REDMOND, J., GUYER, J., CONROY, G. C., AND SEIDLER, H. (2000) Early hominid brain evolution: A new look at old endocasts.

MOORE, J. (1996) Savanna chimpanzees, referential models and the last common ancestor. In: W. C. McGrew, L. F. Marchant and T. Nishida (eds) *Great Ape Societies*. Cambridge: Cambridge University Press, pages 275–292.

STANFORD, C. B. (1996) The hunting ecology of wild chimpanzees: Implications for the evolutionary ecology of Pliocene hominids. *American Anthropologist* **98**:96–113.

WRANGHAM, R. W. (1987) The significance of African apes for reconstructing human social evolution. In: W. G. Kinzey (ed) *The Evolution of Human Behavior: Primate Models*. Albany: State University of New York Press, pages 51–71.

ZIHLMAN, A. L. (1996) Reconstructions reconsidered: Chimpanzee models and human evolution. In: W. C. McGrew, L. F. Marchant and T. Nishida (eds) *Great Ape Societies*. Cambridge: Cambridge University Press, pages 293–304.

EPILOGUE

It is apparent from the descriptions in this book that the more than 200 species of living primates are extremely diverse. Many of them represent highly complicated, social, intelligent, and even beautiful animals. Unfortunately, this text also reveals that a large number of these species are all too often threatened with extinction because of human activities. This is a shame, not only because these animals are fascinating creatures that deserve their rightful places in nature, but also because they are, in a sense, our kin—and kin understand and help each other. In this context, it is fitting that the conclusion to this book includes a powerful story told by Jane Goodall who has become extensively involved with primate conservation, partly because of her research on wild chimpanzees:

> The hero of this story is a human being named Rick Swope who visits the Detroit zoo once a year with his family. One day, as he watched the chimpanzees in their big new enclosure, a fight broke out between two adult males. Jojo, who had been at the zoo for years, was challenged by a younger and stronger newcomer, and Jojo lost. In his fear he fled into the moat which was brand new, and Jojo did not understand water. He had gotten over the barrier erected to prevent the chimpanzees from falling in—for they cannot swim—and the group of visitors and staff that happened to be there watched in horror as Jojo began to drown. He went under once, twice, three times. Rick Swope could bear it no longer. He jumped in to try to save the chimp, despite onlookers yelling at him about the danger. He managed to get Jojo's deadweight over his shoulder, and then crossed the barrier and pushed Jojo onto the bank of the island. Rick held him there—the bank was very steep and if he were to let go Jojo would slide back into the water—even when the other chimps charged toward him, screaming in excitement. Rick held Jojo until he raised his head, took a few staggering steps, and collapsed on more level ground.
>
> The director of the institute called Rick. "That was a brave thing you did. You must have known how dangerous it was. What made you do it?"
>
> "Well, I looked into his eyes. And it was like looking into the eyes of a man. And the message was, 'Won't anybody help me?'"
>
> Rick Swope risked his life to save a chimpanzee, a nonhuman being who sent a message that a human could understand. Now it is up to the rest of us to join in too. (Goodall, 1998:2185)

Another reason for working hard to ensure the survival of nonhuman primates is that, as primates ourselves, we can learn much about our basic nature from them, and this may help us to strengthen the odds for our own survival. Thus, using the methods discussed in this text, primatologists can determine which human traits are likely to have been inherited from primate ancestors, and which appear to be unique. For example, humans seem to share a proclivity for dominance hierarchies and aggression with terrestrial Old World monkeys and (perhaps more telling) our close chimpanzee cousin, *Pan troglodytes*. Fortunately, we also share their tendency (documented especially by Frans de Waal) to reconcile after aggressive interactions. Under some circumstances, humans even manifest a fondness, reminiscent of bonobos (*Pan paniscus*), for "making love, not war." It therefore seems likely that the human propensities for making both war and peace not only are driven culturally but also may be partly the result of evolutionary "baggage." If so, it is best to be aware of these conflicting aspects of our primate nature so that we can consciously control the future of our species. But how can we do this? The answer lies in the features that are special to *Homo sapiens*, namely, enlarged language-bearing, reasoning brains. What a step in the right direction it would be if the world's alpha males and females (read political leaders) became educated about primate dominance interactions and then consciously used their high statuses to help keep our species peaceful and therefore viable!

GLOSSARY

adapids a now extinct family of primates that lived during the Eocene and may have been related to the ancestors of living lemurs and lorises

adaptive radiation one species giving rise to numerous species occupying different niches

ad-libitum or ad-lib sampling the recording of interesting behaviors as they occur rather than according to a preset sampling schedule

alliances a group of two or more individuals that may gang up on another individual

allomothering aunting behavior in which females other than the mother help care for an infant; term may include parenting behavior by males (also called alloparenting)

alpha male the top-ranking male in a social group

altruistic describes behaviors that potentially increase the reproductive fitness of recipients, but decrease that of the actors

anthropoids the suborder of primates that includes monkeys, apes, and humans, also called simians

anthropomorphize to attribute human-like characteristics to another species or individual

arboreal tree living

arboreal theory theory that most of the traits that characterize primates evolved as an initial adaptation to life in the trees

basal primates first primates to arise from mammals that were ancestral to all other primates

bilophodont molars that have two parallel enamel ridges, typical of Old World monkeys

biological species concept the premise that members of two different species cannot successfully interbreed

biomass the total weight of a given species per square kilometer

bipedalism walking on two legs

brachiation arm-over-arm locomotion through arboreal habitats that is used by gibbons and some of the New World monkeys

canines sharp-pointed tooth between the incisors and premolars of primates

catarrhines Old World anthropoids including monkeys, apes, and humans

363

cathemeral   active during parts of the day and night

Cenozoic   the present geological era that began about 65 million years ago; also known as the Age of Mammals

cerebral cortex   outside part of the brain that forms the basis for higher functions in mammals; also called neocortex or cerebral hemispheres

clades   branch of an evolutionary tree that consists of all the species descended from one common ancestor

consortships   periods of time (frequently several days) during which an adult male and female remain together and mate

conspecifics   individuals that belong to the same species

continental drift   slow movement of continents and their underlying crustal plates over the earth's surface

convergent evolution   similar adaptations to similar environments that occur independently in animals that live apart and are extremely remotely related, e.g., various mammals in North America and Australia; see **parallel evolution**

cooperative polyandry   a mating system in which one female has more than one regular mating partner that helps rear her offspring, found especially in New World monkeys

cranial capacity   volume of the braincase (traditionally measured by filling the skull with small seeds), which is used as an approximation of brain size

crepuscular   active at twilight

cryptic feature   trait that serves to conceal an animal, e.g., camouflage or lack of movement

cursorial   physically built for running

dental arcade   the arch formed by the row of teeth in the jaw

dental formula   denotes numbers of each kind of teeth in one-fourth of the jaw; e.g., 2.1.2.3. is the human formula that indicates (beginning center front) 2 incisors, 1 canine, 2 premolars, and 3 molars; may also be written as a ratio of upper to lower teeth, e.g., 2.1.2.3./2.1.2.3.

derived   refers to traits that are specialized or advanced, as opposed to primitive characteristics

dispersed polygyny   mating system where relatively large ranges of males overlap smaller ranges of numerous females that the males visit and mate with

diurnal   active during the day, sleeping at night

dominance   the ability of one individual to intimidate or defeat another in aggressive or competitive interactions

dominance hierarchy   pecking order of a social group in which relatively higher-ranked individuals have greater access to resources

duetting   the ritualized exchange of melodic calls between males and females of some species, e.g., gibbons

ecological species concept   the premise that two species have separated because they have adapted to different environments, but that interbreeding might still be physically possible

ecology   the relationship between species and their environments

encephalized   having more brains than expected for an average mammal of the same body size

endocasts   short for endocranial casts; casts of the interior of skulls that reproduce the size, shape, and some of the details of the outside surface of the brain that were imprinted on the walls of the braincase during life

entomophagy   eating of insects

estrus   hormonally driven state of sexual receptivity or "heat" in females, during which time they become fertile and may conceive

expensive-tissue hypothesis   concept that bigger brains evolved in conjunction with decreased gut size in primates

extinct   describes species that lived in the past but not the present

extra-pair copulations   also known as EPCs or sneak copulations; matings that occur between a paired individual and another that is not that individual's partner

exudates   saps, gums, and resins produced by trees

female-bonded kin groups   cooperative groups of related females, usually living in their natal groups

female choice   a form of intersexual selection in which females select their mates based on individual preferences

female defense polygyny   mating pattern in which one adult male aggressively attempts to monopolize reproductive access to a group of receptive females, e.g., in a harem

fission-fusion   refers to groups that periodically divide (fission) into smaller parties, and join together into larger groups (fusion)

focal animal sampling   observing one or more animals that have been targeted for study (e.g., adult females), in a predetermined order for a given length of time

folivores   animals that eat mostly leaves

fovea   central area of the retina in haplorhines that increases visual acuity

frontal sinuses   spaces contained within the substance of the frontal bones in the general region above the eyes

frugivores   animals that eat primarily fruit

gallery forest   forests found along rivers and streams

genus   taxonomic category that includes one or more species

gestation   the process of carrying developing offspring in the uterus during pregnancy

gracile australopithecines   common name for *Australopithecus africanus*, the least rugged looking of the early hominids from South Africa; see **robust australopithecines**

grade   group of animals that manifest the same general level of adaptation, e.g., lemur grade or monkey grade

graminivorous   describes animals with a grass-eating diet

granivores   animals that eat mostly grains and seeds

gregarious   fond of the company of others; refers to highly social individuals or species

grooming   the process of cleaning the hair or skin of oneself or another animal

gummivores   animals that eat mostly saps and gums

habitat   refers to the specific environment in which a primate lives within its range

habituate   the process of desensitizing primates to the presence of human observers

hallux   big toe

haplorhini   suborder of primates that contains tarsiers, monkeys, apes, and humans

home ranges   the areas within which groups of primates live

Hominidae   the family that contains living humans and their ancestral bipedal relatives

hominoids   members of the superfamily that contains apes and hominids

hyoid bone   a small curved bone above the Adam's apple at the front of the neck

incisors   the front teeth that are used to bite into and tear food in primates

inclusive fitness   the sum of a primate's individual reproductive fitness and that added by altruistic acts directed toward kin

infanticide   the killing of infants

insectivores   animals that eat mostly insects

instantaneous or scan sampling   recording the current behavior of every animal in a group (or subset of the animals in the group) as rapidly as possible

intermembral index   length of the humerus plus length of the radius times 100/length of the femur plus length of the tibia; the lower the intermembral index, the longer the hind limbs

intersexual selection   presence of a trait in one sex (usually males) because it has been preferred by the other, e.g., beautiful tails of male peacocks

intrasexual selection   results from competition between members of one sex (usually males) for sexual access to a member of the opposite sex, and often associated with selection for fighting anatomies such as large canines and bodies

investment   the physiological cost associated with enhancing reproductive fitness, e.g., parental investment

ischial callosities   hairless, cornified pads of skin that cushion the buttocks in Old World monkeys and gibbons; used as sitting pads

Kay's threshold   the observation that purely insectivorous primates usually weigh less than 500 g, which separates them from larger, noninsectivorous primates

kin selection   an increase in fitness of relatives that carry copies of an individual's genes; an individual may increase inclusive fitness by behaving altruistically toward kin

knuckle walking   form of terrestrial locomotion used by chimpanzees and gorillas in which the weight of the forearms is placed on the outsides of the tips of the fingers and knuckles rather than the palms of the hand

K-selection   occurs when mortality is low, life stages are prolonged, and investment of one or both parents in a few number of offpring is high

K-T extinction event   episode between the Cretaceous and Tertiary periods during which 75% of all species became extinct, possibly due to the impact of asteroids or comets on Earth

lactation   production of milk in nursing mothers

living fossils, living links   see **structural ancestor**

Madagascar   island off of the east coast of Africa where lemurs live

matrilineal   linked by descent through females

megafauna   giant animals

Mesozoic era   the geological era from about 225 to 65 million years ago, during which dinosaurs flourished

mitochondrial DNA (mtDNA)   genetic material found in the mitochondria of cells that is particularly useful for molecular evolutionary studies

molars   the broad back teeth that crush and grind food

molecular clocks   concept used to determine dates of evolutionary divergences by comparing the similarities and differences in genes (or their products) among living species, based on the assumption that certain genes mutate at a fairly constant rate

monogamous   having only one mate

monotypic   a genus with only one species; or a species that is not further divided into subspecies

mutations   changes in the structure of genes

natal groups   the groups into which individuals are born

natural selection   theory (formulated by Darwin and Wallace) that attributes changes in gene frequencies between generations to differential reproduction (survival) of fit individuals in nature

neocortex   outside surface of the brain; cerebral hemispheres

neotropical   from tropical regions of Central and South America

nocturnal   active at night, sleeping during the day

omnivore   animals that eat a variety of foods, including plants and animals

omomyids   a now extinct family of primates that lived during the Eocene epoch and may have given rise to tarsiers and anthropoids

opposable thumb   a thumb with a padded surface at the end that may be rotated and brought into contact with the end pad of at least one other digit

orthograde   completely upright posture

pair-bonded   describes two individuals that have strong ties; frequently refers to mated pairs of primates

paleontologists   scientists that specialize in the study of fossils

parallel evolution   evolution of similarities that occurs independently in two groups isolated from each other, e.g., New and Old World monkeys; see **convergent evolution**

patrilineal   descended through the male line

percussive foraging   locating and collecting food such as insects by tapping or drilling on the branches that house them, used by aye-ayes and woodpeckers; see **tap scanning**

philopatry one sex or the other remains in its natal group for life; literally means "love of country"

pleiotropic effects multiple effects of a single gene that may superficially appear to be unrelated

plesiadapiforms primitive Paleocene fossils thought by some to represent the oldest known primates

polyandrous describes females that mate with more than one non-polygynous male

polygamous mating patterns in which both sexes mate with multiple partners

polygynous refers to males that mate with many females

polytypic refers to one species (or genus) containing populations that differ in their appearances, i.e., many types

postcranium part of the skeleton below the skull

power grip forceful grasp using the palm and fingers (e.g., removing a lid)

preadaptive describes a structure that could be changed by natural selection for a new or modified function

precision grip skillful grasp using the tips of the thumb and adjacent fingers (e.g., holding a pen)

premolars teeth between the canines and molars, bicuspids

primary forest forest that is basically undisturbed as opposed to regenerating or **secondary forest**

primitive describes a feature, fossil, or organism that is very generalized, i.e., lacks advanced or specialized traits

prosimians the suborder of primates that includes lemurs, lorises, and tarsiers

quadrumanous literally, four handed, i.e., having feet that function like hands

quadrupedalism moving on all four limbs

reciprocal altruism alternate giving and receiving of altruistic acts over time between two individuals

relative brain size brain weight divided by body weight

reproductive fitness measure of an individual's production of viable offspring

reproductive strategies behavioral mechanisms for increasing the probability that an animal's genes (or copies thereof in relatives) will survive in future generations

residence patterns typical social groups based on ratios of adult females to adult males and associated with particular mating systems

robust australopithecines common name for the rugged-skulled early hominids from South (*Australopithecus* or *Paranthropus robustus*) and East (*P. or A. aethiopicus and boisei*) Africa; see **gracile australopithecines**

r-selection occurs when mortality is high, life is short, and parental investment in large batches or litters of offspring is low

savannas grassland plains with scattered clumps of trees and sources of water and seasonal rainfall

secondary forest   forest that has been disturbed and is regenerating

sexual dichromatism   differences in coat color between the sexes

sexual dimorphism   differences in size and shape of the body and teeth of the two sexes, with male primates usually exhibiting the largest anatomies

sexual selection   a form of natural selection that focuses on matings; usually entails male-male competition (intrasexual selection) or female choice (intersexual selection)

silverbacks   dominant adult male gorillas that have silvery-gray regions of hair across the middle of their backs

simians   anthropoids (monkeys, apes, and hominids)

simian shelf   thickening inside the lower edge of the front of the mandible in apes

socioecology   a subfield of evolutionary biology that focuses on the relationships between social behavior and ecological factors such as food availability, predator pressures, and interactions with other species

species   A population in which individuals are capable of interbreeding and producing viable offspring

sperm competition   intrasexual selection in which males with larger testes and greater volumes of sperm production have greater success at impregnating females that engage in matings with multiple partners

stereoscopic vision   three-dimensional depth perception that is a function of two eyes with overlapping visual fields

stink-fights   fights engaged in by male ring-tailed lemurs, involving rubbing the tail with secretions and flicking it at the opponent

strepsirhini   suborder of primates that includes lemurs and lorises

structural ancestor   a living species that is believed to resemble its ancestor or the common ancestor of it and other species

subfossils   incompletely fossilized specimens that are relatively young compared to fossils

sulcal pattern   the pattern of grooves (sulci) that separate a brain's convolutions

supraorbital tori   thick ridge of bone that crosses the face above the eyes in some primates

symbiotic   refers to a relationship that is mutually beneficial to the partners

sympatric   living in the same geographical area

tapetum lucidum   special light-reflecting area of the retina, seen in the eyes of many nocturnal animals

tap scanning   tapping the finger along a wood surface to locate larvae within a branch; used by aye-ayes; see **percussive foraging**

taxonomic classification   a hierarchical scheme that groups taxa based on their evolutionary relationships and/or physical resemblances

terrestrial   spending most of the waking hours on the ground

territorial   species that aggressively defend their home ranges from conspecific intruders

toilet claws claws usually located on second toes of feet; used for grooming in prosimians

tooth comb formed by projection of lower front teeth in prosimians; used to clean coat; also called dental comb

troops another word for groups that usually refers to multimale, multifemale groups or all-male groups of monkeys and sometimes prosimians (e.g., ring-tailed lemurs)

understory the lower levels of vines and shrubs in the forests

urine washing practice seen in some prosimians and New World monkeys of urinating on hands while rubbing them together

vertical clinging and leaping a form of locomotion in which the animal clings to and leaps between vertical supports

visual predation hypothesis theory that most of the traits that characterize primates evolved as an initial adaptation to preying on insects rather than to life in the trees

Y-5 pattern pattern formed by five cusps on the lower molars of hominoids

zygomatics cheekbones, located below the eyes and toward the sides of the face

BIBLIOGRAPHY

AIELLO, L. C. AND P. WHEELER (1995) The expensive-tissue hypothesis. *Current Anthropology* **36**:199–221.

ALEXANDER, J. P. (1992) Alas, poor *Notharctus*. *Natural History* **101**(8):55–59.

ALTMANN, J. (1974) Observational study of behavior: Sampling methods. *Behaviour* **49**:227–265.

ALTMANN, J. (1992) Leading ladies. *Natural History* **101**(2):48–49.

ANDERSON, M. (1971) A watched potto never grows: A chronicle of the prenatal and first months of a *Perodicticus potto*. *Discovery* **6**:89–98.

ASFAW, B., WHITE, T., LOVEJOY, O., LATIMER, B., SIMPSON, S. AND G. SUWA (1999) *Australopithecus garhi:* A new species of early hominid from Ethiopia. Science **284**: 629–635.

AYRES, J. M. (1990) Scarlet faces of the Amazon. *Natural History*, **99**(2):33–41.

BADRIAN, N. AND R. K. MALENKY (1984) Feeding ecology of *Pan paniscus* in the Lomako Forest, Zaire. In: R. L. Susman (ed) *The Pygmy Chimpanzee.* New York: Plenum Press, pages 275–299.

BAKER, A. J. (1991) Evolution of the social system of the golden lion tamarin (*Leontopithecus rosalia*): Mating system, group dynamics and cooperative breeding. Ph.D. dissertation, University of Maryland, College Park.

BARD, K. A. (1995) Sensorimotor cognition in young feral orangutans (*Pongo pygmaeus*). *Primates* **36**:297–321.

BARNETT, A. A. AND D. BRANDON-JONES (1997) The ecology, biogeography and conservation of the uakaris, *Cacajao* (Pitheciinae). *Folia Primatologica* **68**:223–235.

BARROSO, C. M., SCHNEIDER, H., SCHNEIDER, M. P. C., SAMPAIO, L., HARADA, M. L., CZELUSNIAK, J. AND M. GOODMAN (1997) Update on the phylogenetic systematics of New World monkeys: Further DNA evidence for placing the pygmy marmoset (*Cebuella*) within the genus *Callithrix*. *International Journal of Primatology* **18**:651–674.

BAUERS, S. (1995) No fight for 'drumsticks' when there's 6 of them. *The Philadelphia Inquirer*. June 29. 1995. page C-1.

BEARDER, S. (1987) Lorises, bushbabies, and tarsiers: Diverse societies in solitary foragers. In: B. B. Smuts, D. L. Cheney, R. M. Seyfarth, R. W. Wrangham and T. T. Struhsaker (eds) *Primate Societies*. Chicago: University of Chicago Press, pages 11–24.

BEARDER, S. (1995) Natural calls of the wild. *Natural History* **104**(8):48–57.

BEARDER, S. AND R. D. MARTIN (1979) The social organization of a nocturnal primate revealed by radio tracking. In: C. J. Amlaner and D. W. Macdonald (eds) *A Handbook on Biotelemetry and Radio Tracking*. Oxford: Pergamon Press, pages 633–648.

BECK, B. B. (1980) *Animal Tool Behavior: The Use and Manufacture of Tools by Animals*. New York: Garland STPM Press.

BENEFIT, B. R. AND M. L. MCCROSSIN (1997) Earliest known Old World monkey skull. *Nature* **388**:368–371.

BENNETT, E. L. AND A. G. DAVIES (1994) The ecology of Asian Colobines. In: A. G. Davies and J. F. Oates (eds) *Colobine Monkeys: Their Ecology, Behaviour and Evolution*. Cambridge: Cambridge University Press, pages 129–171.

BERCOVITCH, F. B. (1991) Mate selection, consortship formation, and reproductive tactics in adult female savanna baboons. *Primates* **32**:437–452.

BERGER, L. (1998) Redrawing our family tree? *National Geographic* **194**:90–99.

BLEISCH, W. G. AND N. CHEN (1991) Ecology and behavior of wild black-crested gibbons (*Hylobates concolor*) in China with a reconsideration of evidence for polygyny. *Primates* **32**:539–548.

BLOCH, J. I., FISHER, D. C., GINGERICH, P. D., GUNNELL, G. F., SIMONS, E. L. AND M. D. UHEN (1997) Cladistic analysis and anthropoid origins. *Science* **278**:2134–2135.

BOESCH, C. (1996) Social grouping in Taï chimpanzees. In: W. C. McGrew, L. F. Marchant and T. Nishida (eds) *Great Ape Societies*. Cambridge: Cambridge University Press, pages 101–113.

BOESCH, C. AND H. BOESCH-ACHERMANN (1991) Dim forest, bright chimps. *Natural History* **100**(9):50–57.

BOINSKI, S. (1992) Monkeys with inflated sex appeal. *Natural History*, **101**(7):42–49.

BOINSKI, S. (1995) *Saimiri. Illustrated Monographs of Living Primates*. The Netherlands: Institut voor Ontwikkelingsopdrachten.

BOYD, R. AND J. B. SILK (1997) *How Humans Evolved* (Part Two: Primate Behavior and Ecology). New York: W. W. Norton & Company.

BOYSEN, S. T. (1996) "More is less": The elicitation of rule-governed resource distribution in chimpanzees. In: A. E. Russon, K. H. Bard and S. T. Parker (eds) *Reaching into Thought: The Minds of the Great Apes*. Cambridge: Cambridge University Press, pages 177–189.

BRANDON, R. N. (1996) *Concepts and Methods in Evolutionary Biology*. Cambridge: Cambridge University Press.

BROCKELMAN, W. Y. (1984) Social behaviour of gibbons: Introduction. In: H. Preuschoft, D. Chivers, W. Brockelman and N. Creel (eds) *The Lesser Apes*. Edinburgh: Edinburgh University Press, pages 285–290.

BROCKELMAN, W. Y. AND D. J. CHIVERS (1984) Gibbon conservation: Looking to the future. In: H. Preuschoft, D. Chivers, W. Brockelman, and N. Creel (eds) *The Lesser Apes*. Edinburgh: Edinburgh University Press, pages 3–12.

BROCKELMAN, W. Y., REICHARD, U., TRESUVON, U. AND J. J. RAEMAEKERS (1998) Dispersal, pair formation and social structure in gibbons (*Hylobates lar*). *Behavioral Ecology and Sociobiology* **42**:329–339.

BRUNET, M., BEAUVILAIN, A., COPPENS, Y., HEINTZ, E., MOUTAYE, A. H. E. AND D. PILBEAM (1995) The first australopithecine 2,500 kilometres west of the Rift Valley (Chad). *Nature* **378**:273–275.

BUESCHING, C. D., HEISTERMANN, M., HODGES, J. K. AND E. ZIMMERMANN (1998) Multimodal oestrus advertisement in a small nocturnal prosimian, *Microcebus murinus*. *Folia Primatologica* **69**(suppl 1):295–308.

BURTON, F. B. (1995) *The Multimedia Guide to the Non-human Primates.* Scarborough, Ontario: Prentice Hall Canada.

BUSSE, C. D. AND T. P. GORDON (1984) Infant carrying by adult male mangabeys (*Cercocebus atys*). *American Journal of Primatology* **6**:133–141.

BUTYNSKI, T. M. (1988) Guenon birth seasons and correlates with rainfall and food. In: A. Gautier-Hion, F. Bourliere, J.-P. Gautier and J. Kingdon (eds) *A Primate Radiation: Evolutionary Biology of the African Guenons.* Cambridge: Cambridge University Press, pages 284–322.

BYRNE, W. W. AND A. WHITEN. (1988) *Machiavellian Intelligence.* Oxford: Oxford University Press.

CARTMILL, M. (1974) Rethinking primate origins. *Science* **184**:436–443.

CHAPMAN, C. A. AND R. W. WRANGHAM (1993) Range use of the forest chimpanzees of Kibale: Implications for the understanding of chimpanzee social organization. *American Journal of Primatology* **31**:263–273.

CHARLES-DOMINIQUE, P. (1984) Bush babies, lorises and pottos. In: D. Macdonald (ed) *All the World's Animals: Primates.* New York: Torstar, pages 36–41.

CHEMNICK, L. AND O. RYDER (1994) Cytological and molecular divergence of orangutan subspecies. In: J. J. Ogden et al. (eds) *Proceedings of the International Conference on Orangutans: The Neglected Ape.* San Diego: Zoological Society of San Diego, pages 74–78.

CHENEY, D., SEYFARTH, R., SMUTS, B. AND R. WRANGHAM (1987) The study of primate societies. In: B. B. Smuts, D. L. Cheney, R. M. Seyfarth, R. W. Wrangham and T. T. Struhsaker (eds) *Primate Societies.* Chicago: University of Chicago Press, pages 1–8.

CHENEY, D. L. AND R. M. SEYFARTH (1980) Vocal recognition in free-ranging vervet monkeys. *Animal Behaviour* **38**:362–367.

CHENEY, D. L. AND R. M. SEYFARTH (1990) *How Monkeys See the World.* Chicago: University of Chicago Press.

CHISM, J. (1986) Development and mother-infant relations among captive patas monkeys. *International Journal of Primatology* **7**:49–81.

CHISM, J., OLSON, D. K. AND T. E. ROWELL (1983) Diurnal births and perinatal behavior among wild patas monkeys: Evidence of an adaptive pattern. *International Journal of Primatology* **4**:167–184.

CHISM, J. AND T. E. ROWELL (1988) The natural history of patas monkeys. In: A. Gautier-Hion, F. Bourliere, J.-P. Gautier and J. Kingdon (eds) *A Primate Radiation: Evolutionary Biology of the African Guenons.* Cambridge: Cambridge University Press, pages 412–438.

CHIVERS, D. J. (1972) The siamang and the gibbon in the Malay Peninsula. In: D. Rumbaugh (ed) *Gibbon and Siamang*, Vol. 1. Basel: S. Karger, pages 103–135.

CHIVERS, D. J. (1984) Gibbons. In: D. Macdonald (ed) *All the World's Animals: Primates.* New York: Torstar, pages 118–123.

CHIVERS, D. J. (1994) Functional anatomy of the gastrointestinal tract. In: A. G. Davies and J. F. Oates (eds) *Colobine Monkeys: Their Ecology, Behaviour and Evolution.* Cambridge: Cambridge University Press, pages 205–227.

CLUTTON-BROCK, T. H. (1984) Primates. In: D. Macdonald (ed) *All the World's Animals: Primates.* New York: Torstar, pages 10–21.

COIMBRA-FILHO, A. F. AND R. A. MITTERMEIER (1978) Tree-gouging, exudate-eating, and the "short-tusked" condition in *Callithrix* and *Cebuella.*

In: D. G. Kleiman (ed) *The Biology and Conservation of the Callitrichidae.* Washington, D.C.: Smithsonian Institution Press, pages 105–115.

CONROY, G. C. (1990) *Primate Evolution.* New York: W. W. Norton.

CORBIN, G. D. AND J. SCHMID (1995) Insect secretions determine habitat use patterns by a female lesser mouse lemur (*Microcebus murinus*). *American Journal of Primatology* **37**:317–324.

CORDS, M. (1988) Mating systems of forest guenons: A preliminary review. In: A. Gautier-Hion, F. Bourliere, J.-P. Gautier and J. Kingdon (eds) *A Primate Radiation: Evolutionary Biology of the African Guenons.* Cambridge: Cambridge University Press, pages 323–339.

CROCKETT, C. M. AND J. F. EISENBERG (1987) Howlers: Variations in group size and demography. In: B. B. Smuts, D. L. Cheney, R. M. Seyfarth, R. W. Wrangham and T. T. Struhsaker (eds) *Primate Societies.* Chicago: University of Chicago Press, pages 54–68.

CULOTTA, E. (1992) A new take on anthropoid origins. *Science* **256**:1516–1517.

CULOTTA, E. (1995) Many suspects to blame in Madagascar extinctions. *Science* **268**:1568–1569.

CURTIS, D. J. (1995) Functional anatomy of the trunk musculature in the slow loris (*Nycticebus coucang*). *American Journal of Physical Anthropology* **97**:367–379.

DART, R. A. (1925) Australopithecus africanus: The man-ape of South Africa. *Nature* **115**:195–199.

DAVIES, A. G. AND J. F. OATES (1994) *Colobine Monkeys: Their Ecology, Behaviour and Evolution.* Cambridge: Cambridge University Press.

DE WAAL, F. (1989) *Peacemaking among Primates.* Cambridge, MA: Harvard University Press.

DE WAAL, F. (1997a) Bonobo dialogues. *Natural History* **106**(5):22–25.

DE WAAL, F. (1997b) *The Forgotten Ape.* Berkeley: University of California Press.

DE WAAL, F. (1998) *Chimpanzee Politics.* Baltimore: Johns Hopkins University Press.

DIETZ, J. M. AND A. J. BAKER (1993) Polygyny and female reproductive success in golden lion tamarins (*Leontopithecus rosalia*). *Animal Behaviour* **46**:1067–1078.

DIETZ, J. M., PERES, C. A. AND L. PINDER (1998) Foraging ecology and use of space in wild golden lion tamarins (*Leontopithecus rosalia*). *American Journal of Primatology* **41**:289–305.

DIGBY, L. J. (1995) Social organization in a wild population of *Callithrix jacchus*: II. Intragroup social behavior. *Primates* **36**:361–375.

DIGBY, L. J. AND C. E. BARRETO (1996) Activity and ranging patterns in common marmosets (*Callithrix jacchus*). In: M. A. Norconk, A. L. Rosenberger and P. A. Garber (eds) *Adaptive Radiations of Neotropical Primates.* New York: Plenum Press, pages 173–185.

DIXSON, A. F. (1981) *The Natural History of the Gorilla.* London: Weidenfeld and Nicolson.

DIXON, A. F. (1994) Reproductive biology of the owl monkey. In: J. F. Baer, R. E. Weller and I. Kakoma (eds) *Aotus:* The Owl Monkey. San Diego: Academic Press, pages 113–132.

DORAN, D. M. (1996) Comparative positional behavior of the African apes. In: W. C. McGrew, L. F. Marchant and T. Nishida (eds) *Great Ape Societies.* Cambridge: Cambridge University Press, pages 213–224.

DUCKER, S., REICHLER, S. AND S. PREUSCHOFT (1998) The functional differentiation of affiliative facial displays in *Theropithecus gelada*. *Folia Primatologica* **69**:199–200.

DUNBAR, R. I. M. (1987) Habitat quality, population dynamics, and group composition in colobus monkeys (*Colobus guereza*). *International Journal of Primatology* **8**:299–329.

DUNBAR, R. I. M. (1993) Coevolution of neocortical size, group size and language in humans (target article). *Behavioral and Brain Sciences* **16**:681–735.

DUNBAR, R. I. M. AND U. BOSE (1991) Adaptation to grass-eating in gelada baboons. *Primates* **32**:1–7.

EIMERL, S. AND I. DEVORE (1965) *The Primates*. New York: Time-Life Books.

ELLEFSON, J. O. (1974) A natural history of white-handed gibbons. In: D. Rumbaugh (ed) *Gibbon and Siamang*, Vol. 3. Basel: S. Karger, pages 1–136.

ERICKSON, C. J. (1994) Tap-scanning and extractive foraging in aye-ayes, *Daubentonia madagascariensis*. *Folia Primatologica* **62**:125–135.

ERWIN, J. (1994) Travels and travails in south Sulawesi. *Sulawesi Primate Newsletter* 2(2).

ESTRADA, A. (1982) Survey and census of howler monkeys (*Alouatta palliata*) in the rain forest of "Los Tuxtlas," Veracruz, Mexico. *American Journal of Primatology* **2**:363–372.

FA, J. E. (1984) Baby care in Barbary macaques. In: D. Macdonald (ed) *All the World's Animals: Primates*. New York: Torstar, pages 92–93.

FA, J. E. AND R. LIND (1996) Population management and viability of the Gibraltar Barbary macaques. In: J. E. Fa and D. G. Lindburg (eds) *Evolution and Ecology of Macaque Societies*. Cambridge: Cambridge University Press, pages 235–262.

FALK, D. (1981) Sulcal patterns of fossil *Theropithecus* baboons: Phylogenetic and functional implications. *International Journal of Primatology* **2**:57–69.

FALK, D. (1983) Cerebral cortices of East African early hominids. *Science* **221**:1072–1074.

FALK, D., REDMOND, J., GUYER, J., CONROY, G. C., RECHEIS, W., WEBER, G. W. AND H. SEIDLER (2000) Early hominid brain evolution: A new look at old endocasts. In press.

FEDIGAN, L. AND L. M. FEDIGAN (1988) *Cercopithecus aethiops*: A review of field studies. In: A. Gautier-Hion, F. Bourliere, J.-P. Gautier and J. Kingdon (eds) *A Primate Radiation: Evolutionary Biology of the African Guenons*. Cambridge: Cambridge University Press, pages 389–411.

FEDIGAN, L. M., ROSENBERGER, A. L., BOINSKI, S., NORCONK, M. A. AND P. A. GARBER (1996) Critical issues in cebine evolution and behavior. In: M. A. Norconk, A. L. Rosenberger and P. A. Garber (eds) *Adaptive Radiations of Neotropical Primates*. New York: Plenum Press, pages 219–228.

FEDIGAN, L. M. AND S. ZOHAR (1997) Sex differences in mortality of Japanese macaques: Twenty-one years of data from the Arashiyama West population. *American Journal of Physical Anthropology* **102**:161–175.

FEISTNER, A. T. C. AND E. J. STERLING (1994) The aye-aye: Madagascar's most puzzling primate. *Folia Primatologica* **62**.

FLEAGLE, J. G. (1994) Primate locomotion and posture. In: S. Jones, R. Martin and D. Pilbeam (eds) *The Cambridge Encyclopedia of Human Evolution*. Cambridge: Cambridge University Press, pages 75–79.

FONTAINE, R. (1981) Uakaris, genus *Cacajao*. In: A. F. Coimbra-Filho and R.

Mittermeier (eds) *Ecology and Behavior of Neotropical Primates*. Rio de Janeiro: Academia Brasileira de Ciencias, pages 443–493.

FORD, S. M. (1994) Taxonomy and distribution of the owl monkey. In: J. F. Baer, R. E. Weller and I. Kakoma (eds) *Aotus:* The Owl Monkey. San Diego: Academic Press, pages 1–57.

FORD, S. M. AND L. C. DAVIS (1992) Systematics and body size: Implications for feeding adaptations in New World monkeys. *American Journal of Physical Anthropology* **88**:415–468.

FOSSEY, D. (1983) *Gorillas in the Mist*. Boston: Houghton Mifflin.

GAGNEUX, P., WOODRUFF, D. S. AND C. BOESCH (1997) Furtive mating in female chimpanzees. *Nature* **387**:358–359.

GALDIKAS, B. (1995) *Reflections of Eden: My Life with the Orangutans of Borneo*. New York: Little, Brown/Gollancz.

GALDIKAS, B. M. F. (1984) Adult female sociality among wild orangutans at Tanjung Puting Reserve. In: M. F. Small (ed) *Female Primates: Studies by Women Primatologists*. New York: Alan R. Liss, pages 217–235.

GALEF, B. (1990) Tradition in animals: Field observations and laboratory analyses. In: M. Bekoff and D. Jamieson (eds) *Interpretation and Explanation in the Study of Animal Behavior*, Vol. I. Boulder: Westview Press, pages 74–95.

GALLUP, G. (1970) Chimpanzee: Self-recognition. *Science* **167**:86–87.

GARBER, P. A. (1993) Feeding ecology and behaviour of the genus *Saguinus*. In: A. B. Rylands (ed) *Marmosets and Tararins: Systematics, Behaviour, and Ecology*. Oxford: Oxford University Press, pages 273–295.

GARBER, P. A. (1994) Phylogenetic approach to the study of tamarin and marmoset social systems. *American Journal of Primatology* **34**:199–220.

GARBER, P. A., ROSENBERGER, A. L. AND M. A. NORCONK (1996) Marmoset misconceptions. In: M. A. Norconk, A. L. Rosenberger and P. A. Garber (eds) *Adaptive Radiations of Neotropical Primates*. New York: Plenum Press, pages 87–95.

GARDNER, R. A. AND B. T. GARDNER (1969) Teaching sign language to a chimpanzee. *Science* **165**:664–672.

GAUTIER-HION, A. (1988) Polyspecific associations among forest guenons: Ecological, behavioural and evolutionary aspects. In: A. Gautier-Hion, F. Bourliere, J.-P. Gautier and J. Kingdon (eds) *A Primate Radiation: Evolutionary Biology of the African Guenons*. Cambridge: Cambridge University Press, pages 452–476.

GAUTIER-HION, A., BOURLIERE, F., GAUTIER, J.-P. AND J. KINGDON (1988) *A Primate Radiation: Evolutionary Biology of the African Guenons*. Cambridge: Cambridge University Press.

GEBO, D. L., MACLATCHY, L., KITYO, R., DEINO, A., KINGSTON, J. AND D. PILBEAM (1997) A hominoid genus from the early Miocene of Uganda. *Science* **276**:401–404.

GIBBONS, A. AND E. CULOTTA (1997) Miocene primates go ape. *Science* **276**:355–356.

GILBERT, B. (1996) New ideas in the air at the National Zoo. *Smithsonian* **27** (June):32–43.

GLANDER, K. E. (1994) Morphometrics and growth in captive aye-ayes (*Daubentonia madagascariensis*). *Folia Primatologica* **62**:108–114.

GLANDER, K. E. (1996) *The Howling Monkeys of La Pacifica*. Durham: Duke University Primate Center.

GOLDIZEN, A. W. (1987) Tamarins and marmosets: Communal care of off-spring. In: B. B. Smuts, D. L. Cheney, R. M. Seyfarth, R. W. Wrangham and T. T. Struhsaker (eds) *Primate Societies*. Chicago: University of Chicago Press, pages 34–43.

GOLDIZEN, A. W. AND J. TERBORGH (1986) Cooperative polyandry and helping behavior in saddle-backed tamarins (*Saguinus fuscicollis*). In: J. G. Else and P. C. Lee (eds) *Primate Ecology and Conservation*. Cambridge: Cambridge University Press, pages 191–198.

GOLDMAN, E. N. AND J. LOY (1997) Longitudinal study of dominance relations among captive patas monkeys. *American Journal of Primatology* **42**:41–51.

GONDER, M. K., OATES, J. F., DISOTELL, T. R., FORSTNER, M. R. J., MORALES, J. C. AND D. J. MELNICK (1997) A new west African chimpanzee subspecies? *Nature* **388**:337.

GOODALL, J. (1990) *Through a Window*. Boston: Houghton Mifflin.

GOODALL, J. (1998) Learning from the chimpanzees: A message humans can understand. *Science* **282**:2184–2185.

GOODMAN, M., PORTER, C. A., CZELUSNIAK, J., PAGE, S. L., SCHNEIDER, H., SHOSHANI, J., GUNNELL, G. AND C. P. GROVES (1998) Toward a phylogenetic classification of primates based on DNA evidence complemented by fossil evidence. *Molecular Phylogenetic Evolution* **9**:585–598.

GOODMAN, S. M., O'CONNOR, S. AND O. LANGRAND (1993) A review of predation on lemurs: Implications for the evolution of social behavior in small, nocturnal primates. In: P. M. Kappeler and J. U. Ganzhorn (eds) *Lemur Social Systems and Their Ecological Basis*. New York: Plenum Press, pages 51–66.

GREEN, G. M. G. AND R. W. SUSSMAN (1990) Deforestation history of the eastern rain forests of Madagascar from satellite images. *Science* **248**:212–215.

GREEN, S. K. (1978) *Gentle Gorilla: The Story of Patty Cake*. New York: Richard Marek Publishers.

GROVES, C. P. (1993) Order Primates. In: D. E. Wilson and D. M. Reader (eds) *Mammalian Species of the World: A Taxonomic and Geographic Reference*, 2nd ed. Washington, D.C.: Smithsonian Institution Press, pages 243–277.

GURSKY, S. L. (1994) Infant care in the spectral tarsier (*Tarsius spectrum*) Sulawesi, Indonesia. *International Journal of Primatology* **15**:843–853.

GURSKY, S. L. (1996) Group size and composition in the spectral tarsier, *Tarsius spectrum*: Implications for social organization. *Tropical Biodiversity* **3**:57–68.

GUST, D. A. AND T. P. GORDON (1991) Male age and reproductive behaviour in sooty mangabeys, *Cercocebus torquatus atys*. *Animal Behaviour* **41**:277–283.

GUST, D. A. AND T. P. GORDON (1994) The absence of a matrilineally based dominance system in sooty mangabeys, *Cercocebus torquatus atys*. *Animal Behaviour* **47**:589–594.

GUST, D. A., MCCASTER, T., GORDON, T. P., GERGITS, W. F., CASNA, N. J. AND H. M. MCCLURE (1998) Paternity in sooty mangabeys. *International Journal of Primatology* **19**:83–94.

HAIMOFF, E. H. (1986) Convergence in the duetting of monogamous Old World primates. *Journal of Human Evolution* **15**:51–59.

HALL, K. R. L. AND I. DeVORE (1965) Baboon social behavior. In: I. DeVore (ed) *Primate Behavior: Field Studies of Monkeys and Apes.* New York: Holt Rinehart and Winston, pages 53–110.

HAMILTON, W. D. (1964) The genetical evolution of social behaviour: I and II. *Journal of Theoretical Biology* 7:1–52.

HARCOURT, A. H. (1984) Gorilla. In: D. Macdonald (ed) *All the World's Animals: Primates.* New York: Torstar, pages 136–143.

HARDING, R. S. O. AND D. K. OLSON (1986) Patterns of mating among male patas monkeys (*Erythrocebus patas*) in Kenya. *American Journal of Primatology* 11:343–358.

HARTWIG, W. C., ROSENBERGER, A. L., GARBER, P. A. AND M. A. NORCONK (1996) On atelines. In: M. A. Norconk, A. L. Rosenberger and P. A. Garber (eds) *Adaptive Radiations of Neotropical Primates.* New York: Plenum Press, pages 427–431.

HARVEY, P. H., MARTIN, R. D. AND T. H. CLUTTON-BROCK (1987) Life histories in comparative perspective. In: B. B. Smuts, D. L. Cheney, R. M. Seyfarth, R. W. Wrangham and T. T. Struhsaker (eds) *Primate Societies.* Chicago: University of Chicago Press, pages 181–196.

HAUSFATER, G. (1975) *Dominance and Reproduction in Baboons (Papio cynocephalus). A Quantitative Analysis.* Basel: Karger.

HERSHKOVITZ, P. (1977) *Living New World Monkeys (Platyrrhini),* Vol. 1. Chicago: University of Chicago Press.

HIKAMI, K., HSEGAWA, Y. AND T. MATSUZAWA (1990) Social transmission of food preferences in Japanese monkeys (*Macaca fuscata*) after mere exposure or aversion training. *Journal of Comparative Psychology* 104:233–237.

HILL, D. A. AND N. OKAYASU (1996) Determinants of dominance among female macaques: Nepotism, demography and danger. In: J. E. Fa and D. G. Lindburg (eds) *Evolution and Ecology of Macaque Societies.* Cambridge: Cambridge University Press, pages 459–472.

HILL, O. (1953) *Primates: Comparative Anatomy and Taxonomy. I—Strepsirhini.* Edinburgh: At the University Press.

HOAGE, R. J. (1978) Parental care in *Leontopithecus rosalia rosalia*: Sex and age differences in carrying behavior and the role of prior experience. In: D. G. Kleiman (ed) *The Biology and Conservation of the Callitrichidae.* Washington, D.C.: Smithsonian Institution Press, pages 293–305.

HOELZER, G. A. AND D. J. MELNICK (1996) Evolutionary relationships of the macaques. In: J. E. Fa and D. G. Lindburg (eds) *Evolution and Ecology of Macaque Societies.* Cambridge: Cambridge University Press, pages 3–19.

HOROVITZ, I., ZARDOYA, R. AND A. MEYER (1998) Platyrrhine systematics: A simultaneous analysis of molecular and morphological data. *American Journal of Physical Anthropology* 106:261–281.

HORR, D. A. (1977) Orangutan maturation: Growing up in a female world. In: S. Chevalier-Skolnikoff and F. Poirier (eds) *Primate Biosocial Development.* New York: Garland Press, pages 289–321.

HOSHINO, J. (1985) Feeding ecology of mandrills (*Mandrillus sphinx*) in Campo Animal Reserve, Cameroon. *Primates* 26:248–273.

HRDY, S. B. (1977) *The Langurs of Abu: Female and Male Strategies of Reproduction.* Cambridge, MA: Harvard University Press.

HRDY, S. B. (1981) *The Woman That Never Evolved.* Cambridge, MA: Harvard University Press.

HRDY, S. B., JANSON, C. AND C. VAN SCHAIK (1994/1995) Infanticide: Let's not throw out the baby with the bath water. *Evolutionary Anthropology* 3:151–154.

HUFFMAN, M. A. AND M. SEIFU (1989) Observations on the illness and consumption of a possibly medicinal plant *Vernonia amygdalina* (del.) by a wild chimpanzee in the Mahale Mountains National Park, Tanzania. *Primates* **30**:51–63.

INGMANSON, E. J. (1996) Tool-using behavior in wild *Pan paniscus*: Social and ecological considerations. In: A. E. Russon, K. A. Bard and S. T. Parker (eds) *Reaching into Thought: The Minds of the Great Apes.* Cambridge: Cambridge University Press, pages 190–210.

IZAWA, K. AND A. MIZUNO (1977) Palm-fruit cracking behavior of wild black-capped capuchin (*Cebus apella*). *Primates* **18**:773–792.

JOLLY, A. (1988) Madagascar's lemurs: On the edge of survival. *National Geographic* **174**:132–161.

JOLLY, A. (1998) Pair-bonding, female aggression and the evolution of lemur societies. *Folia Primatologica* **69**(suppl 1):1–13.

JUNGERS, W. L., GODFREY, L. R., SIMONS, E. L., AND P. S. CHATRATH (1995) Subfossil *Indri indri* from the Ankarana Massif of northern Madagascar. *American Journal of Physical Anthropology* **97**:357–366.

JUNGERS, W. L., GODFREY, L. R., SIMONS, E. L., CHATRATH, P. S. AND B. RAKOTOSAMIMANANA (1991) Phylogenetic and functional affinities of *Babakotia* (primates), a fossil lemur from northern Madagascar. *Proceedings of the National Academy of Science of the United States of America* **88**:9082–9086.

JUNGERS, W. L. AND R. L. SUSMAN (1984) Body size and skeletal allometry in African apes. In: R. L. Susman (ed) *The Pygmy Chimpanzee: Evolutionary Biology and Behavior.* New York: Plenum Press, pages 131–177.

KANO, T. (1990) The bonobos' peaceable kingdom. *Natural History* **99**(11):62–71.

KANO, T. (1992) *The Last Ape: Pygmy Chimpanzee Behavior and Ecology.* Stanford: Stanford University Press.

KAPLAN, G. AND L. ROGERS (1994) *Orang-Utans in Borneo.* Armidale, NSW, Australia: University of New England Press.

KAPPELER, P. M. (1993) Sexual selection and lemur social systems. In: P. M. Kappeler, and J. U. Ganzhorn (eds) *Lemur Social Systems and Their Ecological Basis.* New York: Plenum Press, pages 223–240.

KAPPELMAN, J. (1996) The evolution of body mass and relative brain size in fossil hominids. *Journal of Human Evolution* **30**:243–276.

KAWAI, M. (1965) Newly-acquired pre-cultural behavior of the natural troop of Japanese monkeys on Koshima islet. *Primates* **6**:1–30.

KAWAMURA, S. (1958) Matriarchal social order in the Minoo-B troop: A study on the rank system of Japanese monkeys. *Primates* **1**:149–156.

KAY, R. F., ROSS, C. AND B. A. WILLIAMS. (1997) Anthropoid origins. *Science* **275**:797–804.

KAY, R. F., ROSS, C., WILLIAMS, B. A. AND D. JOHNSON (1997) Response to Bloch et al. (Cladistic analysis and anthropoid origins). *Science* **278**:2135–2136.

KINGDON, J. (1988) What are face patterns and do they contribute to reproductive isolation in guenons? In: A. Gautier-Hion, F. Bourliere, J.-P. Gautier and J. Kingdon (eds) *A Primate Radiation: Evolutionary Biology of the African Guenons.* Cambridge: Cambridge University Press, pages 227–245.

KINZEY, W. G. (1997) Synopsis of New World primates (16 genera). In: W. G. Kinzey (ed) *New World Primates: Ecology, Evolution, and Behavior.* New York: Aldine de Gruyter, pages 169–305.

KIRKPATRICK, R. C. (1997) Search for the snub-nosed monkey. *Natural History* **106**(4):42–47.

KLEIMAN, D. G., BECK, B. B., DIETZ, J. M. AND L. A. DIETZ (1991) Costs of re-introduction and criteria for success: Accounting and accountability in the golden lion tamarin conservation program. *Symposium of the Zoological Society of London* **62**:125–142.

KNOTT, C. (1998) Orangutans in the wild. *National Geographic* **194**:30–57.

KOHLER, M. AND S. MOYA-SOLA (1997) Fossil muzzles and other puzzles. *Nature* **388**:327–328.

KUMMER, H. (1968) *Social Organization of Hamadryas Baboons*. Chicago: University of Chicago Press.

KUMMER, H. (1995) *In Quest of the Sacred Baboon: A Scientist's Journey*. Princeton: Princeton University Press.

LANCASTER, J. B. (1984) Introduction. In: M. F. Small (ed) *Female Primates: Studies by Women Primatologists*. New York: Alan R. Liss, pages 1–10.

LE GROS CLARK, W. E. (1959) *The Antecedents of Man*. New York: Harper and Row.

LEIGHTON, D. R. (1987) Gibbons: Territoriality and monogamy. In: B. B. Smuts, D. L. Cheney, R. M. Seyfarth, R. W. Wrangham and T. T. Struhsaker (eds) *Primate Societies*. Chicago: University of Chicago Press, pages 135–145.

LEIGHTON, M. (1993) Modeling dietary selectivity by Bornean orangutans: Evidence for integration of multiple criteria in fruit selection. *International Journal of Primatology* **14**:257–313.

LOIREAU, J.-N. AND A. GAUTIER-HION (1988) Olfactory marking behaviour in guenons and its implications. In: A. Gautier-Hion, F. Bourliere, J.-P. Gautier and J. Kingdon (eds) *A Primate Radiation: Evolutionary Biology of the African Guenons*. Cambridge: Cambridge University Press, pages 246–253.

LOY, J., ARGO, B., NESTELL, G., VALLETT, S. AND G. WANAMAKER (1993) A reanalysis of patas monkeys' "grimace and gecker" display and a discussion of their lack of formal dominance. *International Journal of Primatology* **14**:879–893.

MACDONALD, D. (1984) *All the World's Animals: Primates*. New York: Torstar Books.

MACDONALD, S. E. (1994) Gorillas' (*Gorilla gorilla gorilla*) spatial memory in a foraging task. *Journal of Comparative Psychology* **108**:107–113.

MACKINNON, J. (1978) *The Ape within Us*. London: Collins.

MACKINNON, J. AND K. MACKINNON (1977) The formation of a new gibbon group. *Primates* **18**:701–708.

MACKINNON, J. AND K. MACKINNON (1980) The behavior of wild spectral tarsiers. *International Journal of Primatology* **1**:361–379.

MACNEILAGE, P. F., STUDDERT-KENNEDY, M. G. AND B. LINDBLOM (1987) Primate handedness reconsidered. *Behavioral and Brain Sciences* **10**:247–303.

MACPHEE, R. D. E. AND P. A. MARX (1997) The 40,000-year plague: Humans, hyperdisease, and first-contact extinctions. In: S. M. Goodman and B. D. Patterson (eds) *Natural Change and Human Impact in Madagascar*. Washington, D.C.: Smithsonian Institution Press, pages 169–217.

MAESTRIPIERI, D. (1998) The evolution of male-infant interactions in the tribe Papionini (Primates: Cercopithecidae). *Folia Primatologica* **69**:247–251.

MARTIN, R. (1973) A review of the behaviour and ecology of the lesser mouse lemur (*Microcebus murinus* J. F. Miller 1777). In: R. Michale and J. Crook (eds) *Comparative Ecology and Behaviour of Primates*. London: Academic Press, pages 1–68.

MARTIN, R. D. (1987) Long night for owl monkeys. *Nature* **326**:639–640.

MARTIN, R. D. (1993) Primate origins: Plugging the gaps. *Nature* **363**:223–234.

MATSUZAWA, T. (1996) Chimpanzee intelligence in nature and in captivity: isomorphism of symbol use and tool use. In: W. C. McGrew, L. F. Marchant and T. Nishida (eds) *Great Ape Societies*. Cambridge: Cambridge University Press, pages 196–209.

McCROSSIN, M. L., BENEFIT, B. R., GITAU, S. N., PALMER, A. K. AND K. T. BLUE (1998) Fossil evidence for the origins of terrestriality among Old World higher primates. In: E. Strasser (ed) *Primate Locomotion*. New York: Plenum Press, pages 353–396.

McGREW, W. C. (1992) *Chimpanzee Material Culture: Implications for Human Evolution*. Cambridge: Cambridge University Press.

McGREW, W. C. (1994a) Cultural implications of differences between populations of free-ranging chimpanzees in Africa. In: R. A. Gardner et al. (eds) *The Ethnological Roots of Culture*. The Netherlands: Kluwer Academic Publishers, pages 61–79.

McGREW, W. C. (1994b) Tools compared: The material of culture. In: R. W. Wrangham, W. C. McGrew, F. B. M. de Waal and P. G. Heltne (eds) *Chimpanzee Cultures*. Cambridge, MA: Harvard University Press, pages 25–89.

McGREW, W. C. (1998) Culture in nonhuman primates? *Annual Review of Anthropology* **27**:301–328.

McGREW, W. C., HAM, R. M., WHITE, L. J. T., TUTIN, C. E. G. AND M. FERNANDEZ (1997) Why don't chimpanzees in Gabon crack nuts? *International Journal of Primatology* **18**:353–374.

McGREW, W. C. AND L. F. MARCHANT (1992) Chimpanzees, tools, and termites: Hand preference or handedness? *Current Anthropology* **33**:114–119.

McGREW, W. C. AND L. F. MARCHANT (1997) Current issues in and meta-analysis of the behavioral laterality of hand function in nonhuman primates. *Yearbook of Physical Anthropology* **40**:201–232.

McGREW, W. C., MARCHANT, L. F. AND T. NISHIDA (1996) *Great Ape Societies*. Cambridge: Cambridge University Press.

McHENRY, H. (1992) How big were early hominids? *Evolutionary Anthropology* **1**:15–25.

MEHLMAN, P. T. (1989) Comparative density, demography, and ranging behavior of Barbary macaques (*Macaca sylvanus*) in marginal and prime conifer habitats. *International Journal of Primatology* **10**:269–292.

MELNICK, D. J. AND M. C. PEARL (1987) Cercopithecines in multimale groups: Genetic diversity and population structure. In: B. B. Smuts, D. L. Cheney, R. M. Seyfarth, R. W. Wrangham and T. T. Struhsaker (eds) *Primate Societies*. Chicago: University of Chicago Press, pages 121–134.

MENARD, N. AND D. VALLET (1996) Demography and ecology of Barbary macaques (*Macaca sylvanus*) in two different habitats. In: J. E. Fa and D. G. Lindburg (eds) *Evolution and Ecology of Macaque Societies*. Cambridge: Cambridge University Press, pages 106–131.

MILES, H. L. W. (1990) The cognitive foundations for reference in a signing orangutan. In: S. T. Parker and K. R. Gibson (eds) *'Language' and Intelligence in Monkeys and Apes: Comparative Developmental Perspectives*. Cambridge: Cambridge University Press, pages 511–539.

MILES, L. H., MITCHELL, R. W. AND S. E. HARPER (1996) Simon says: The development of imitation in an enculturated orangutan. In: A. E. Russon, K. A. Bard and S. T. Parker (eds) *Reaching into Thought: The Minds of the Great Apes*. Cambridge: Cambridge University Press, pages 278–299.

MILTON, K. (1981) Diversity of plant foods in tropical forests as a stimulus to mental developments in primates. *American Anthropologist* **83**:534–548.

MILTON, K. (1986) Ecological background and conservation priorities for woolly spider monkeys (*Brachyteles arachnoides*). In: K. Benirschke (ed) *Primates: The Road to Self-sustaining Populations*. New York: Springer-Verlag, pages 241–250.

MILTON, K. (1993) Diet and primate evolution. *Scientific American* **269** (August):86–93.

MITANI, J. C. (1984) The behavioral regulation of monogamy in gibbons (*Hylobates muelleri*). *Behavioral Ecology and Sociobiology* **15**:225–229.

MITANI, J. C. (1985) Sexual selection and adult male orangutan long calls. *Animal Behaviour* **33**:272–283.

MITANI, J. C. (1996) Comparative studies of African ape vocal behavior. In: W. C. McGrew, L. F. Marchant and T. Nishida (eds) *Great Ape Societies*. Cambridge: Cambridge University Press, pages 241–254.

MITANI, J. C., GRETHER, G. F., RODMAN, P. S. AND D. PRIATNA (1991) Associations among wild orang-utans: Sociality, passive aggregations or chance? *Animal Behavior* **42**:33–46.

MITTERMEIER, R. A. (1986) Primate conservation priorities in the neotropical region. In: K. Benirschke (ed) *Primates: The Road to Self-sustaining Populations*. New York: Springer-Verlag, pages 221–240.

MITTERMEIER, R. A., TATTERSALL, I., KONSTANT, W. R., MEYERS, D. M. AND R. B. MAST (1994) *Lemurs of Madagascar*. Washington, D.C.: Conservation International.

MOORE, J. (1984) Female transfer in primates. *International Journal of Primatology* **5**:537–589.

MOORE, J. (1992) "Savanna" chimpanzees. In: T. Nishida, W. C. McGrew, P. Marler, M. Pickford and F. B. M. de Waal (eds) *Topics in Primatology*, Vol. 1. *Human Origins.* Tokyo: University of Tokyo Press, pages 99–118.

MOORE, J. (1996) Savanna chimpanzees, referential models and the last common ancestor. In: W. C. McGrew, L. F. Marchant and T. Nishida (eds) *Great Ape Societies*. Cambridge: Cambridge University Press, pages 275–292.

MORBECK, M. E. AND A. L. ZIHLMAN (1988) Body composition and limb proportions. In: J. H. Swartz (ed) *Orang-utan Biology*. Oxford: Oxford University Press, pages 285–297.

MORBECK, M. E. AND A. L. ZIHLMAN (1989) Body size and proportions in chimpanzees, with special reference to *Pan troglodytes schweinfurthii* from Gombe National Park, Tanzania. *Primates* **30**:369–382.

MORELL, V. (1994) Will primate genetics split one gorilla into two? *Science* **265**:1661.

MORELL, V. (1998) A new look at monogamy. *Science* **281**:1982–1983.

MORIN, P. A., MOORE, J. J., CHAKRABORTY, R., JIN, L., GOODALL, J. AND D. S. WOODRUFF (1994) Kin selection, social structure, gene flow, and the evolution of chimpanzees. *Science* **265**:1193–1201.

MOYNIHAN, M. (1976) Notes on the ecology and behavior of the pygmy marmoset (*Cebuella pygmaea*) in amazonian Colombia. In: R. W. Thorington and P. G. Heltne (eds) *Neotropical Primates: Field Studies and Conservation*. Washington, D.C.: National Academy of Sciences, pages 79–84.

MUROYAMA, Y. (1995) Developmental changes in mother-offspring grooming in Japanese macaques. *American Journal of Primatology* **37**:57–64.

NADLER, R. D., GALDIKAS, B. F. M., SHEERAN, L. K. AND N. ROSEN (eds) (1995) *The Neglected Ape*. New York: Plenum Press.

NAKAGAWA, N. (1998) Indiscriminate response to infant calls in wild patas monkeys (*Erythrocebus patas*). *Folia Primatologica* **69**:93–99.

NAKAMICHI, M., ITOIGAWA, N., IMAKAWA, S. AND S. MACHIDA (1995) Dominance relations among adult females in a free-ranging group of Japanese monkeys at Katsuyama. *American Journal of Primatology* **37**:241–251.

NAKAMICHI, M., KATO, E., KOJIMA, Y. AND N. ITOIGAWA (1998) Carrying and washing of grass roots by free-ranging Japanese macaques at Katsuyama. *Folia Primatologica* **69**:35–40.

NAPIER, J. R. (1960) Studies of the hands of living primates. *Proceedings of the Zoological Society of London* **134**:647–657.

NAPIER, J. R. AND P. H. NAPIER (1994) *The Natural History of the Primates*. Cambridge, MA: MIT Press.

NAPIER, J. R. AND A. C. WALKER (1967) Vertical clinging and leaping—A newly recognized category of locomotor behavior of primates. *Folia Primatologica* **6**:204–219.

NASH, L. T. (1998) Vertical clingers and sleepers: Seasonal influences on the activities and substrate use of *Lepilemur leucopus* at Beza Mahafaly Special Reserve, Madagascar. *Folia Primatologica* **69**(suppl 1):204–217.

NASH, L. T., BEARDER, S. K. AND T. R. OLSON (1989) Synopsis of *Galago* species characteristics. *International Journal of Primatology* **10**:57–80.

NEWTON, P. N. (1988) The variable social organization of Hanuman langurs (*Presbytis entellus*), infanticide, and the monopolization of females. *International Journal of Primatology* **9**:59–77.

NEWTON, P. N. AND R. I. M. DUNBAR (1994) Colobine monkey society. In A. G. Davies and J. F. Oates (eds) *Colobine Monkeys: Their Ecology, Behaviour and Evolution*. Cambridge: Cambridge University Press, pages 311–346.

NIEMITZ, C. (1984) *Biology of Tarsiers*. New York: Gustav Fischer Verlag.

NIEMITZ, C., NIETSCH, A., WARTER, S. AND Y. RUMPIER (1991) *Tarsius dianae*: A new primate species from Central Sulawesi (Indonesia). *Folia Primatologica* **56**:105–116.

NIETSCH, A. AND S. THEILE (1998) Significance of vocal duetting in a polygynous primate species. *Folia Primatologica* **69**:236.

NISHIDA, T. (1990) *The Chimpanzees of the Mahale Mountains: Sexual and Life History Strategies*. Tokyo: University of Tokyo Press.

NISHIDA, T. AND M. HIRAIWA-HASEGAWA (1987) Chimpanzees and bonobos: Cooperative relationships among males. In: B. B. Smuts, D. L. Cheney, R. M. Seyfarth, R. W. Wrangham and T. T. Struhsaker (eds) *Primate Societies*. Chicago: University of Chicago Press, pages 165–177.

NISHIDA, T. AND K. HOSAKA (1996) Coalition strategies among adult male chimpanzees of the Mahale Mountains, Tanzania. In: W. C. McGrew, L. F. Marchant and T. Nishida (eds) *Great Ape Societies*. Cambridge: Cambridge University Press, pages 114–134.

NORCONK, M. A., ROSENBERGER, A. I. AND P. A. GARBER (1996) *Adaptive Radiations of Neotropical Primates*. New York: Plenum Press.

OATES, J. (1994) Natural history of African colobines. In: A. G. Davies and J. F. Oates (eds) *Colobine Monkeys: Their Ecology, Behaviour and Evolution*. Cambridge: Cambridge University Press, pages 75–128.

OATES, J. F. AND A. G. DAVIES (1994) What are the colobines? In: A. G. Davies and J. F. Oates (eds) *Colobine Monkeys: Their Ecology, Behaviour and Evolution.* Cambridge: Cambridge University Press, pages 1–9.

OATES, J. F., DAVIES, A. G. AND E. DELSON (1994) The diversity of living colobines. In: A. G. Davies and J. F. Oates (eds) *Colobine Monkeys: Their Ecology, Behaviour and Evolution.* Cambridge: Cambridge University Press, pages 45–73.

OATES, J. F., WHITESIDES, G. H., DAVIES, A. G., WATERMAN, P. G., GREEN, S. M., DASILVA, G. L. AND S. MOLE (1990) Determinants of variation in tropical forest primate biomass: New evidence from West Africa. *Ecology* **71**:328–343.

OHSAWA, H., INOUE, M. AND O. TAKENAKA (1993) Mating strategy and reproductive success of male patas monkeys (*Erythrocebus patas*). *Primates* **34**:533–544.

PACKER, C., COLLINS, D. A., SINDIMWO, A. AND J. GOODALL (1995) Reproductive constraints on aggressive competition in female baboons. *Nature* **373**:60–63.

PALOMBIT, R. A. (1993) Lethal territorial aggression in a white-handed gibbon. *American Journal of Primatology* **31**:311–318.

PALOMBIT, R. A. (1994a) Dynamic pair bonds in hylobatids: implications regarding monogamous social systems. *Behaviour* **128**:65–101.

PALOMBIT, R. A. (1994b) Extra-pair copulations in a monogamous ape. *Animal Behavior* **47**:721–723.

PALOMBIT, R. A. (1995) Longitudinal patterns of reproduction in wild female siamang (*Hylobates syndactylus*) and white-handed gibbons (*Hylobates lar*). *International Journal of Primatology* **16**:739–760.

PALOMBIT, R. A. (1996) Pair bonds in monogamous apes: A comparison of the siamang *Hylobates syndactylus* and the white-handed gibbon *Hylobates lar.* *Behaviour* **133**:321–356.

PARISH, A. R. (1996) Female relationships in bonobos (*Pan paniscus*). *Human Nature* **7**:61–96.

PERRET, M. (1992) Environmental and social determinants of sexual function in the male lesser mouse lemur (*Microcebus murinus*). *Folia Primatologica* **59**:1–25.

PETTER, J. J. (1977) The aye-aye. In: P. Rainier and G. H. Bourne (eds) *Primate Conservation.* New York: Academic Press, pages 37–57.

POLLOCK, J. I. (1977) The ecology and sociology of feeding in *Indri indri.* In: T. H. Clutton-Brock (ed) *Primate Ecology.* London: Academic Press, pages 37–69.

POLLOCK, J. I. (1979) Female dominance in *Indri indri. Folia Primatologica* **31**:143–164.

POLLOCK, J. I. (1986) The song of the indris (*Indri indri;* Primates: Lemuroidea): Natural history, form, and function. *International Journal of Primatology* **7**:225–264.

POOK, A. G. AND G. POOK (1981) A field study of the socio-ecology of the Goeldi's monkey (*Callimico goeldii*) in northern Bolivia. *Folia Primatologica* **35**:288–312.

PUSEY, A., WILLIAMS, J. AND J. GOODALL (1997) The influence of dominance rank on the reproductive success of female chimpanzees. *Science* **277**:828–831.

RADINSKY, L. (1975) Primate brain evolution. *American Scientist* **63**:656–663.

RAJPUROHIT, L. S. AND S. M. MOHNOT (1991) The process of weaning in Hanuman langurs *Presbytis entellus entellus. Primates* **32**:213–218.

RAMIREZ, M. F., FREESE, C. H. AND J. REVILA (1978) Feeding ecology of the pygmy marmoset, *Cebuella pygmaea*, in northeastern Peru. In: D. G. Kleiman (ed) *The Biology and Conservation of the Callitrichidae*. Washington, D.C.: Smithsonian Institution Press, pages 91–104.

RATAJSZCZAK, R. (1998) Taxonomy, distribution, and status of the lesser slow loris *Nycticebus pygmaeus* and their implications for captive management. *Folia Primatologica* **69**(suppl 1): 171–174.

REICHARD, U. (1995) Extra-pair copulations in a monogamous gibbon (*Hylobates lar*). *Ethology* **100**:99–112.

REICHARD, U. AND V. SOMMER (1997) Group encounters in wild gibbons (*Hylobates lar*): Agonism, affiliation, and the concept of infanticide. *Behaviour* **134**:1135–1174.

REICHLER, S., DUCKER, S. AND S. PREUSCHOFT (1998) Agonistic behaviour and dominance style in gelada baboons (*Theropithecus gelada*). *Folia Primatologica* **69**:200

REMIS, M. (1995) Effects of body size and social context on the arboreal activities of lowland gorillas in the Central African Republic. *American Journal of Physical Anthropology* **97**:413–433.

RICHARD, A. F. (1987) Malagasy prosimians: Female dominance. In: B. B. Smuts, D. L. Cheney, R. M. Seyfarth, R. W. Wrangham and T. T. Struhsaker (eds) *Primate Societies*. Chicago: University of Chicago Press, pages 25–33.

RICHARD, A. F. AND R. W. SUSSMAN (1987) Framework for primate conservation in Madagascar. In: C. W. Marsh and R. A. Mittermeier (eds) *Primate Conservation in the Tropical Rain Forest*. New York: Alan R. Liss, pages 329–341.

RIJKSEN, H. D. (1995) The neglected ape? In: R. D. Nadler, B. F. M. Galdikas, L. K. Sheeran and N. Rosen (eds) *The Neglected Ape*. New York: Plenum Press, pages 13–21.

ROBBINS, M. M. (1995) A demographic analysis of male life history and social structure of mountain gorillas. *Behaviour* **132**:21–48.

ROBINSON, J. G. AND C. H. JANSON (1987) Capuchins, squirrel monkeys, and atelines: Socioecological convergence with Old World primates. In: B. B. Smuts, D. L. Cheney, R. M. Seyfarth, R. W. Wrangham and T. T. Struhsaker (eds) *Primate Societies*. Chicago: University of Chicago Press, pages 69–82.

ROBINSON, J. G., WRIGHT, P. C. AND W. G. KINZEY (1987) Monogamous cebids and their relatives: Intergroup calls and spacing. In: B. B. Smuts, D. L. Cheney, R. M. Seyfarth, R. W. Wrangham and T. T. Struhsaker (eds) *Primate Societies*. Chicago: University of Chicago Press, pages 44–53.

RODMAN, P. S. (1988) Diversity and consistency in ecology and behavior. In: J. H. Schwartz (ed) *Orang-utan Biology*. New York: Oxford University Press, pages 31–51.

RODMAN, P. S. AND J. C. MITANI (1987) Orangutans: Sexual dimorphism in a solitary species. In: B. B. Smuts, D. L. Cheney, R. M. Seyfarth, R. W. Wrangham and T. T. Struhsaker (eds) *Primate Societies*. Chicago: University of Chicago Press, pages 146–154.

ROGERS, L. J. AND G. KAPLAN (1993) Koehler and tool use in orang-utans. *International Journal of Comparative Psychology* **6**:234–241.

ROGERS, L. J. AND G. KAPLAN (1994) A new form of tool use by orang-utans in Sabah, East Malaysia. *Folia Primatologica* **63**:50–52.

ROGERS, L. J. AND G. KAPLAN (1996) Hand preferences and other lateral biases in rehabilitant orang-utans, *Pongo pygmaeus pygmaeus*. *Animal Behavior* **51**:13–25.

ROGERS, L. J. AND G. KAPLAN (1998) Orangutans. In: G. Greenberg and M. M. Haraway (eds) *Comparative Psychology: A Handbook*. New York: Garland Publishing, pages 465–472.

ROGERS, M. E., ABERNETHY, K. A., FONTAINE, B., WICKINGS, E. J., WHITE, L. J. T. AND C. E. G. TUTIN (1996) Ten days in the life of a mandrill horde in the Lope Reserve, Gabon. *American Journal of Primatology* **40**:297–313.

ROMER, A. S. (1967) Major steps in vertebrate evolution. *Science* **158**:1629–1637.

ROSEN, S. I. (1974) *Introduction to the Primates*. New Jersey: Prentice Hall.

ROSENBERGER, A. L. (1992a) Evolution of feeding niches in New World monkeys. *American Journal of Physical Anthropology* **88**:525–562.

ROSENBERGER, A. L. (1992b) Evolution of New World monkeys. In: S. Jones, R. Martin and D. Pilbeam (eds) *The Cambridge Encyclopedia of Human Evolution*. Cambridge: Cambridge University Press, pages 209–216.

ROSENBERGER, A. L., NORCONK, M. A. AND P. A. GARBER (1996) New perspectives on the pitheciines. In: M. A. Norconk, A. L. Rosenberger and P. A. Garber (eds) *Adaptive Radiations of Neotropical Primates*. New York: Plenum Press, pages 329–333.

ROWE, N. (1996) *The Pictorial Guide to the Living Primates*. New York: Pogonias Press.

ROWELL, T. E. (1984) Guenons, macaques and baboons. In: D. Macdonald (ed) *All the World's Animals: Primates*. New York: Torstar, pages 74–85.

RUITER, J. R. DE (1996) Infanticide counter-strategies. *Evolutionary Anthropology* **5**:5.

RUSSON, A. E. (1996) Imitation in everyday use: Matching and rehearsal in the spontaneous imitation of rehabilitant orangutans (*Pongo pygmaeus*). In: A. E. Russon, K. A. Bard and S. T. Parker (eds) *Reaching into Thought: The Minds of the Great Apes*. Cambridge: Cambridge University Press, pages 152–176.

RUVOLO, M. (1997) Molecular phylogeny of the hominoids: Inferences from multiple independent DNA sequence data sets. *Molecular Biology and Evolution* **14**:248–265.

RYLANDS, A. B. (1993) The ecology of the lion tamarins, *Leontopithecus*: Some intrageneric differences and comparisons with other callitrichids. In: A. B. Rylands (ed) *Marmosets and Tamarins: Systematics, Behaviour, and Ecology*. Oxford: Oxford University Press, pages 296–313.

RYLANDS, A. B. AND D. S. DE FARIA (1993) Habitats, feeding ecology, and home range size in the genus *Callithrix*. In: A. B. Rylands (ed) *Marmosets and Tamarins: Systematics, Behaviour, and Ecology*. Oxford: Oxford University Press, pages 262–272.

SAGAN, C. (1994) *Pale Blue Dot: A Vision of the Human Future in Space*. New York: Random House.

SAKURA, O. (1994) Factors affecting party size and composition of chimpanzees (*Pan troglodytes verus*) at Bossou, Guinea. *International Journal of Primatology* **15**:167–183.

SAUTHER, M. (1989) Antipredator behavior in troops of free-ranging *Lemur catta* at Beza Mahafaly Special Reserve, Madagascar. *International Journal of Primatology* **10**:595–606.

SAUTHER, M. (1991) Reproductive behavior of free-ranging *Lemur catta* at Beza Mahafaly Special Reserve, Madagascar. *American Journal of Physical Anthropology* **84**:463–477.

SAUTHER, M. (1993) Resource competition in wild populations of ring-tailed lemurs (*Lemur catta*): Implications for female dominance. In: P. Kappeler and J. Ganzhorn (eds) *Lemur Social Systems and Their Ecological Basis.* New York: Plenum Press, pages 135–152.

SAUTHER, M. (1994) Wild plant use by pregnant and lactating ring-tailed lemurs, with implications for early hominid foraging. In: N. Etkin (ed) *Eating on the Wild Side.* Tucson: University of Arizona Press, pages 240–256.

SAUTHER, M. (1998) Interplay of phenology and reproduction in ring-tailed lemurs: Implications for ring-tailed lemur conservation. *Folia Primatologica* **69**(suppl 1):309–320.

SAUTHER, M. AND R. SUSSMAN (1993) A new interpretation of the social organization and mating system of the ring-tailed lemur (*Lemur catta*). In: P. Kappeler and J. Ganzhorn (eds) *Lemur Social Systems and Their Ecological Basis.* New York: Plenum Press, pages 111–121 .

SAVAGE-RUMBAUGH, S. AND R. LEWIN (1994) *Kanzi: The Ape at the Brink of the Human Mind.* New York: John Wiley and Sons.

SCHALLER, G. B. (1963) *The Mountain Gorilla.* Chicago: University of Chicago Press.

SCHINO, G., AURELI, F., D'AMATO, F. R., D'ANTONI, M., PANDOLFI, N. AND A. TROISI (1993) Infant kidnapping and co-mothering in Japanese macaques. *American Journal of Primatology* **30**:257–262.

SCHNEIDER, H. AND A. L. ROSENBERGER (1996) Molecules, morphology, and platyrrhine systematics. In: M. A. Norconk, A. L. Rosenberger and P. A. Garber (eds) *Adaptive Radiations of Neotropical Primates.* New York: Plenum Press, pages 3–19.

SCHULTZ, A. H. (1969) *The Life of Primates.* New York: Universe Books.

SCHULTZ, A. H. (1973) The skeleton of the Hylobatidae and other observations on their morphology. In: D. Rumbaugh (ed) *Gibbon and Siamang,* Vol. 2. Basel: S. Karger, pages 1–54.

SCHWARTZ, J. (1988) *Orang-utan Biology.* Oxford: Oxford University Press.

SCHWARTZ, J. (1996) *Pseudopotto martini:* a new genus and species of extant lorisiform primate. *Anthropological Papers of the American Museum of Natural History* **78**:1–14.

SEMENDEFERI, K. AND H. DAMASIO (1999) Brain size in living hominoids. *Journal of Human Evolution* (in press).

SENUT, B. AND C. TARDIEU (1985) Functional aspects of Plio-Pleistocene hominid limb bones: Implication for taxonomy and phylogeny. In: E. Delson (ed) *Ancestors: The Hard Evidence.* New York: Alan R. Liss, pages 193–201.

SICOTTE, P. (1993) Inter-group encounters and female transfer in mountain gorillas: Influence of group composition on male behavior. *American Journal of Primatology* **30**:21–36.

SIGG, H., STOLBA, A., ABEGGLEN, J.-J. AND V. DASSER (1982) Life history of hamadryas baboons: Physical development, infant mortality, reproductive parameters and family relationships. *Primates* **23**:473–487.

SIMONS, E. L., BURNEY, D. A., CHATRATH, P. S., GODFREY, L. R., JUNGERS, W. L. AND B. RAKOTOSAMIMANANA (1995) AMS ^{14}C dates for extinct lemurs from caves in the Ankarana Massif, northern Madagascar. *Quarterly Research* **43**:249–254.

SIMPSON, G. G., PITTENDRIGHT, C. S. AND L. H. TIFFANY (1957) *Life, an Introduction to Biology.* New York: Harcourt, Brace.

SMITH, A. G. SMITH, D. G. AND B. M. FUNNELL (1994) *Atlas of Mesozoic and Cenozoic Coastlines.* Cambridge: Cambridge University Press.

SMITH, D. G., KANTHASWAMY, S. AND J. L. WAGNER (1998) Paternity and female choice in two groups of hamadryas baboons. *American Journal of Primatology* **45**:209.

SMUTS, B. (1987) What are friends for? *Natural History* **96**(2):36–44.

SMUTS, B. B. AND J. M. WATANABE (1990) Social relationships and ritualized greetings in adult male baboons (*Papio cynocephalus anubis*). *International Journal of Primatology* **11**:147–172.

SOINI, P. (1988) The pygmy marmoset, genus *Cebuella*. In: R. A. Mittermeier, A. B. Rylands, A. F. Coimbra-Filho and G. A. B. da Fonseca (eds) *Ecology and Behavior of Neotropical Primates*, Vol. 2. Washington, D.C.: World Wildlife Fund, pages 79–129.

SOINI, P. (1993) The ecology of the pygmy marmoset, *Cebuella pygmaea*: Some comparisons with two sympatric tamarins. In: A. B. Rylands (ed) *Marmosets and Tamarins: Systematics, Behaviour, and Ecology*. Oxford: Oxford University Press, pages 257–261.

SOMMER, V., SRIVASTAVA, A. AND C. BORRIES (1992) Cycles, sexuality, and conception in free-ranging langurs (*Presbytis entellus*). *American Journal of Primatology* **28**:1–27.

STAFFORD, B. J., ROSENBERGER, A. L., BAKER, A. J., BECK, B. B., DIETZ, J. M. AND D. G. KLEIMAN (1996) Locomotion of golden lion tamarins (*Leontopithecus rosalia*). In: M. A. Norconk, A. L. Rosenberger and P. A. Garber (eds) *Adaptive Radiations of Neotropical Primates*. New York: Plenum Press, pages 111–132.

STAFFORD, D. K., MILLIKEN, G. W. AND J. P. WARD (1990) Lateral bias in feeding and brachiation in *Hylobates*. *Primates* **31**:407–414.

STAMMBACH, E. (1987) Desert, forest and montaine baboons: Multilevel-societies. In B. B. Smuts, D. L. Cheney, R. M. Seyfarth, R. W. Wrangham and T. T. Struhsaker (eds) *Primate Societies*. Chicago: University of Chicago Press, pages 112–120.

STANFORD, C. B. (1995a) Chimpanzee hunting behavior and human evolution. *American Scientist* **83**:256–261.

STANFORD, C. B. (1995b) To catch a colobus. *Natural History* **104**(1):48–55.

STANFORD, C. B. (1996) The hunting ecology of wild chimpanzees: Implications for the evolutionary ecology of Pliocene hominids. *American Anthropologist* **98**:96–113.

STANFORD, C. B., WALLIS, J., MATAMA, H. AND J. GOODALL (1984) Patterns of predation by chimpanzees on red colobus monkeys in Gombe National Park, 1982–1991. *American Journal of Physical Anthropology* **94**:213–228.

STERLING, E. J. (1994a) Evidence for nonseasonal reproduction in wild aye-ayes (*Daubentonia madagascariensis*). *Folia Primatologica* **62**:46–53.

STERLING, E. J. (1994b) Taxonomy and distribution of *Daubentonia*: A historical perspective. *Folia Primatologica* **62**:8–13.

STEWART, C-B. AND T. R. DISOTELL (1998) Genes, geology, and geography: A synthetic hypothesis of catarrhine primate evolution. *Current Biology* **8**:R582-R588.

STEWART, K. J. AND A. H. HARCOURT (1987) Gorillas: Variation in female relationships. In: B. B. Smuts, D. L. Cheney, R. M. Seyfarth, R. W. Wrangham and T. T. Struhsaker (eds) *Primate Societies*. Chicago: University of Chicago Press, pages 155–164.

STRIER, K. B. (1993) Menu for a monkey. *Natural History*, **102**(3):34–43.

STRIER, K. B. (1996) Reproductive ecology of female muriquis (*Brachyteles arachnoides*). In: M. A. Norconk, A. L. Rosenberger and P. A. Garber (eds) *Adaptive Radiations of Neotropical Primates*. New York: Plenum Press, pages 511–532.

STRUHSAKER, T. T. AND L. LELAND (1987) Colobines: Infanticide by adult males. In B. B. Smuts, D. L. Cheney, R. M. Seyfarth, R. W. Wrangham and T. T. Struhsaker (eds) *Primate Societies*. Chicago: University of Chicago Press, pages 83–97.

STRUM, S. C. (1987) *Almost Human: A Journey into the World of Baboons*. New York: Random House.

SUAREZ, S. D. AND G. G. GALLUP, JR. (1981) Self-recognition in chimpanzees and orangutans, but not gorillas. *Journal of Human Evolution* **10**:175–188.

SUGIYAMA, Y. (1994) Age-specific birth rate and lifetime reproductive success of chimpanzees at Bossou, Guinea. *American Journal of Primatology* **32**:311–318.

SUSSMAN, R. (1992) Male life history and intergroup mobility among ring-tailed lemurs (*Lemur catta*). *International Journal of Primatology* **13**:394–413.

SUSSMAN, R. W., CHEVERUD, J. M. AND T. Q. BARTLETT (1994/95) Infant killing as an evolutionary strategy: Reality or myth? *Evolutionary Anthropology* **3**:149–151.

SUSSMAN, R. W. AND P. A. GARBER (1987) A new interpretation of the social organization and mating system of the Callitrichidae. *International Journal of Primatology* **8**:73–92.

SUSSMAN, R. W. AND W. G. KINZEY (1984) The ecological role of the Callitrichidae: A review. *American Journal of Physical Anthropology* **64**:419–449.

TAKAHATA, Y., IHOBE, H. AND G. IDANI (1996) Comparing populations of chimpanzees and bonobos: Do females exhibit proceptivity or receptivity? In: W. C. McGrew, L. F. Marchant and T. Nishida (eds) *Great Ape Societies*. Cambridge: Cambridge University Press, pages 146–155.

TATTERSALL, I. (1982) *The Primates of Madagascar*. New York: Columbia University Press.

TATTERSALL, I. (1993) Madagascar's lemurs. *Scientific American* **268**(January):110–117.

TENAZA, R. R. (1989a) Female sexual swellings in the Asian colobine *Simias concolor*. *American Journal of Primatology* **17**:81–86.

TENAZA, R. R. (1989b) Intergroup calls of male pig-tailed langurs (*Simias concolor*). *Primates* **30**:199–206.

TERBORGH, J. AND M. STERN (1987) The surreptitious life of the saddle-backed tamarin. *American Scientist* **75**:260–269.

THALMANN, UL, GEISSMANN, T., SIMONA, A. AND T. MUTSCHLER (1993) The indris of Anjanaharibe-Sud, northeastern Madagascar. *International Journal of Primatology* **14**:357–381.

THIERRY, B. (1990) Feedback loop between kinship and dominance: The macaque model. *Journal of Theoretical Biology* **145**:511–521.

TOKIDA, E., TANAKA, I., TAKEFUSHI, H. AND T. HAGIWARA (1994) Tool-using in Japanese macaques: Use of stones to obtain fruit from a pipe. *Animal Behaviour* **47**:1023–1030.

TOMASELLO, M. AND J. CALL (1994) Social cognition of monkeys and apes. *Yearbook of Physical Anthropology* **37**:273–305.

TRIVERS, R. L. (1972) Parental investment and sexual selection. In: B. H. Campbell (ed) *Sexual Selection and the Descent of Man*. Chicago: Aldine, pages 136–179.

TSUMORI, A. (1967) Newly acquired behavior and social interactions of Japanese monkeys. In: S. A. Altmann (ed) *Social Communication among Primates*. Chicago: University of Chicago Press, pages 207–219.

TSUMORI, A., KAWAI, M. AND R. MOTOYOSHI (1965) Delayed response of wild Japanese monkeys by the sand-digging test (I)—Case of the Koshima troop. *Primates* **6**:195–212.

TUTIN, C. E. G. (1996) Ranging and social structure of lowland gorillas in the Lopé Reserve, Gabon. In: W. C. McGrew, L. F. Marchant and T. Nishida (eds) *Great Ape Societies*. Cambridge: Cambridge University Press, pages 58–70.

TUTIN, C. E. G. AND M. FERNANDEZ (1993) Composition of the diet of chimpanzees and comparisons with that of sympatric lowland gorillas in the Lopé Reserve, Gabon. *American Journal of Primatology* **30**:195–211.

TUTIN, C. E. G., HAM, R. M., WHITE, L. J. T. AND M. J. S. HARRISON (1997) The primate community of the Lopé Reserve, Gabon: Diets, responses to fruit scarcity, and effects on biomass. *American Journal of Primatology* **42**:1–24.

TUTTLE, R. H. (1986) *Apes of the World*. Park Ridge, NJ: Noyes Publications.

UTAMI, S. S. AND J. A. R. A. M. VAN HOOFF (1997) Meat eating by adult female orangutans (*Pongo pygmaeus abelii*). *American Journal of Primatology* **43**:159–165.

VAN DEN AUDENAERDE, D. F. E. T. (1984) The Tervuren Museum and the pygmy chimpanzee. In: R. L. Susman (ed) *The Pygmy Chimpanzee*. New York: Plenum Press, pages 3–11.

VAN HOOFF, J. A. R. A. M. (1990) Macaques and allies. In S. P. Parker (ed) *Grzimek's Encyclopedia of Mammals*, Vol. 2. New York: McGraw-Hill, pages 208–285.

VAN SCHAIK, C., FOX, E. AND A. SITOMPUL (1996) Manufacture and use of tools in wild Sumatran orangutans. *Naturwissenschaften* **83**:186–188.

VAN SCHAIK, C. P. AND J. A. R. A. M. VAN HOOFF (1996) Toward an understanding of the orangutan's social system. In: W. C. McGrew, L. F. Marchant and T. Nishida (eds) *Great Ape Societies*. Cambridge: Cambridge University Press, pages 3–15.

VERVAECKE, H. AND L. VAN ELSACKER (1992) Hybrids between common chimpanzees and pygmy chimpanzees in captivity. *Mammalia* **56**:667–669.

VISALBERGHI, E. (1993) Capuchin monkeys: A window into tool use in apes and humans. In: K. R. Gibson and T. Ingold (eds) *Tools, Language and Cognition in Human Evolution*. Cambridge: Cambridge University Press, pages 138–150.

VISALBERGHI, E. AND D. M. FRAGASZY (1990) Do monkeys ape? In: S. T. Parker and K. R. Gibson (eds) *"Language" and Intelligence in Monkeys and Apes*. Cambridge: Cambridge University Press, pages 247–273.

WALKER, A. AND M. TEAFORD (1989) The hunt for *Proconsul*. *Scientific American* **260**(January):76–82.

WATANABE, K. (1981) Variations in group composition and population density of the two sympatric Mentawaian leaf-monkeys. *Primates* **22**:145–160.

WATTS, D. P. (1984) Composition and variability of mountain gorilla diets in the central Virungas. *American Journal of Primatology* **7**:323–356.

WATTS, D. P. (1989) Infanticide in mountain gorillas: New cases and a reconsideration of the evidence. *Ethology* **81**:1–18.

WATTS, D. P. (1991) Mountain gorilla reproduction and sexual behavior. *American Journal of Primatology* **24**:211–225.

WATTS, D. P. (1994) Agonistic relationships between female mountain gorillas (*Gorilla gorilla beringei*). *Behavioral Ecology and Sociobiology* **34**:347–358.

WATTS, D. P. (1996) Comparative socio-ecology of gorillas. In: W. C. McGrew, L. F. Marchant and T. Nishida (eds) *Great Ape Societies*. Cambridge: Cambridge University Press, pages 16–28.

WEINER, J. (1994) *The Beak of the Finch*. New York: A. A. Knopf.

WHEELER, P. E. (1988) Stand tall and stay cool. *New Scientist* **12**(May):62–65.

WICKINGS, E. J. AND A. F. DIXSON (1992) Testicular function, secondary sexual development, and social status in male mandrills (*Mandrillus sphinx*). *Physiology and Behavior* **52**:909–916.

WILLIAMSON, E. A., TUTIN, C. E. G. AND M. FERNANDEZ (1988) Western lowland gorillas feeding in streams and on savannas. *Primate Report* **19**:29–34.

WINN, R. M. (1994) Preliminary study of the sexual behaviour of three aye-ayes (*Daubentonia madagascariensis*) in captivity. *Folia Primatologica* **62**:63–73.

WOLFE, L. D. (1984) Japanese macaque female sexual behavior: A comparison of Arashiyama East and West. In: M. Small (ed) *Female Primates: Studies by Women Primatologists*. New York: Alan R. Liss, pages 141–157.

WRANGHAM, R. W. (1984) Chimpanzees. In: D. Macdonald (ed) *All the World's Animals: Primates*. New York: Torstar, pages 126–131.

WRANGHAM, R. W. (1987) The significance of African apes for reconstructing human social evolution. In: W. G. Kinzey (ed) *The Evolution of Human Behavior: Primate Models*. Albany: State University of New York Press, pages 51–71.

WRANGHAM, R. W. (1997) Subtle, secret female chimpanzees. *Science* **277**:774–775.

WRANGHAM, R. W., McGREW, W. C., DE WAAL, F. B. M. AND P. G. HELTNE (1994) *Chimpanzee Cultures*. Cambridge, MA: Harvard University Press.

WRIGHT, P. (1992) Primate ecology, rain forest conservation, and economic development: Building a national park in Madagascar. *Evolutionary Anthropology* **1**:25–33.

WRIGHT, P. C. (1994a) Night watch on the Amazon. *Natural History* **103**(5):44–51.

WRIGHT, P. C. (1994b) The behavior and ecology of the owl monkey. In: J. F. Baer, R. E. Weller and I. Kakoma (eds) *Aotus. The Owl Monkey*. New York: Academic Press, pages 97–112.

YAMAGIWA, J., MARUHASHI, T., YUMOTO, T. AND N. MWANZA (1996) Dietary and ranging overlap in sympatric gorillas and chimpanzees in Kahuzi-Biega National Park, Zaire. In: W. C. McGrew, L. F. Marchant and T. Nishida (eds) *Great Ape Societies*. Cambridge: Cambridge University Press, pages 82–98.

YEAGER, C. P. (1990) Notes on the sexual behavior of the proboscis monkey (*Nasalis larvatus*). *American Journal of Primatology* **21**:223–227.

YEAGER, C. P. (1991) Proboscis monkey (*Nasalis larvatus*) social organization: Intergroup patterns of association. *American Journal of Primatology* **23**:73–86.

ZELLER, A. C. (1987) Communication by sight and smell. In: B. B. Smuts, D. L. Cheney, R. M. Seyfarth, R. W. Wrangaham and T. T. Struhsaker (eds) *Primate Societies*. Chicago: University of Chicago Press, pages 433–439.

ZIHLMAN, A. L. (1996a) Looking back in anger. *Nature* **384**:35–36.

ZIHLMAN, A. L. (1996b) Reconstructions reconsidered: Chimpanzee models and human evolution. In: W. C. McGrew, L. F. Marchant and T. Nishida (eds) *Great Ape Societies*. Cambridge: Cambridge University Press, pages 293–304.

ZIHLMAN, A. L. (1997) Natural history of apes: Life-history features in females and males. In: M. E. Morbeck, A. Galloway and A. L. Zihlman (eds) *The Evolving Female: A Life-History Perspective*. Princeton: Princeton University Press, pages 86–103.

ZILLES, K. AND G. REHKAMPER (1988) The brain, with special reference to the telencephalon. In: J. H. Schwartz (ed) *Orang-utan Biology*. New York: Oxford University Press, pages 157–176.

ZIMMERMANN E. (1989) Aspects of reproduction and behavioral and vocal development in Senegal bushbabies. *International Journal of Primatology* **10**:1–16.

ZINNER, D. (1998) Male take-overs and sexual swellings in hamadryas baboons (*Papio hamadryas*). *Folia Primatologica* **69**:240.

ZUCKER, E. L. (1994) Severity of agonism of free-ranging patas monkeys differs according to the composition of dyads. *Aggressive Behavior* **20**:315–323.

CREDITS

Original drawings provided by John Guyer.

INTRODUCTION

i-1(a) Underwood & Underwood/CORBIS. i-1(b) Noel Rowe. i-2 From Figure 5.5, p. 156, in R. Boyd and J. Silk, 1997, *How Humans Evolved*, W. W. Norton. i-4 Redrawn from Figure 3 in Noel Rowe, 1996, *The Pictorial Guide to the Living Primates*, Pogonias Press. i-7 From Figure 3.7, p. 93, in G. Conroy, 1997, *Reconstructing Human Origins*, W. W. Norton. i-8 Modified from Figure 3.5c, p. 90, in G. Conroy, 1997, *Reconstructing Human Origins*, W. W. Norton.

CHAPTER 1

1.1 From Figure 36.18, p. 855, in C. McFadden, 1996 *Biology: An Exploration of Life*, W. W. Norton. B-1 Modified from Figure 1.10, p. 37, in G. Conroy, 1990, *Primate Evolution*, W. W. Norton. 1.2 Illustration by Lukrezia Bieler in R. Martin, 1993, Primate origins: Plugging the gaps, *Nature* 363:223–234. Reproduced by permission. 1.3 Redrawn from p. 32 in A. G. Smith, D. G. Smith, and B. M. Funnell, 1994, *Atlas of Mesozoic and Cenozoic Coastlines*, Cambridge University Press. 1.4 Redrawn from p. 30 in A. G. Smith, D. G. Smith, and B. M. Funnell, 1994, *Atlas of Mesozoic and Cenozoic Coastlines*, Cambridge University Press. 1.5 Carnegie Museum of Natural History. 1.6 Redrawn from p. 28 in A. G. Smith, D. G. Smith, and B. M. Funnell, 1994, *Atlas of Mesozoic and Cenozoic Coastlines*, Cambridge University Press. 1.7 From Figure 12.5 in J. G. Fleagle, 1988, *Primate Adaptation and Evolution*, Academic Press, San Diego. Reprinted by permission of Academic Press and the author. N-1 modified from RADINSKY and THE PRIMATE BRAIN. 1.9 From Figure 2 in C-B. Reprinted from *Current Biology 8*, Stewart and T. R. Disotell, Primate evolution: In and out of Africa, R582–588, 1998, with kind permission from Elsevier Science Ltd, Middlesex House, 34–42 Cleveland Street, London W1P 6LE, UK. 1.10 Redrawn from p. 26 in A. G. Smith, D. G. Smith, and B. M. Funnell, 1994, *Atlas of Mesozoic and Cenozoic Coastlines*, Cambridge University Press. 1.13 Redrawn from p. 24 in A. G. Smith, D. G. Smith, and B. M. Funnell, 1994, *Atlas of Mesozoic and Cenozoic Coastlines*, Cambridge University Press. 1.14 Modified from p. 47 in Le Gros Clark, W. E., 1959, *The Antecedents of Man*. Harper & Row. 1.15 Modified from Figure 7, p. 260 in C. Stanford 1995, Chimpanzee hunting behavior and human evolution, *American Scientist* 83:256–261. 1.16 Noel Rowe.

CHAPTER 2

2.1 Tim Laman. 2.2 Loomis Dean, Life Magazine © Time Inc. 2.5 Noel Rowe. 2.6 From Figure 7.33, p. 235, in R. Boyd and J. Silk, 1997, *How Humans Evolved*, W. W. Norton. 2.7 From Figure 7.45 in R. Boyd and J. Silk, 1997,

How Humans Evolved, W. W. Norton. 2.8 Noel Rowe. 2.9 Nicole Duplaix/Peter Arnold, Inc. 2.10 From Figure 7.10a, p. 215, in R. Boyd and J. Silk, 1997, *How Humans Evolved*, W. W. Norton. 2.11 Mike Turco. 2.12 Kennan Ward/CORBIS. 2.14 From Figure 6.6, p. 190, in R. Boyd and J. Silk, 1997 *How Humans Evolved*, W. W. Norton. 2.15 Yann Arthus-Bertrand/CORBIS. 2.16 Joseph Popp/Anthro-Photo. 2.17 Tom Brakefield/CORBIS.

CHAPTER 3

3.3 Redrawn from Figure 5 in W. E. Le Gros Clark, 1996, *History of the Primates*, University of Chicago Press. 3.4 Adapted from Figure 4.4, p. 148, in G. Conroy, 1990, *Primate Evolution*, W. W. Norton. N-3 Modified from Figure 3.19, p. 133, in G. Conroy, 1990, *Primate Evolution*, W. W. Norton. 3.6 Clem Haagner, ABPL/CORBIS, 3.8 David Haring/Duke University Primate Center. 3.9 H. Sprankel/WRPRC AV Archives. 3.11 David Haring/Duke University Primate Center. 3.12 Redrawn from Figure 39, p. 159, in W. C. Osman Hill, 1953, *Primates: Comparative Anatomy and Taxonomy. I—Strepsirhini*, Edinburgh University Press. 3.14 David Haring/Duke University Primate Center.

CHAPTER 4

4.1 Drawing by Stephen Nash/Conservation International. B-4 Illustration by Patricia Wynne in I. Tattersall, 1993, Madagascar's lemurs, *Scientific American* 268:110–117. 4.2 Courtesy of Michelle Sauther. 4.4 Courtesy of Michelle Sauther. 4.5 Nick Garbutt/BBC Natural History Unit. 4.6 Peter Oxford/BBC Natural History Unit. 4.9(a) Alan Walker/Anthro-Photo. 4.9(b) David Haring/Duke University Primate Center. 4.12 Wolfgang Kaehler/CORBIS. 4.14 David Haring/Duke University Primate Center. 4.15 From Figures 186 and 188 in W. C. Osman Hill, 1953, *Primates: Comparative Anatomy and Taxonomy. I—Strepsirhini*, Edinburgh University Press. 4.17 From Figure 6 in R. M. Winn, 1994, Preliminary study of the sexual behavior of three aye-ayes (*Daubentonia madagascariensis*) in captivity. E. J. Sterling, 1994, Taxonomy and distribution of *Daubentonia*: A historical perspective. *Folia Primatologica* 62:63–73. Reproduced with permission of S. Karger AG, Basel. F-4 David Sanders/DS Photography. N-4 Figure from p. 60 in D. Falk, 1992, *Braindance*, Henry Holt & Company.

CHAPTER 5

B-5 From Figure 19-7, p. 471, in *Life: An Introduction to Biology*, © 1957 by George Gaylord Simpson, Colin S. Pittendrigh, and Lewis H. Tiffany, and renewed 1985 by Anne R. Simpson, Joan Simpson Burns, Ralph Tiffany, Helen Vishniac, and Elizabeth Leonie S. Wurr, reprinted by permission of Harcourt, Inc. 5.3 Figure from p. 211 in S. Jones, R. Martin, and D. Pilbeam, 1992, *The Cambridge Encyclopedia of Human Evolution*, Reprinted with the permission of Cambridge University Press. 5.5 Redrawn from Figure 1, p. 420 in R. W. Sussman and W. G. Kinzey, 1984, The ecological role of the Callitrichidae: A review. *American Journal of Physical Anthropology* 64:419–449. Reprinted by permission of Wiley-Liss, Inc., a division of John Wiley & Sons, Inc. 5.6 Gerard Lacz/Peter Arnold, Inc. 5.7 Noel Rowe. 5.9 Mitsuaki Iwago/Minden Pictures. 5.11 Noel Rowe. 5.13(a) Courtesy of Paul Beaver/Amazonia Expeditions. 5.15 Kevin Schafer/CORBIS.

CHAPTER 6

6.1 From Figure 8 and 9 in A. L. Rosenberger, 1992, Evolution of New World monkeys, pages 209–216 in S. Jones, R. Martin, and D. Pilbeam, *The Encyclopedia of Human Evolution*, Cambridge University Press. Reprinted by permission of A. L. Rosenberger. 6.2 From Figure 3, p. 419, in S. Ford and L. C. Davis, 1992, Systematics and body size: Implications for feeding adaptations in New World monkeys, *American Journal of Physical An-*

thropology 88:415–468. Reprinted by permission of Wiley-Liss, Inc., a division of John Wiley & Sons, Inc. 6.5 Roland Seitre/Peter Arnold, Inc. 6.8 Kevin Schafer/CORBIS. B-6 and 6.10 Luiz Claudio Marigo. 6.11 L. C. Marigo/WRPRC AV Archives. 6.12 Redrawn from map by Joe LeMonnier in J. M. Ayres, 1990, Scarlet faces of the Amazon, *Natural History* 99(2):33–41. Reprinted by permission of Joe LeMonnier. N-6 Noel Rowe. 6.13 D. J. Chivers/Anthro-Photo. 6.14 R. Fontaine/WRPRC AV Archives. 6.17 L. C. Marigo/WRPRC AV Archives.

CHAPTER 7

7.1 Noel Rowe. 7.2 From Figure 1.1, p. 4, in A. G. Davies and J. F. Oates (eds), 1992, *Colobine Monkeys: Their Ecology, Behaviour and Evolution*. Reprinted with the permission of Cambridge University Press. 7.3 The Purcell Team/CORBIS. 7.5 and 7.7 W. Scott McGraw. 7.9 Kevin Schafer/CORBIS. 7.11 Courtesy of Craig Kirkpatrick. 7.13 Irven DeVore/Anthro-Photo. 7.15 From Figure 11.3, p. 321 in P. N. Newton and R. I. M. Dunbar, 1994, Colobine monkey society. In A. G. Davies and J. F. Oates (eds) *Colobine Monkeys: Their Ecology, Behaviour and Evolution*. Reprinted with the permission of Cambridge University Press. 7.16 Mike Turco.

CHAPTER 8

8.4 W. Scott McGraw. 8.5 K. G. Preston-Mafham/Premaphotos Wildlife. 8.7 Noel Rowe. 8.9 From Figure 21.6, p. 426, in J. Chism and T. E. Rowell, 1988, The natural history of patas monkeys. In A. Gautier-Hion, F. Bourliere, J-P. Gautier and J. Kingdon (eds.) *A Primate Radiation: Evolutionary Biology of the African Guenons*. Reprinted with the permission of Cambridge University Press. 8.10 Hans Kummer/WRPRC AV Archives. 8.13 and N-8(a) Irwin Bernstein/WRPRC AV Archives. 8.17 W. Scott McGraw.

CHAPTER 9

9.1 Richard T. Nowitz/CORBIS. 9.2a-b Noel Rowe. 9.4 Roger Ressmeyer/CORBIS. 9.5 Yann Arthus-Bertrand/CORBIS. 9.6 Kennosuke Tsuda/Nature Production. 9.7 A. Tsumori, M. Kawai, and R. Motoyoshi, 1965, Delayed response of wild Japanese monkeys by the sand-digging test (1)—Case of the Koshima troop, *Primates* 6:195–212. 9.9 Wolfgang Kaehler/CORBIS. 9.10 Nancy Nicolson/Anthro-Photo. 9.11 and 9.12 Irven DeVore/Anthro-Photo.

CHAPTER 10

10.2 From Figure 17, p. 51 in A. H. Schultz, 1969, *The Life of Primates*, Weidenfeld & Nicolson, London. 10.3 Gerry Ellis/Minden Pictures. 10.4 D. Robert Franz/Wildlife Collection. 10.5 Galen Rowell/CORBIS. 10.6 Figure from p. 120 in D. McDonald (ed), 1984, *The Primates, Torstar*. Reprinted by permission of Andromedia Oxford Ltd., England. 10.7 Noel Rowe. 10.8 Bertrand Deputte/WRPRC AV Archive. 10.9 Nina Leen/Life Magazine. 10.10 and 10.11 Gerry Ellis/ENP Images. 10.12 Gerard Lacz/Peter Arnold, Inc. 10.13 From p. 212 in Noel Rowe, 1996, *The Pictorial Guide to the Living Primates*, Pogonias Press. 10.14 Noel Rowe.

CHAPTER 11

11.1 Redrawn from Figure 1.1, p. 4, in C. P. Van Schaik, and J. A. R. A. M. Van Hooff. 1996, toward an understanding of the orangutan's social system, pages 3–15 in W. C. McGrew, L. F. Marchant, and T. Nishida (eds.) *Great Ape Societies*, Cambridge University Press. 11.2 From Plate 7 in A. H. Schultz, 1969, *The Life of Primates*, Weidenfeld & Nicolson, London. 11.3 Tim Laman. 11.4 Noel Rowe. 11.5(left) Martin Harvey/Wildlife Collection. 11.5(right) Noel Rowe. 11.6 Mike Turco. 11.7 Konrad Wothe/Minden Pictures. 11.8 Timothy Laman/NGS Image Collection. 11.9 Bates Littlehales/NGS Image Collection. 11.10 Mike Turco. N-11 Redrawn from illustration by Tom Dunne on p. 537, *American Scientist* vol. 80, Nov–Dec. 11.13 H. D. Rijksen.

CHAPTER 12

12.1 Modified from Figure 5, p. 17 in A. F. Dixson, 1981, *The Natural History of the Gorilla*, Weidenfeld & Nicolson, London. 12.2 Museum of Modern Art Film Stills Archive. 12.4 Frans Lanting/Minden Pictures. 12.5 Peter G. Veit. 12.6 From Plate 5 in A. H. Schultz, 1969, *The Life of Primates*, Weidenfeld & Nicolson, London. 12.7 Gerry Ellis/ENP Images. 12.8 Noel Rowe. 12.11 A. Veder and B. Webber/WRPRC AV Archives. 12.12 Peter G. Veit. 12.13 Dr. Ronald H. Cohn/The Gorilla Foundation.

CHAPTER 13

13.1 From Figure 5.19, p. 176, in R. Boyd and J. Silk, 1997, *How Humans Evolved*, W. W. Norton. 13.2 IFA/Peter Arnold, Inc. 13.3(a) Frans Lanting/Minden Pictures. 13.3(b) Richard Wrangham/Anthro-Photo. 13.4 Frans Lanting/Minden Pictures. B-13 Peter Menzel Photography. 13.12 Amy Parish/Anthro-Photo. 13.13 Steve Winter/Language Research Center.

CHAPTER 14

14.2 Adapted from Figure 4.1, p. 130, in G. Conroy, 1997, *Reconstructing Human Origins*, W. W. Norton. N-14 John Guyer. 14.4 From Figure 5.20, p. 213, in G. Conroy, 1997, *Reconstructing Human Origins*, W. W. Norton. 14.5 From Figure 4.12c, p. 161, in G. Conroy, 1997, *Reconstructing Human Origins*, W. W. Norton. 14.7 Redrawn from *Human Evolution Coloring Book* by Adrienne Zihlman. Copyright © 1982 by Coloring Concepts, Inc. Reprinted by permission of HarperCollins Publishers, Inc. 14.9 From Figure 5.32c, p. 241, in G. Conroy, 1997, *Reconstructing Human Origins*, W. W. Norton.

Every effort has been made to contact the copyright holders of each of the selections. Rights holders of any selections not credited should contact W. W. Norton & Company, Inc., 500 Fifth Avenue, New York, NY 10110, in order for a correction to be made in the next reprinting of our work.

INDEX